John S

Mathematics Education Library

VOLUME 18

The titles published in this series are listed at the end of this volume.

APPROACHES TO ALGEBRA

Perspectives for Research and Teaching

Edited by

NADINE BEDNARZ,
CAROLYN KIERAN
and
LESLEY LEE

Département de Mathématiques,
Université du Québec à Montréal

KLUWER ACADEMIC PUBLISHERS
DORDRECHT / BOSTON / LONDON

A C.I.P. Catalogue record for this book is available from the Library of Congress.

ISBN 0-7923-4145-7 (HB)
ISBN 0-7923-4168-6 (PB)

Published by Kluwer Academic Publishers,
P.O. Box 17, 3300 AA Dordrecht, The Netherlands.

Kluwer Academic Publishers incorporates
the publishing programmes of
D. Reidel, Martinus Nijhoff, Dr W. Junk and MTP Press.

Sold and distributed in the U.S.A. and Canada
by Kluwer Academic Publishers,
101 Philip Drive, Norwell, MA 02061, U.S.A.

In all other countries, sold and distributed
by Kluwer Academic Publishers Group,
P.O. Box 322, 3300 AH Dordrecht, The Netherlands.

Printed on acid-free paper

Printed in the Netherlands

TABLE OF CONTENTS

PART IV. A MODELING PERSPECTIVE ON THE INTRODUCTION OF ALGEBRA

PART VI. SYNTHESIS AND DIRECTIONS FOR FUTURE RESEARCH

ACKNOWLEDGMENTS

No production can move forward without a network of relationships in which alliances and complicity play a part. The present work is no exception. Obviously, its conceptual and empirical viability may be ascribed to the participating authors, but it is also attributable to the contributions that several other persons made to the form which it ultimately assumed.

We are particularly grateful to the *Comité des publications de l'Université du Québec à Montréal,* which supported this publication by means of a grant. We also wish to express our sincere appreciation to the Social Sciences and Humanities Research Council of Canada and the *Programme d'aide aux chercheur-e-s et aux créateurs-trices de l'Université du Québec à Montréal* which, thanks to their financial support of an international research colloquium, brought together all of the contributors to the various chapters of this book.

Finally, without the expertise of Pat McQuatty-Lytle, who lent a hand with the language corrections, and Céline Beaulieu and Jocelyne Deschênes, who helped out with the word processing, we would have been unable to present the following reflection on the learning and teaching of algebra.

Nadine Bednarz
Carolyn Kieran
Lesley Lee

INTRODUCTION

CHAPTER 1

APPROACHES TO ALGEBRA: PERSPECTIVES FOR RESEARCH AND TEACHING

NADINE BEDNARZ, CAROLYN KIERAN, LESLEY LEE

The introduction of school algebra can take many different directions: the rules for transforming and solving equations (to which current teaching often reduces algebra), the solving of specific problems or classes of problems (which has played an important role historically in the development of algebra and its teaching), the generalization of laws governing numbers (a very strong focus in certain curricula), the more recent introduction of the concepts of variable and function (which appeared much later historically and which occupy a position of growing importance in some programs), and the study of algebraic structures (which marked the school curriculum in the 1960s under the influence of modern mathematics).

Thus, different options for introducing algebra have appeared over time, options that are again coming to the fore in new programs throughout the international community. These curricular options determine to a large extent the algebraic conceptions students develop, as has been shown in studies on the meanings that students ascribe to algebraic symbols and notation, the poor strategic decisions they make, and, in a more general way, the relationship they continue to maintain with algebra several years after having been introduced to it.[1] Many studies have shown the difficulties of students at several school levels with respect to the concepts focused on in the various ways of introducing algebra: equation solving, the manipulation of algebraic expressions, problem solving, and the handling of fundamental concepts such as that of variable.[2] Analysis of these difficulties has brought to light certain notions or interpretations that students develop with respect to the use of letters, notations, and written conventions associated with certain concepts, such as equality (Kieran, 1981) and the nature of mathematical "answers" (Collis, 1974), which in fact interfere with the very process of algebraic knowledge construction. Some of these studies point to important conceptual changes that students must make in negotiating the passage to algebra and its development.

These findings suggest numerous questions regarding the beginnings of algebraic learning and the appropriateness of associated didactic interventions. For example, what are the kinds of situations that are conducive to the emergence and development of algebraic understanding and to giving sense to fundamental algebraic concepts and to their use?

In the international community, the teaching and learning of algebra have recently received a great deal of research interest, and different approaches aimed at making this learning meaningful for students have been proposed: generalization of numerical and geometric patterns and of the laws governing numerical relations, problem solving, equation solving aided by the use of concrete models, introduction of functional situations, and the modeling of physical and mathematical phenomena.

One can associate these approaches, and their underlying theoretical foundations, with different conceptions (explicit or more often implicit) of algebra: the study of a

N. Bednarz et al. (eds.), Approaches to Algebra, 3-12.

language and its syntax; the study of solving procedures for certain classes of problems, algebra being here conceived of not only as a tool for solving specific problems but also as a tool for expressing general solutions; the study of regularities governing numerical relations, a conception of algebra that centers on generalization and that can be widened by adding components of proof and validation (Margolinas, 1991); and the study of relations among quantities that vary. The different approaches to introducing algebra that have been put forward by the international community appear, in the light of this analysis, to be related to the conceptions of algebra held by their authors.

According to Bell (1988) and Vergnaud (1989), only an equilibrium among the different components and the various situations that make them meaningful can enable students to deeply understand the pertinence of algebra, its structure, the significance of fundamental algebraic concepts such as variable, and the use of algebraic reasoning (knowing to apply it in contexts where algebraic manipulations can be fruitful).

Now what do we know of each of these approaches, of their contributions, of the difficulties they elicit on the part of students? What do we know of the passage between these diverse components that are essential to the teaching of algebra, for example, how students negotiate the passage between algebra as a tool for generalization and algebra as a problem-solving tool or algebra as a tool for modeling in functional situations. Is the passage made as easily as some might think or does it imply important conceptual adjustments, notably with respect to the notion of variable?

In order to better understand the emergence and development of algebraic knowledge in these different contexts, the articulation of these different components, and the ways in which the introduction of algebra can have a bearing upon this development, this book looks more closely at some of the options that aim at giving meaning to algebra. These options, which are at the forefront of contemporary research and curriculum efforts on the introduction of secondary school algebra, are as follows: generalizing patterns (numeric and geometric) and the laws governing numerical relations (Arzarello, 1991a; Mason, Graham, Pimm, & Gowar, 1985), solving of specific problems or classes of problems (Puig & Cerdán, 1990; Filloy & Rubio, 1991; Rubio, 1990, 1994), modeling of physical phenomena (Chevallard, 1989), and focusing on the concepts of variable and function (Confrey, 1992a; Garançon, Kieran, & Boileau, 1990; Heid & Zbiek, 1993; Kieran, 1994).

Issuing from an international colloquium on "Research Perspectives on the Emergence and Development of Algebraic Thought," held in Montréal in May 1993 and organized jointly by the *Centre Interdisciplinaire de Recherche sur l'Apprentissage et le Développement en Éducation* (CIRADE) and the Mathematics Department of the Université du Québec à Montréal, the contributions in this book shed light upon the different possible approaches to the teaching of school algebra. Contributors were asked to address the following questions. What are the essential characteristics of algebraic thinking that are developed in each of these approaches (or what would we like students of algebra to know)? On what thought processes can this learning rely (or what can algebra learning build on)? What thought processes must this learning work against (or what are the obstacles in algebra learning)? What situations might promote the emergence and development of algebraic thinking in students?

The different chapters of this book provide a discussion of various theoretical positions taken by researchers in the field of the learning and teaching of algebra, and of the several possible choices to be made in its introduction and development at the secondary level. They offer an in-depth reflection on important characteristics of algebraic thought, on the difficulties students encounter in moving into algebra, and on situations that might facilitate its development. They present a selection of studies that examine the emergence and development of algebra from different perspectives (generalization, problem solving, modeling, functions). The analysis of research proposed in this book--on these different components of algebra and on the theoretical and epistemological foundations on which they rest--is strengthened by a double perspective: didactic and historical (via the history of the development of algebra itself).

1. HISTORICAL PERSPECTIVES IN THE DEVELOPMENT OF ALGEBRA

The history of algebra is rich on several counts. For example, it allows us to better appreciate the complexity of algebraic concepts and the breakthroughs that occurred during its construction. The historical study of the notion of variable is rather instructive in this regard (Janvier, Charbonneau, & René de Cotret, 1989). The history of negative numbers, which is at the heart of the development of algebra, is equally rich in examples of epistemological obstacles (Gallardo, 1994; Pycior, 1984) that allow us to be aware of the difficulties inherent in the new demands being made in the exercise of mathematical reasoning. Historical analysis permits us to better understand the advances, oppositions, and steps backward in the evolution of a knowledge system. Thus, the study of phenomena observable in the history of mathematics, phenomena involving a rupture or change in models and languages, is important for our understanding of students' difficulties.

History also affords us, as we shall see later, a framework for analyzing those situations that led to the creation of algebraic knowledge. The use of these situations requires, however, that the researcher adjust for current conditions of knowledge production and for the constraints under which such a learning experience takes place (Brousseau, 1989).

The glimpse that we take of the historical evolution of algebra, in the three chapters of Part I of this book, permits us to make explicit some of the key ideas in the development of algebra and provides us with indicators of change that signaled the transition to a mode of algebraic thinking. Thus, the contributions of Charbonneau and of Radford, which are both based on historical analysis of the emergence of algebra, yield evidence of the links that unite the development of algebra to that of geometry. Certain geometric approaches to which they refer seem to be a possible precursor to the emergence of analytic thinking in the learning of algebra. With respect to the present tendencies in primary school mathematics and the problems posed by the arithmetic-algebra transition, the above analysis has the advantage of re-situating in a broader context the questions related to the emergence of algebra and of demonstrating the non-negligible contribution of geometry to this evolution. This geometric perspective is found to a certain extent in those approaches to algebra that focus on the generalization of geometric patterns and on the construction of geometric formulas.

The three chapters of this first part of the book thus point to key elements in the evolution of algebraic thought, elements that are taken up at least partially in the didactic analyses of the later chapters. The analytic approach is one of these essential characteristics (history shows us that access to the analytic mode of problem solving, in contrast to the synthetic that was used prior to this, took time to be introduced). This major conceptual change marks for students too the passage from an arithmetic mode of reasoning in problem solving to an algebraic one (Bednarz & Janvier, this volume).

The theory of proportionality also plays, according to this historical analysis, a central role in the development of algebra (the importance of proportional reasoning, which is shown by Radford in both the numerical approaches used and in the method of false position, appears to be a possible key entry point to the concept of unknown in the solving of algebraic problems). The actual conditions of knowledge production in school have reduced to a considerable degree the importance accorded to proportional reasoning in mathematics programs. Nevertheless, the contribution of numerical approaches remains a possible access route in the passage to algebra (in this regard, see the chapters by Rojano and by Kieran et al. in this volume). The historical analysis in the first three chapters shows, thus, the contribution of geometric and numerical approaches to the emergence of algebra and points to the potential of certain prior knowledge on the part of students in the passage to algebra.

Finally, the role of symbolism in this evolution is emphasized in the historical analysis of the Rojano chapter, which shows the direct links between the creation of algebraic language and the development of algebraic thought. Thus, the geometric significance of symbols and the related barriers of homogeneity served as the point of departure for creating new symbols (the coefficients) whose initial function was to compensate for the dimensions of the terms containing the unknown. This creation of new symbols was not, however, disconnected, as is the case today for algebra students, from those situations that provided meaning for the symbols. Such historical analysis, as pointed out by Rojano, can remind us of the potential learning risks of presenting symbolic manipulation as an object of knowledge in itself:

> The well known separation that students tend to make between algebraic manipulation and its use in problem solving may originate in an educational approach based on this oversimplified vision of algebra, which hides the significance of its origins and the semantic background of its grammar. (Rojano, Part I of this volume)

According to Rojano, the potential of algebraic syntax cannot be appreciated by students until they have reached the limits of their prior mathematical knowledge.

Radford's chapter raises, on the other hand, important questions regarding the learning of fundamental concepts of algebra, such as the unknown and variable, by showing that the conceptions underlying the notion of variable are related to situations whose aims and intentions are essentially different from those related to the concept of unknown (establishing relations between numbers vs. solving problems). This historical analysis leads to the didactic question of the articulation between these two essential components of algebra learning--between an approach emphasizing generalization and the construction of formulas where the symbolism takes on the sense of generalized number and an approach focusing on problem solving where the symbols represent unknowns. This reflection is taken up in several subsequent contributions. We return now to the other chapters and their perspectives on different approaches to algebra.

2. A GENERALIZATION PERSPECTIVE ON THE INTRODUCTION OF ALGEBRA

Several attempts at introducing algebra through generalizing activities (numeric or geometric patterns, laws governing numbers, etc.) have recently come to light particularly in England (Bell, 1988; Mason et al., 1985) and have given rise to investigations of the knowledge and processes that sustain such approaches (MacGregor & Stacey, 1993; Orton & Orton, 1994; Stacey, 1989). The various chapters in the second part of this book allow us to re-examine the issues involved in a generalization approach to the introduction of algebra.

A scrutiny of established school practice involving generalization in algebra reveals that often, starting from geometric figures or numeric sequences, the emphasis is on the construction of tables of values from which a closed-form formula is extracted and checked with one or two examples. This approach, as Mason demonstrates in his chapter, in effect short-circuits all the richness of the process of generalization:

> I suggest that students remain unaware of the generality in a formula they conjecture because, in most mathematical topics, teachers collude with them to keep their attention focused on the technique in particular cases, and not on questions about the domain of applicability of the technique. (Mason, Part II of this volume)

Thus, one of the keys to generalization is to be found in the pedagogical approach that frames this process in the classroom. Mason suggests several possible investigative approaches that can lead to students' construction of formulas: visualization; manipulation of the figure on which the generalizing process is based, thereby facilitating the construction of the formula; formulating a recursive rule which shows how to construct the following terms from preceding ones; and finding a pattern which leads directly to a formula.

The remarks of the various contributors indicate a certain number of difficulties inherent in the introduction of algebra through a generalization approach. Different reactions to questions on number patterns are reported, revealing difficulties encountered by students at different levels.

For example, one of the major problems Lee found in a teaching experiment involving adults, was not that of "seeing a pattern" but, in her words, that of "finding an algebraically useful" pattern. Fixing on an irrelevant or unhelpful pattern turns out to be an obstacle to students in elaborating formulas that account for a general procedure or a relationship between the quantities.

These difficulties also raise the issue of the role of teachers who are called upon to manage this pedagogical situation. How can they be alert to patterns, reasoning, students' symbolizations? Can they recognize the pertinence and the validity of these various visualizations if they themselves are fixed on a certain pattern that eclipses others?

Thus, Lee reminds us that a certain flexibility seems to be an entry point for this process of generalization:

> Perceptual agility seemed to be a key (in this process of generalization in algebra): being able to see several patterns and willing to abandon those that do not prove useful. (Lee, this volume)

But can we limit the difficulties involved in this generalization process to a purely perceptual level? Generalization, seen in the historical light of Radford's chapter (in Part II), also depends on the elements (numbers, configurations) that are the object of this generalization. Moreover, generalization is situated within a certain theory that allows for its justification. Thus, it depends on the knowledge and intentions of the observer.

Another important question raised by Radford is linked to the very nature of the algebraic concepts this approach admits. At the conceptual level, important differences need to be examined (these are discussed later in this volume by Janvier), for example, formula construction appears to be built upon the concept of generalized number (Küchemann, 1981). Can the latter be seen as a pre-concept to that of variable? And how does the eventual junction with the very different concept of unknown come about?

The process of generalization as an approach to algebra appears intimately related to that of justification (Lee & Wheeler, 1989; Radford, Part II of this volume). This cannot be said about the process of problem solving that finds its logical basis in its analytical nature. Is it possible to link these two approaches? What do we know about the latter approach?

3. A PROBLEM-SOLVING PERSPECTIVE ON THE INTRODUCTION OF ALGEBRA

The important role played by problem solving in the development of algebra and its teaching over the centuries (Chevallard, 1989) cannot be denied. This essential component of algebraic thinking merits, therefore, to be examined in depth. Consequently, the contributions of the various authors to Part III, from the double perspective of history (Charbonneau & Lefebvre) and didactics (Bednarz & Janvier; Rojano; Bell), attempt to clarify what is involved in the problem solving component, the possible forms such an approach could take, and the difficulties to which it gives rise. The historical analysis raises questions about the very notion of a problem and the role problems have played (within the theory or theory-in-the-making). The historical study of the works of Diophantus, Al-Khwarizmi, Cardano, and Viète situates problem solving in a wider than usual context by revealing the scope of what is meant by a problem and the evolution of the contexts in which these problems arose. The diversity of words used to denote a problem and their use within theory indicate a variety of underlying conceptions. These conceptions range from a simple application of rules to a specific solution calling for certain skills, and from a specific solution to a more general solution for which the set of rules used and problems to be solved are expanded.

From the same perspective, Bell also locates problem solving in a wider context which includes various components that all deserve a place in the curriculum (forming and solving equations, generalization, working with functions and formulas). These diverse components are illustrated by showing how an open exploration of problems can form the basis of an algebra course, with the accent nevertheless on what we usually associate with problem solving, the forming and solving of equations.

From a narrower perspective, that of solving word problems, the Bednarz and Janvier chapter looks at the fundamental issue of the arithmetic/algebra transition. They analyze this transition, not from the viewpoint of the symbolism used in the equation, but more from the angle of an analysis of the problems and reasoning used by pupils. The symbolic framework involving a relational calculus, which was developed by the research team, not only allows one to examine the arithmetic-algebra passage but, beyond this, permits one to account for the relative difficulty of different types of problems given in algebra and to anticipate student difficulties (Bednarz & Janvier, 1994). Moreover, studying the continuities and discontinuities that can be found in students' reasoning helps to uncover those situations that force the limits of students' previous mathematical knowledge and that might motivate an eventual passage into algebra or an evolution within algebra itself. The conceptual changes required by the passage into and the continued development within algebra can also be identified using this framework. It is, however, the various profiles of arithmetic reasoning, reasoning which is constructed by students before any introduction to algebra, that is the focus of attention in this analysis.

This consideration of previously elaborated problem-solving methods is a preoccupation of Rojano's teaching approach as well: "The students' own strategies are incorporated as the necessary starting point for the learning of systematic and general methods of problem solving." These are put into play in a spreadsheet computer environment that helps students symbolize their informal procedures for the problem. The realization of a relationship among the unknowns, the choice of an unknown, and the work on the unknown to find its numerical value distinguish the approach in this environment from the informal strategy of numeric trials on which it is based.

Commentary chapters on the various contributions presented in this third part of the book go back to certain questions linking this approach with some of the other components of algebraic thinking. The link with generalization is raised by Mason: What do we know at the historical level, for example, about this transition from the particular to the general (Did not Diophantus and others already work in general terms through the particular manifestations of problems?). The place of symbolic language and its manipulation in such an approach is raised by Bell, who reminds us of the important role symbolism plays in the development of algebraic thinking. Thus, all problem-solving approaches, according to Bell, ought to take into account the interactions that are woven between the formulation of equations and their manipulation in the solving process. The question of the arithmetic-algebra transition is at the heart of the debate in Wheeler's commentary. He wonders if it would not be possible to take a completely different approach to algebra: Is there a reason why we couldn't teach algebra to high school students as something totally new and different from what they have learned before?

Other approaches for introducing algebra that might fit the paradigm put forward by Wheeler, such as modeling and a functional approach, are advanced in the next two parts of the book.

4. A MODELING PERSPECTIVE ON THE INTRODUCTION OF ALGEBRA

The two principal chapters (by Nemirovsky and Janvier) in this fourth part of the book clearly point out that the early recognition of the pertinence and validity of the new mathematical object, algebra, requires putting students in a situation that allows them to construct meaning for various representations (graphs, equations, etc.) and to use them with a certain flexibility in the description and interpretation of physical phenomena or world events.

For Nemirovsky, this modeling process relies on a certain verbalization which gives meaning to the symbolism that is gradually developed by the student. The symbolism is based on real or imaginary actions (displacement of a car, growth of a plant, etc.) and appears as the expression of a certain continuous variation and not as the result of isolated measurements. The use of "mathematical narratives" appears to be a rich context for describing the way things change while linking this description to a certain form of symbolic expression, as the author clearly shows in the following extract:

> It is through the construction of mathematical narratives--our interpretation of "modeling"--that children's experiences with change, that is, with the different ways in which change occurs, become the subject of mathematical generalization. (Nemirovsky, Part IV of this volume)

The contribution of modeling to the introduction to algebra is also addressed by Janvier, who shows the opportunities offered by the approach in developing the student's sense of variable. The crucial point in this modeling process, according to Janvier, is the formulation phase that results in the creation of the model (symbolic expression, graph, table of values, etc.) on the basis of hypotheses. In this approach which, like the functional approaches, favors an introduction to algebra via the notion of variable, one must nevertheless be aware of the many hidden pitfalls that await the student. These difficulties are clearly shown by the author through an analysis of the meanings students give to symbolic representations and various notations on which this modeling is based. The distinctions he makes among formula, equation, unknown, variable, indeterminate value, and polyvalent noun in a symbolic expression help us to better situate the various approaches to algebra, reminding us that these distinctions cannot be determined a priori but that they are essentially linked to the student's activity.

This approach to algebra gives rise to a new field of research questions, which Heid points out in her commentary chapter. This research cannot be limited to clarifying what students think of the symbols and the way they use them. It involves above all knowing a little more about the way students create and use mathematical models and the way they manage the contradictory information related to the analysis of the frequently complex phenomena that are the objects of modeling.

Such questions apply as well to the so called functional approach to algebra, which also aims at constructing a certain meaning for the notion of variable, but does not necessarily subscribe to the same point of view as is found in the modeling approach to algebra.

5. A FUNCTIONAL PERSPECTIVE ON THE INTRODUCTION OF ALGEBRA

With the arrival of computers and the possibilities offered by recent technological developments, the concepts of variable and function are increasingly present in certain approaches used in introducing algebra at the secondary level. The projects that are described in this part of the book (Heid; Kieran, Boileau, & Garançon) are examples of this. However, the introductory approaches in both cases refer to rather different conceptions of what is meant by a functional approach.

In the first (Heid), *Computer Intensive Algebra* is a curricular development that emphasizes the exploration of functions in "real world situations". These real world settings are in some ways the initial pretext motivating the introduction of an entire family of functions (linear, quadratic, exponential, etc.) and the study of their properties. This is done through the elaboration and interpretation of certain mathematical models (graphs, tables, equations, inequalities, etc.) that allow one to deal with these real world situations.

In the second case (Kieran, Boileau, & Garançon), the introduction of algebra in the *CARAPACE* environment takes place in a problem-solving context and aims at finding, for a given problem, an expression for the general functional relationship between the various quantities in the problem. Contrary to the classical tendency in problem solving which emphasizes the setting up and solving of an equation, this approach aims at a much more general procedure for dealing with the problem, the solution being based on the previously established functional relationship.

In Heid, the approach requires an ability to deal with real world situations using several representations (graphs, numerical, symbolic) and thinking with these. In Kieran et al., the approach is based on numeric trials and requires the establishment of a general functional representation expressed in natural language. The resulting procedural representation then becomes the gateway to tables of values, Cartesian graphs, and symbolic expressions.

Several studies referred to briefly in Heid's chapter show the contributions of the *Computer Intensive Algebra* curriculum to students' understanding of functions and to the development of their problem-solving abilities and the manipulation of algebraic expressions. Kieran et al. describe six of their studies involving students ranging in age from 12 to 15 years. The findings of these six studies reveal the usefulness of both numeric trials and natural language as a bridge to developing a conceptual understanding of the processes used in representing and solving problems, and also point out some of the difficulties students meet in going from natural language to standard algebraic expression. These two research programs, each in its own way, demonstrate how students can evolve in their understanding of fundamental algebraic concepts without previously acquiring symbol manipulation skills.

Thus, these chapters help to situate the functional approach and the various concepts involved in it with respect to the other approaches analyzed in this book. The functional approach, Heid tells us, is not unlike the modeling approach in that it focuses on the elaboration and interpretation of certain mathematical models, those that relate to particular "real" situations. Her approach differs, however, in its focus on the introduction of particular families of functions as opposed to a modeling approach that is more open in its choice of problems and models. The functional approach developed in *CARAPACE* (Kieran et al.,) can be likened to a problem-solving approach to algebra in which a particular generalization is the goal (accounting for a general procedure).

In his commentary chapter, Nemirovsky questions the contribution of "realistic situations" to the learning of algebra. Here his analysis tries to get at the very notion of "real world problems" as presented by Heid (e.g., where is the reality in a real problem for the student? what makes it real? for whom?) and their appropriateness for the development of certain algebraic concepts (such as the integers and their operations, or the laws of exponents).

Regarding the functional approach described in the Kieran et al. chapter, Nemirovsky raises questions related to the nature of the graphical representation favored in the *CARAPACE* environment (a set of discrete points rather than a continuous curve). He suggests that such a representation, which features a point-wise transition to Cartesian graphs, may implicitly promote a concept of function that emphasizes the correspondence between pairs of elements. A "variational approach to functions," according to Nemirovsky, aims at developing a different notion of the concept of function, one that focuses on the relationship between quantities that vary. This fundamental distinction that Nemirovsky raises concerning the concepts of function and variable is not far from earlier concerns expressed by Janvier and Radford in this book: What kind of understanding of the concepts of variable and function can develop using a point-wise approach? Can it be an understanding of the variational nature of functions? Is there a need for both types of understanding and what are their interactions?

The reflections and discussions presented in this book do not pretend to deal once and for all with the many questions related to the introduction of algebra, as can be clearly understood from Wheeler's concluding chapter to this book. Rather they are an invitation to all those interested in the learning of algebra to examine the foundations and the effects of various educational practices that attempt to give meaning to this learning. This book opens new avenues both for reflection and action and as such constitutes an illustration of the richness of current work in this area.

NOTES

1 See, for example, the following studies on: the meanings students ascribe to algebraic symbols (Bednarz & Janvier, 1995; Booth, 1984; Küchemann, 1981), poor strategic decisions (Lee & Wheeler, 1989; Margolinas, 1991), the continued relationship with algebra (Schmidt, 1994).

2 For analyses of various student difficulties, the reader might refer to the following works on: equation solving (Filloy & Rojano, 1984a, 1989; Herscovics & Linchevski, 1991; Kieran, 1989, 1992), the manipulation of algebraic expressions (Booth, 1984; Sleeman, 1986), problem solving (Clement, 1982; Kaput & Sims-Knight, 1983; Lochhead & Mestre, 1988), the handling of variables (Küchemann, 1981).

PART I

HISTORICAL PERSPECTIVES IN THE DEVELOPMENT OF ALGEBRA

CHAPTER 2

FROM EUCLID TO DESCARTES: ALGEBRA AND ITS RELATION TO GEOMETRY

LOUIS CHARBONNEAU

Algebra is not solely the product of the evolution of arithmetic. It owes much to geometry. Some algebraic reasoning is present in Greek geometry. Geometric analysis, as well as the theory of proportions, played an important role in the development of algebra in the Renaissance. Until Viète's algebraic revolution at the end of the 16th century, geometry was a means to prove algebraic rules, and, likewise, algebra was a means to solve some geometrical problems. In this chapter, I discuss some of the relations which, from Euclid to Descartes, bound algebra to geometry.

1. INTRODUCTION

It is difficult to characterize algebraic thinking. The historian, Michael Mahoney, who knows well the history of algebra from 1550 to 1670, has proposed that an algebraic way of thinking involves:[1]

> 1) Operational symbolism. 2) The preoccupation with mathematical relations rather than with mathematical objects, which relations determine the structures constituting the subject-matter of modern algebra. The algebraic mode of thinking is based, then, on relational rather than on predicate logic. 3) Freedom from any ontological questions and commitments and, connected with this, abstractness rather than intuitiveness.

Of course, this is a point of view based on our modern view of algebra. But as we will see, it applies also to algebra of the 17th century. This is why I will use it as a reference to situate the different examples that I will use in my chapter.

Let's begin at the end. Since Descartes is the most recent mathematician whom I am going to discuss, here is a quotation from *La Géométrie* published in 1637:[2]

> Any problem in geometry can easily be reduced to such terms that a knowledge of the lengths of certain straight lines is sufficient for its construction. . .
>
> Finally, so that we may be sure to remember the names of these lines, a separate list should always be made as often as names are assigned or changed, for example, we may write, AB = 1, that is AB is equal to 1; GH = a, BD = b, and so on.
>
> If, then, we wish to solve any problem, we first suppose the solution already effected, and give names to all the lines that seem needful for its construction--to those that are unknown as well as to those that are known. Then, making no distinction between known and unknown lines, we must unravel the difficulty in any way that shows most naturally the relations between these lines, until we find it possible to express a single quantity in two ways. This will constitute an equation, since the terms of one of these two expressions are together equal to the terms of the other.

In this quotation, Descartes describes his method, now called analytical geometry, to solve geometrical problems with the help of algebra. Descartes' algebra

N. Bednarz et al. (eds.), Approaches to Algebra, 15-37.

has some of the characteristics found in Mahoney's description of algebraic thinking. The symbolism is central for Descartes. He represents all significant lines by letters. The importance of "mathematical relations" may best be seen in the last paragraph of the excerpt. There he says explicitly that one has to show "the relations between lines." The relations he has in mind are those which lead to equations, the privileged relation of algebra. But the importance of relations goes further. Descartes begins his paragraph by stating that "we suppose the solution [of the problem] already effected." Therefore, he puts himself clearly within the analytical program being unfolded at that time. (We will come back to this question of the analytical approach further on in this chapter). It is because of that program that relations are of prime importance. "Making no distinction between known and unknown lines" indicates clearly that it is just the relations between lines which matter, not the status of lines. Those relations may take different forms: equation and proportion being the main forms of relation. Of course, the "mathematical relations" Mahoney has in mind are different in terms of generality from Descartes'. Nevertheless, by putting the focus on relation and away from the usual distinction between known and unknown, Descartes gives a special imprint to his algebra, leading eventually to the algebra of structures. As for "Freedom from any ontological questions and commitments," does it apply to Descartes' algebra? In the context of what we now call analytical geometry, letters represent "lines." Those lines are not the actual lines, but, as Descartes indicates implicitly in the second paragraph, they represent the measures of geometric lines. Therefore, his algebra is an algebra based on the measure of geometrical magnitudes and relations between these measures. In that sense, there are "ontological questions and commitments," but they are circumscribed.

The above discussion of Descartes' algebra shows the importance of certain elements in the evolution of algebraic thinking: 1) analytical thinking and 2) a theory of measure. We will see that other elements are also of prime importance in the history of algebra: 3) the theory of proportion and, foremost, 4) the need for algebra to be a homogeneous theory, that is, a theory self-sufficient in terms of proof.

This chapter is divided into four parts:

- Algebraic Reasoning in Greek Geometry.
- Geometric Proofs of Algebraic Rules.
- Algebraic Solution of Geometric Problems.
- Toward a Homogeneous Algebra.

Some didactic remarks inspired by these historical considerations conclude the chapter.

2. ALGEBRAIC REASONING IN GREEK GEOMETRY

As we know and as we have seen with Descartes, equality is of prime importance in algebra. Equality is not restricted to algebra. It is a relation which one encounters over all mathematics, among others in geometry.

2.1. *Equality in Greek Geometry*

2.1.1. *Equality of magnitude*

2.1.1.1. *Euclid's Common Notions.* At the beginning of the first book of the *Elements,* following the definitions and the postulates, Euclid gives five Common Notions of which the first three are:[3]

1) Things which are equal to the same thing are also equal to one another.
2) If equals be added to equals, the wholes are equal.
3) If equals be subtracted from equals, the remainders are equal.

The three first Common Notions are very familiar to algebraists. The first one is the transitivity of equality. Common Notions 2 and 3 are fundamental to any manipulations of equations. Eleven centuries after Euclid, Al-Khwarizmi names the second one al-jabr, meaning complement, and the third one al-muqabala, meaning balancing.[4]

Euclid uses those Common Notions in an almost algebraic way. For example, the end of the proof of Proposition III-35 is characteristic.[5] The reasoning is straightforward. The presentation here is modified in order to emphasize the similarities with standard algebraic reasoning. The letters $a, a', b, b', c, \ldots, g, h$ are actual segments. Euclid uses rather the standard notation AB, CD, ... corresponding to the letter designating the extremities of the segments. Of course, there is no symbolism used to denote operations and equalities. The reasoning is described by Euclid in full sentences.

Euclid's *Elements* (Prop. III-35):

If in a circle two straight lines cut one another, the rectangle contained by the segments of the one is equal to the rectangle contained by the segments of the other.

Proof (using modern notation, the end of the argument proceeds as follows):
$ab + c^2 = d^2$.
Let the square of e be added: $ab + c^2 + e^2 = d^2 + e^2$,
but $c^2 + e^2 = f^2$ and $d^2 + e^2 = g^2$,
therefore $ab + f^2 = g^2$,
and $g^2 = h^2$,
therefore $ab + f^2 = h^2$,
for the same reason $a'b' + f^2 = h^2$,
but we also proved $ab + f^2 = h^2$,
therefore $a'b' + f^2 = ab + f^2$.
Let the square on f be subtracted from each: $a'b' = ab$.
Q.E.D.

This kind of reasoning is common in primitive trigonometry. For example, Ptolemy uses it often, in Book I, section 10, of his *Almagest,* where he shows how he calculated his table of chords.[6]

2.1.1.2. *Basic algebraic identities (*Euclid's *Elements Book II).* Among historians, it is a general belief that Book II of Euclid's *Elements* has an algebraic tone.[7] The first 10 propositions, out of 14, may be seen as algebraic identities proved geometrically. They were interpreted by historians as tools to eventually solve geometrical problems equivalent to algebraic problems. The first four are quite direct. In modern notation, one may write them as follows:[8]

1) $a(b + c + d + \dots) = ab + ac + ad + \dots,$
2) $(a + b)a + (a + b)b = (a + b)^2,$
3) $(a + b)a = ab + a^2,$
4) $(a + b)^2 = a^2 + b^2 + 2ab.$

The translation into modern notation of the other identities is not as straightforward. Proposition II-5 is enunciated as follows:[9]

If a straight line be cut into equal and unequal segments, the rectangle contained by the unequal segments of the whole together with the square on the straight line between the points of section is equal to the square on the half.

Depending on the way you name the segments involved (see Figure 1), it may be translated at least into the following identities:

5a) $ab + \left(\frac{a+b}{2} - b\right)^2 = \left(\frac{a+b}{2}\right)^2$ [a+b : the "straight line"; a and b : "the two unequal segments"],

5b) $(\alpha + \beta)(\alpha - \beta) + \beta^2 = \alpha^2$ [α : half of the "straight line"; β : "the straight line between the points of section"].

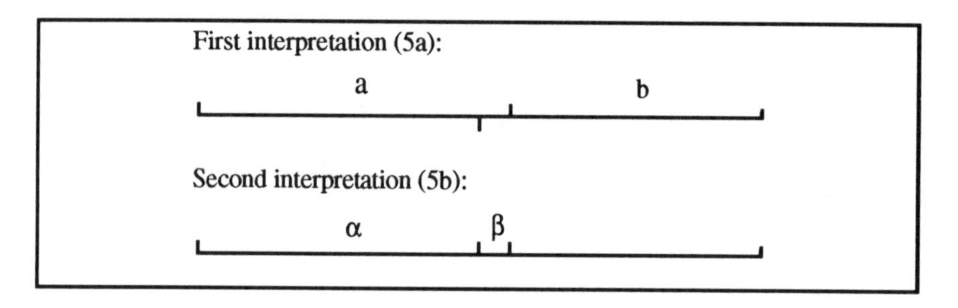

Figure 1.

These identities are general, that is, they are always true, whichever two segments you take. A lot more can be said about those identities.[10] At the least, they certainly respect Descartes' requirement that the same thing be exhibited in two different forms.

The interpretation of Book II as an algebraic book must be tempered by the fact that there are only two propositions that correspond to the resolution of an equation:[11]

Proposition 11: To cut a given straight line so that the rectangle contained by the whole and one of the segments is equal to the square on the remaining segment.

Proposition 14: To construct a square equal to a given rectilineal figure.

Proposition 11 corresponds to the equation $a(a - x) = x^2$ and Proposition 14 to the equation x^2 = A where A is the area of a plane figure. The solution given by Euclid uses Propositions II-6 (for II-11) and II-5 (for II-14); but of course it is a geometrical one. This is why the equation x^2 = A is not exactly equivalent to Proposition II-14. The geometrical problem asks for the construction of a square of a certain area. However, area in Euclid must not be understood as the measure of a plane figure, but simply as a magnitude. Also, the A in the equation above should be seen as a representation of a magnitude of a different nature than x. A is an area and x is a segment. In a geometrical context, one may say that two magnitudes may be equal only if the two magnitudes are of the same nature, in particular of the same dimension.

2.1.1.3. *Conclusion.* The use of equalities in Euclid's *Elements* is characterized by a non-numerical environment. Magnitudes are compared, but not their measures. There are operations on those magnitudes, operations like "putting together" and "obtaining a rectangle with two given sides." We may then speak of an "arithmetic" of magnitudes by analogy to the arithmetic of number.

This "arithmetic" of magnitudes must not be confused with the arithmetic of numbers. For example, there is an important difference between equality of magnitudes and equality of numbers. Numbers have standard representations--the symbols used in forming a system of written numeration, and their names in forming a system of oral numeration. Also, in reasoning involving numbers, the numbers change and the memory of them is lost when an operation transforms two numbers into one which will be used in the rest of the reasoning because the new number, and its representation, has a meaning by itself. Magnitudes do not have such standard representations. When a new magnitude comes from an operation on two magnitudes, the new magnitude has a meaning only in relation with those from which it comes. Even if one represents it by a symbol (like AC for a segment), the symbol does not by itself carry any meaning. To reason with segments, one must continuously refer to the other segments. Compared with the use of numeric symbols, it is more demanding; but in the same way, the need of a general argument is inescapable.

2.1.2. *Equality of ratios: the Eudoxian theory of proportions*

There is another kind of equality which is used heavily by Euclid: the equalities between ratios, in other words, proportions. But the meaning of this equality is more delicate than that of equality of magnitudes.

The definition of ratio, in Book V (Definition 3) of Euclid's *Elements*, gives a clue as to why this is so:[12]

A *ratio* is a sort of relation in respect to size between two magnitudes of the same kind.

It is clearly difficult to say that two ratios are equal if one doesn't know what a ratio is. Therefore, the basic definition is not the above, but Definition 5 which gives a criterion for proving that two ratios are actually equal:[13]

Magnitudes are said to *be in the same ratio*, the first to the second and the third to the fourth, when, if any equimultiples whatever be taken of the first and third, and any equimultiples whatever of the second and fourth, the former equimultiples alike exceed, are alike equal to, or alike fall short of, the latter equimultiples respectively taken in corresponding order.

It is a quite complicated criterion. It means, using modern notation:

If a, b, c, and d are magnitudes (a and b of the same kind of magnitude, and c and d of the same kind of magnitude, but of a different kind than that of a and b), a, b and c, d are of the same ratio , that is, $\frac{a}{b} = \frac{c}{d}$, if and only if

$$\forall\, n, m \in \mathbf{N}, \qquad \begin{aligned} ma > nb &\Rightarrow mc > nd; \\ ma = nb &\Rightarrow mc = nd; \\ ma < nb &\Rightarrow mc < nd. \end{aligned}$$

Here the notion of equality is quite different from the notion of equality between two geometrical figures. This latter equality is formulated in Common Notion 4 of Book I of Euclid's *Elements* :[14]

Things which coincide with one another are equal to one another.

Heath, in a note on this formulation, emphasizes the geometrical nature of this criterion.[15] It corresponds to the intuitive meaning of the equality of two geometrical figures. The equality of two figures is intimately related to the geometrical nature of those figures. This kind of semantic formulation cannot be used with the equality of ratios since ratio is so vaguely defined. It is impossible to graphically illustrate a ratio. As stated in Definition 3 above, a ratio is a relation. Therefore, everything is in our mind. It does not have a natural non symbolic representation. Also, to define a proportion, one has to refer not only to the magnitudes involved, but also to the relation between them. One may say that the equality of ratios is at a second level of abstractness compared with the equality of magnitudes. Reasoning at such a second level is difficult, especially if the magnitudes compared in both ratios are incommensurable.[16] Therefore, a technical criterion must be used here, a criterion using first level relations: to be equal to, to be larger than, to be smaller than. The Euclidian (I shall say Eudoxian) theory of proportions is then less ontological than the geometry of figures. By giving central importance to the criterion of proportionality instead of the meaning of equality, the theory of proportions seems to be the first axiomatic theory which is not based on ontological definitions. It also comes close to fitting the second and third characteristics of algebra given by Mahoney.

However, can we say that the theory of proportion is an algebra? There is a very primitive symbolism, that is, letters to represent magnitudes, like A or B. But the operational apparatus is very weak. There is no addition or multiplication of ratios. The theory of proportions is not an "arithmetic" of proportions. We are far from high school algebra.

There is nevertheless a problem which has an algebraic flavor: To find a fourth proportional, that is, knowing three magnitudes a, b, of the same kind, and c, which may be of a different kind, to find d such that $\frac{a}{b} = \frac{c}{d}$. In the context of the theory of proportions, this is not a simple problem. For example, since a, b and c, d are not necessarily the same kind of magnitudes, one cannot simply use something like $ad = bc$, the products ad and bc being not necessarily defined.[17]

2.1.3. *Apollonius and the Conics*[18]

A special use of equality of magnitudes can be found in Apollonius' *Conics*. It is in some way similar to the use of equations in analytical geometry, but it is also basically different. For example, to characterize a parabola, Apollonius uses a relation that corresponds in modern notation[19] to $x^2 = k \cdot y$ where k is the fourth proportional of A, B and C, and where A, B are two rectangles and C is a segment. C and the sides of those two rectangles are segments which characterize the cone supporting the parabola and the specific location of the plane cutting the cone. Another relation, restricted to the plane containing the parabola, is called the symptom of the parabola: $\frac{x^2}{X^2} = \frac{y}{Y}$ where x, X, y, and Y are as in Figure 2. This relation may be easily derived from the first relation.

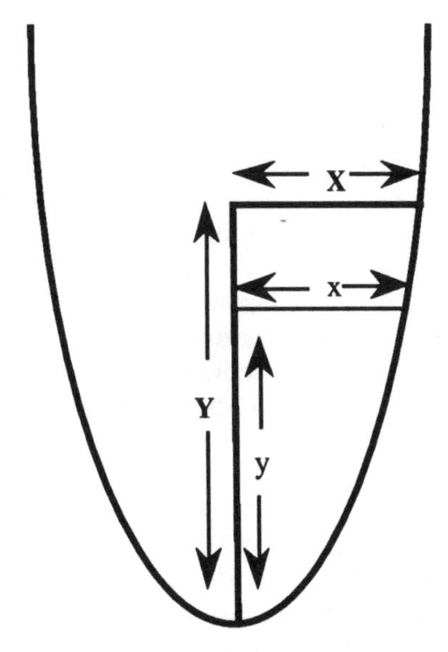

Figure 2.

We may be tempted to see here a Greek non symbolic form of analytic geometry, but the basic relation and the corresponding symptom of the ellipse will dissuade us from making such a premature generalization. The relation which characterizes the ellipse is: $x^2 = y \cdot y' \left(\frac{C \cdot D}{E^2}\right)$ where y' is the complement of y to the axis in y (see Figure 3), and C, D, E are segments characteristic of the cone supporting the ellipse and of the specific location of the plane cutting the cone. The symptom is then

$$\frac{x^2}{X^2} = \frac{y \cdot y'}{Y \cdot Y'}$$

where x, X, y, y', Y, and Y' are as in Figure 3.

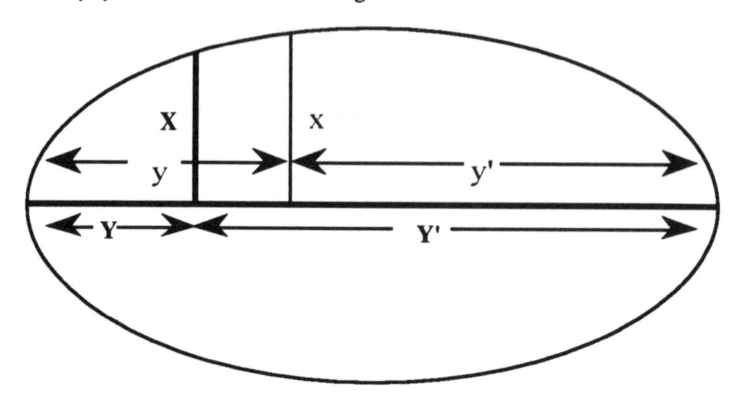

Figure 3.

In a symptom, the equality is the equality of ratios because the characteristic relation, which takes into account only the curve itself and not the cone from which it is derived, is a proportion. The symptom refers also to more than the two classical variables now denoted x and y. "Greek geometry did not seek to reduce the number of unknown quantities or lines in a figure to one or two, as is done in astronomy or in plane Cartesian coordinates, but they sought instead to make as simple as possible the relationships in terms of areas. ... Of Greek geometry one may say that equations are determined by curves, but not that curves are defined by equations."[20] The absence of numbers also pushes toward the use of proportions instead of equalities of magnitudes. It emphasizes the importance of relations, but an algebraic way of thinking is not really present in Apollonius.

2.1.4. *Conclusion*

The use of equalities in Greek geometry has two faces: equalities of magnitudes and equalities of ratios, that is, proportions. Two magnitudes which are equal must be of the same kind: either lines, areas, or volumes. For ratios, however, it is important to notice that this is only partially true. Equality of ratios must, of course, be considered between two ratios, but those ratios could be, on the one hand, a ratio of areas and, on the other hand, a ratio of segments. The ratio goes beyond the nature of the magnitudes compared. The level of abstractness is greater than that of magnitudes particularly if one takes into account that the theory of proportions was put together in the aftermath of the discovery of incommensurable magnitudes.

The difficulty in dealing with ratios may be seen by the fact that Euclid, while giving in his Common Notions the general properties of equality for magnitudes in terms of basic operations, does not introduce similar operations with ratios. By moving to a higher level of abstractness, one focuses more on relations and thus develops a theory which is less ontologically oriented. To ease the manipulation of ratios and proportions, those difficulties will have to be overcome--eventually by extending the notion of measure.

2.2. *Analysis (as Opposed to Synthesis)*

Another element of Greek geometry that plays an important role in algebra is analysis. What is analysis? Here is Pappus' definition:[21]

> Now analysis is the passage from the thing sought, as if it were admitted, through the things which follow in order [from it], to something admitted as the result of synthesis.

By "result of synthesis," Pappus means a proposition which has been proved or which is true. Therefore, analysis may be represented by the following chain of implications, going from the unknown to the known:

$$P_{solved} \implies P_n \implies \ldots \implies P_{true},$$

where P_{solved} is the proposition corresponding to the theorem or the problem to be solved, and P_{true} is the proposition "result of synthesis."

On the other hand, synthesis reverses this path. To prove P_{solved}, one thus goes from the known to the unknown:

$$P_{true} \implies P_1 \implies \ldots \implies P_{solved}.$$

There are two steps involved in solving a geometrical problem analytically. First, the problem is purported to be solved. In other words, the relations between the different objects involved in the problems are considered to be true. This consideration constitutes a hypothesis. Second, this hypothesis having been formulated, one searches for objects which can be constructed or can be arrived at from the objects which are known in the problem. If one of these objets can be constructed, then, from it, one can construct the object which is sought.

Proposition 105 from Book VII of Pappus' *Mathematical Collection* will illustrate this process. Only the skeleton of Pappus' argument without all the justifications is given here. It is a more dynamic presentation than the one Pappus provided. The terminology for the different parts is not actually used by Pappus, but was added by Hankel.[22] It corresponds nevertheless to the general pattern of the use of analysis by the Greeks in the resolution of geometrical problems.

> Proposition 105: Given a circle ABC and two points D, E external to it, to draw straight lines DB, EB from D, E to a point B on the circle such that, if DB, EB produced meet the circle again in C, A, AC shall be parallel to DE.

Analysis:

Hypothesis: Suppose the problem solved, i.e. one has AC parallel to DE, and the tangent at A drawn, meeting ED produced in F (see Figure 4a).

Transformation: Using the hypothesis, one sees that the angles FAB, ACB and BDE are equal and thus that A, B, D, F are concyclic. Therefore (by a generalization of Proposition III-35 of Euclid's *Elements* discussed in Section 2.1.1)

$$EB \cdot EA = ED \cdot EF$$ (see Figure 4b).

The question is then to find a way to construct a point F which satisfies this last relation, by using only what the problem considers as given, that is, without using the fact that FA is tangent to the circle and that AC is parallel to DE. This is done in the second part of analysis, the resolution.

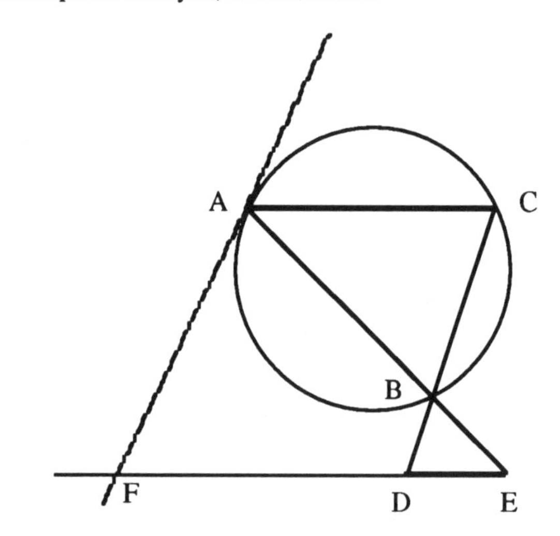

Figure 4a.

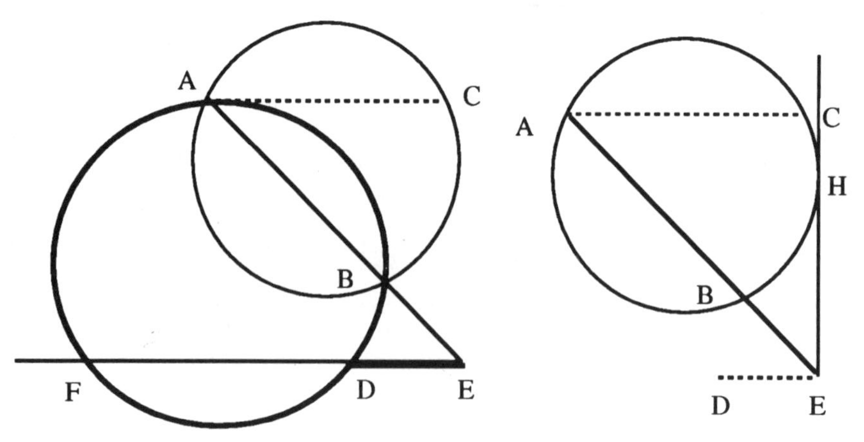

Figure 4b.

Figure 4c.

Resolution: One notices that the "rectangle" (i.e., the product) EB•EA is equal to the square on EH, where H is the point on the circle of the tangent to the circle from E (Figure 4c). Therefore

$$EH \bullet EH = EB \bullet EA = ED \bullet EF.$$

Since ED is given in the problem, in position and in magnitude, one can find EF in position and magnitude. Knowing EF, F may also be found. F is then given.

One may then find A because FA is tangent to the circle.

To establish the truth of the construction found by the analysis, one uses the synthesis. The construction, the first part of the synthesis, comes from the resolution of the analysis. This construction is followed by the demonstration, which corresponds to what one searches for in the problem. It is the reverse of the transformation of the analysis.

Synthesis:

Construction: Take a rectangle contained by ED and by a certain straight line EF equal to the square on the tangent to the circle from E.

From F draw FA touching the circle in A; join ABE and then DB, producing DB to meet the circle at C. Join AC. (Figure 4a)

I say then that AC is parallel to DE.

Demonstration: By construction one has: EH•EH = EB•EA = ED•EF. Therefore A B, D, F are on a circle and then the angles FAB and BDE are equal. But the angles FAE and ACB are equal. Therefore the angles ACB and BDE are equal. Therefore AC is parallel to DE.

Q.E.D.

Solving a problem in words with high school algebra is an analytical process. Writing the equation which corresponds to the problem is actually supposing the problem solved. Finding this equation corresponds to the transformation of analysis in Pappus' discussion above. Solving the equation corresponds to the resolution of the analysis. The synthesis is then simply to replace the unknown by the value found and to verify that one has the equality. It is the demonstration stage of the synthesis. The construction corresponds to showing that the equation is actually a good representation of the problem.

We see that the second part of analysis involves relations between objects. This is the essence of the second characteristic of Mahoney's characterization of algebraic thinking. Furthermore, step one of analysis being an hypothesis, it imposes the development of a certain way of representing the unknown magnitudes considered given in this hypothesis. The need for a form of representation is at the heart of analysis. A figure may play this role in geometrical problems, but in numerical problems, a language of the unknown has to be crafted. As we will notice later, discussions of geometrical analysis by Greek commentators will play an important role in the evolution of Renaissance algebra.

2.3. *Summary*

In Greek geometry, there is an arithmetic of magnitudes in which, in terms of numbers, only integers are involved. This theory of measure is limited to exact measure. Operations on magnitudes cannot be actually numerically calculated, except if those magnitudes are exactly measured by a certain unit. The theory of proportions does not have access to such operations. It cannot be seen as an "arithmetic" of ratios.

Even if Euclidean geometry is done in a highly theoretical context, its axioms are essentially semantic. This is contrary to Mahoney's second characteristic. This cannot be said of the theory of proportions, which is less semantic.

Only synthetic proofs are considered rigorous in Greek geometry. Arithmetic reasoning is also synthetic, going from the known to the unknown.

Finally, analysis is an approach to geometrical problems that has some algebraic characteristics and involves a method for solving problems that is different from the arithmetical approach.

3. GEOMETRIC PROOFS OF ALGEBRAIC RULES

Until the second half of the 19th century, Euclid's *Elements* was considered a model of a mathematical theory. This may be one reason why geometry was used by algebraists as a tool to demonstrate the accuracy of rules otherwise given as numerical algorithms. It may also be that geometry was one way to represent general reasoning without involving specific magnitudes. To go a bit deeper into this, here are three geometric proofs of algebraic rules, the first by Al-Khwarizmi, the other two by Cardano.

3.1. *The Geometric Puzzle of Al-Khwarizmi ($x^2 + 10x = 39$)*

The rule given in words by Al-Khwarizmi (circa 825 A.D.) to solve the equivalent to the equation $x^2 + 10x = 39$ corresponds to the formula:

$$x = \sqrt{\left(\frac{10}{2}\right)^2 + 39} - \frac{10}{2}$$

To prove that this algorithm will always give the (positive) solution of this type of equation, Al-Khwarizmi uses a well known argument, based on the interpretation of the two sides of the equation as areas of rectangles. The left side corresponds to a rectangle formed by a square whose side is x and by a rectangle of base 10 and height x. The right side says that the whole rectangle has an area of 39. Seeing the whole rectangle more or less as a puzzle, Al-Khwarizmi divides the rectangle $10x$ into four smaller rectangles of area $\frac{10}{4}x$ and places them on the four sides of the square x^2 (see Figure 5). The new figure, which has the shape of a cross, still has an area of 39. By completing it to form a larger square, one gets a square with an area of $39 + 4\left(\frac{10}{4}\right)^2$, with the side being $x + 2 \cdot \frac{10}{4}$ or $x + \frac{10}{2}$. The formula above is readily deduced from there.

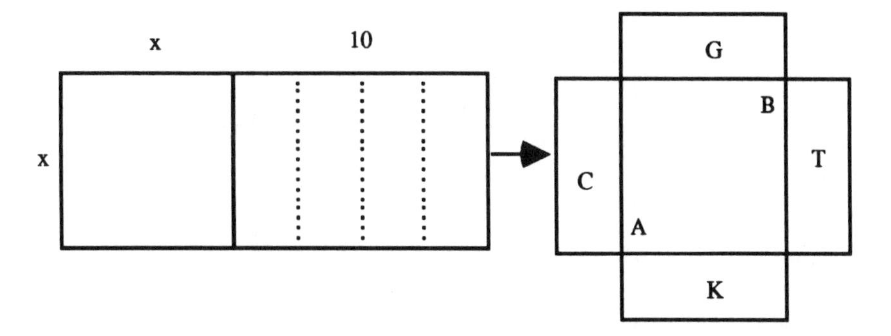

Figure 5. (Al-Khwarizmi draws only the figure on the right)

The proof begins: "The figure to explain this, the sides of which are unknown. It represents the square, the which, or the root of which, you wish to know. This is the Figure AB, each side of which may be considered as one of its roots" (Al-Khwarizmi, 1831, p. 13). Later, he uses the hypothesis that the square AB plus the four rectangles C, G, K, and T are together equal to 39. (This is Al-Khwarizmi's notation, the only difference being that A, B, C, G, T, K are written in Arabic.) In this, the proof of the mathematician of Baghdad has a clear analytical flavor since it is supposed that the equation is satisfied. Therefore this is not a Euclidean proof. There is also another difference with respect to Greek proofs: Numbers are involved. It identifies what is measured (segments, squares and rectangles) with the measure itself (10, x, 39, etc.). Nevertheless, all operations are geometric. Only geometric magnitudes are involved in the reasoning. The measures of those magnitudes are used only to actually calculate. In that sense, one may say the proof is non-numerical and within the "arithmetic" of magnitudes. The generality of the proof proceeds from the generality of this "arithmetic."

3.2. *Cardano's Resolution of* $x^2 = ax + cxy$

In Cardano's *Ars Magna* (1545/1968), Chapter X, one of the problems is, in modern notation: "Find y if x is given and if $x^2 = ax + cxy$."[23]

The rule Cardano gives as a solution of the equation is a rhetorical form of the modern expression $y = (x - a)/c$. He proves it by using Figure 6. In this figure: x = BC = AB, a = DC, y = BF and c is the exact number of times BF goes into BD.

The reasoning is as follows. The square of x (AC) is formed by the rectangle ax (EC) and the rectangle cyx (AD). Then cy (BD) equals x (BC) minus a (DC). Therefore y (BF) is $x - a$ divided by c (c being the number of times y is contained in cy).

As in the case of Al-Khwarizmi's proof, Cardano's proof also has an analytical flavor. It is based on the known fact that two rectangles with the same area and the same base have the same height.

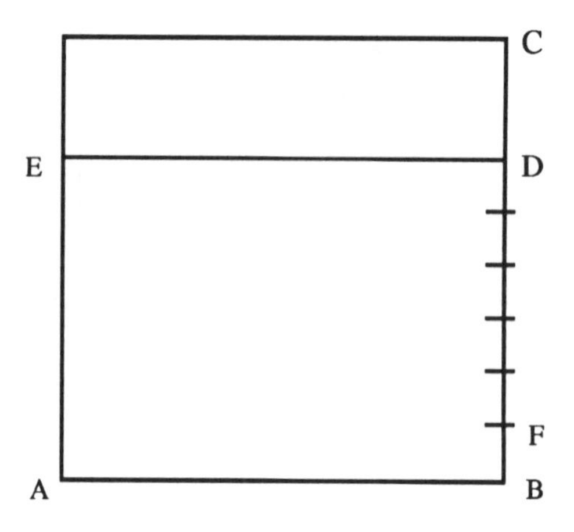

Figure 6.

But the main feature of this proof is the presence of two kinds of multiplication: multiplication as an area and multiplication as a repeated addition.[24] This double kind of multiplication limits the extent of application of the proof because c must be a natural number. But, on the other hand, it allows avoidance of the use of a three dimensional figure. A probably more important consequence of the existence of those two multiplications is that it also prevents the identification of the structure of the "arithmetic" of magnitudes with the structure of the arithmetic of numbers.

3.3. *Cardano's Resolution of* $x^3 + 6x = 20$

Another ambivalence of the relation of a geometric proof with the algebraic rule it proves may be seen in the proof of Cardano's formula for the equation $x^3 + 6x = 20$. This proof,[25] contrary to the two previous examples, is synthetic. It is a proof with a geometrical outlook, but done in a way that is not really geometrical. Cardano bases his reasoning on the identity

$$(u + t)^3 = u^3 + 3tu^2 + 3ut^2 + t^3$$

and on Figure 7 (where I stands for u^3, II stands for $3ut^2$, III stands for $3tu^2$, IV stands for t^3).

It is interesting to see him use this figure only to stabilize the representation of the formula for $(u + t)^3$ which would normally need a three dimensional figure. The figure only reminds the reader that there are four monomials in the development of $(u + t)^3$. Later in the proof, he considers, in an even less geometrical manner, the development of $(a + b)^3$ when b is negative.

This proof once again brings out an ambiguity of the relation between the proof and the proven. But this time, the ambiguity is more in the use of a figure than in the nature of the reasoning. The proof is still under a geometrical influence, but it is not really geometrical. It is not arithmetical either.

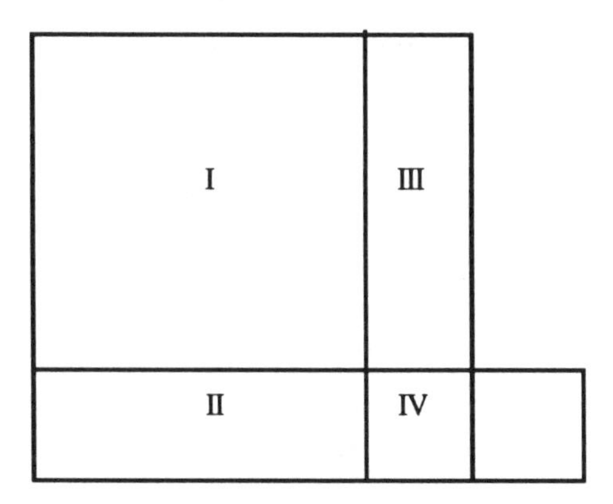

Figure 7. (This figure differs from Cardano's only in the use of I, II, III, IV to identify the quadrilaterals. Cardano uses letters as in Figure 6.)

3.4. *Conclusion*

In the three examples that have just been briefly discussed, the proofs and the content of the rules proven are at different levels, heterogeneous one to the other. The former has a geometrical support and is non-numerical. The latter is seen as purely numerical. The relation between the proof and the rule is based on the assumption that there is a way to measure magnitudes. This theory of measure is intuitive and not clearly defined.

The geometric proof is based on the "arithmetic" of magnitudes and derived its generality from the generality of the reasoning conducted within this "arithmetic." No numbers participate in this generality. The geometric reasoning is general. The precise length of the segments is not important. Two figures are considered to be the same figure if one figure is proportional to the other. It is the relations between different lines which are significant. This recalls the second characteristic of algebra given by Mahoney. But, as seen in the last proof, magnitudes may not always have geometric representations. The use by Cardano of a figure which he doesn't really need and which is not a good geometrical representation indicates that this figure is useful because it probably stabilizes the mental representation of magnitudes. The fact that the reasoning is still general shows that the generality of the "arithmetic" of magnitudes goes further than the generality of geometry.

Geometrical proofs of algebraic rules seem to be different in their nature from purely Euclidean proofs. Euclidean proofs are synthetic, while the others are usually analytical. Those analytical proofs are successions of equivalent propositions. They correspond to a process of exploration. Synthetic proofs need only implications to go from one step to the next. They are unidirectional.

Restrictions due to dimensional considerations limit the generality of geometrical proof. A figure cannot easily represent a situation of dimension higher than 3. As we saw in our second and third examples, a figure of lower dimension than what the problem *a priori* suggests can be used, but at a price.

As for the rules, they are purely numerical. The restriction just mentioned does not apply in a numerical context. But, as we have seen, there is no numerical reasoning.

4. ALGEBRAIC SOLUTION OF GEOMETRIC PROBLEMS

The intuitive theory of measure which allows the use of geometrical arguments to prove numerical rules was very often used to numerically solve geometrical problems. I will limit myself to two examples from the second half of the 15th century.[26]

4.1. *Chuquet*

In 1484, Nicolas Chuquet writes a large and important treatise on arithmetic and on the "rule of first terms," that is, algebra, entitled *La triparty en la science des nombres*. To this manuscript is attached another treatise, *La Géométrie,* in which Chuquet solves many geometrical problems in the tradition of Italian algebraists.[27] The following is one of his classic problems:[28]

Three (equal) circles are included and contiguously enclosed in the circumference of another which has the diameter of 12, which is *ab*. To determine, what are the diameters of each of these contained circles.

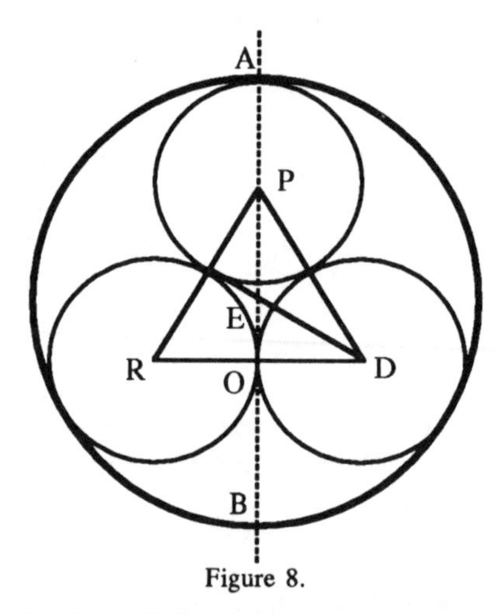

Figure 8.

Chuquet's reasoning is as follows, using our modern notation. Let E be the center of the large circle. The unknown x is chosen as half the side of the triangle formed by the centers of the three inscribed circles. It corresponds to DO or RO in Figure 8. By the Pythagorean theorem, PO is $\sqrt{3}\,x$. Naming E the intersection of the medians, EO is a third of PO, that is, $\frac{\sqrt{3}}{3}x$ or $\frac{x}{\sqrt{3}}$. Using again the Pythagorean theorem, one finds that ED is the square root of $\left(\frac{x}{\sqrt{3}}\right)^2 + x^2$. On the other hand, the radius of the large circle is ED + OD.

Therefore, the equation is

$$\sqrt{\left(\frac{x}{\sqrt{3}}\right)^2 + x^2} + x = \frac{12}{2}.$$

which is equivalent to $\frac{x^2}{3} + 12x = 36$. This is a standard equation with a well known rule for solving it.

The solution is not primarily geometrical, but essentially numerical. As for the geometric proofs of numerical rules, the context is heterogeneous. Here a geometric problem is solved numerically. Actually, the problem is seen from the beginning as a numerical problem. The difficulty is to put forward relations between segments "until we find it possible to express a single quantity in two ways" as Descartes says. It supposes, from the start, an identification between magnitudes and their measures. But this identification is not well defined. It seems to be a matter of course. But, what does the product $\sqrt{3}x$ mean? $\sqrt{3}$ does not correspond to a segment and $\sqrt{3}x$ is then not an area. Can one say that $\sqrt{3}x$ is $\sqrt{3}$ times the measure x of a segment? It is the status of number and its relation to a segment which is at stake here. Regiomontanus, the astronomer, will tackle this question.

4.2. *Regiomontanus*

Regiomontanus' *On the Triangle* (written circa 1464 but published only in 1533) sheds a different light on the relation between geometry and the developing algebra.[29] This book is the first European book in which trigonometry is considered as a branch of mathematics, independent of astronomy. It is a landmark. The problems in this book are of finding the sides of a triangle when some information is given. Algebra (*ars rei et census,* as he says) is not central to this book. Regiomontanus uses it only twice, in Book II, theorems 12 and 23. More interesting for us are his efforts to clarify the relation between numbers (whole numbers) and the measure of a segment. His treatise begins with a definition:[30]

A *quantity* is considered *known* if it is measured by a known or arbitrarily assigned measure a known number of times. One quantity is said to *measure* another when the former is contained in the latter a known number of times or when the former is found in the latter quantity as often as the unit value is found in that known number.

For Regiomontanus geometrical problems are transformed into numerical problems. He is therefore able to remove an ambiguity in the meaning of the operations. For example, multiplication is well defined.

Multiplication of a number produces another number which contains the multiplicand as often as the multiplier contains unity.[31]

The restriction to whole numbers has to be dealt with. Not all segments can be represented solely by whole numbers.

However, since it often happens that the numbers we use to measure our squares are not squares yet we certainly know (as human affairs are knowable) a close approximation, then hereafter we shall use the term "of known quantity" more loosely than we defined it at the start. Therefore, we will consider any quantity to be known whether it is precisely known or is almost equal to a known quantity. I personally think it nicer to know the near-truth than to neglect it completely, for it is worthwhile not only to reach the goal but also to approach close to it. ... But, indeed, even if we are accustomed to take preciseness as the truth, yet enough things are accepted as definition by the reader which are barely approximations of the truth.[32]

This book is based on a theory of approximation. The differentiation between an "arithmetic" of magnitudes and the arithmetic of number is clearly seen. This is not surprising coming from an astronomer writing on trigonometry. It is another conscious step toward a more abstract and non-ontological approach to algebraic manipulation. It puts numbers center stage.

The second half of the 16th century and the beginning of the 17th century is the period when the use of decimal numbers begins to spread in practical and theoretical mathematics. John Napier publishes his first book on logarithms in 1614. Regiomontanus' book is very influential during that period. It stresses the need to go further than intuition in dealing with magnitudes and their measures.

5. TOWARD A HOMOGENEOUS ALGEBRA

We have seen in previous sections algebraic behavior in geometry, mixed behavior in geometrical proofs of numerical rules, and algebraic solutions of geometrical problems. Algebra is then always seen through a numerical and/or a geometrical lens.

When does this change? When does algebra become a branch of mathematics seen as autonomous and self-sufficient, homogeneous?

Viète and Descartes both push in this direction of having a homogeneous algebra, but in two somehow different ways.

5.1. *Viète and the Analytic Art*

In 1691, when he publishes his *Introduction to the Analytic Art,* Viète is the first mathematician to tackle, with great success, this problem of heterogeneity. His inspiration probably comes from the pedagogical principles developed by Peter Ramus, the sixteenth century humanist. For Ramus, knowledge should be organized in subjects which should be intrinsically homogeneous. Algebra, being dependent on arithmetic and on geometry, did not satisfy his principle.

For Viète, giving the algebra a real homogeneity meant extracting it from the realm of arithmetic and geometry, and giving it a status by itself. His first step, following Ramus' interpretation of algebra,[33] is to see those processes as analytical processes. He divides his analytic art into three parts. The first part is the Zetetics, which is a set of rules of manipulation of letters.[34] As Teresa Rojano points out in her chapter (Part I of this volume), those rules "create the systematic context that 'defines' the object to which these rules apply." The object of the Zetetics is to produce symbolic equalities, but not just any equalities, rather those which are in a sense equivalent to a proportion. Why this restriction? Simply because it is through the use of proportions that this analytical and symbolic process can be translated into geometry in such a way that the geometrical output is purely geometrical. The need for homogeneity is here omnipresent. The Zetetics, and its *logistica speciosa*, has its own aims, its own mode of demonstration. It is seen as complete by itself. It does not depend on arithmetic or geometry, yet it has to be useful in arithmetic and geometry. But the translation of zetetical results in arithmetical or geometrical terms is not part of the Zetetics; it belongs to the third part of analysis, the Exegetics.

The second part of the analytic art, the Poristics, is aimed at assuring that the Zetetics process leads to a conclusive translation. In other words, seen in an analytical perspective as opposed to a synthetical one, the Poristics makes sure that the chain of implications of analysis can be reversed so as to obtain a synthetic demonstration.[35]

The third part of the analytic art is precisely the translation of the Zetetical analysis into geometrical terms (the third part is then called Geometrical Exegetics) or arithmetical terms (the third part is then called Numerical Exegetics). As stated above, this translation is a way to homogenize algebra, and, as a matter of fact, geometry and arithmetic as well. The main tool of this translation in geometrical terms is the theory of proportions. It allows Viète to resolve geometrical problems without passing through a transformation of the geometrical problem into a numerical problem. But this solution will be synthetical and entirely geometrical. The analytical and symbolical process which led him to the geometrical solution will have vanished from the proposed geometrical solution.

Two last remarks on Viète. First, the central role of proportions has a price: All equations must be homogeneous, that is, all monomials have the same degree. Therefore, his algebra is not far from being simply numerical. It is an algebra based solely on an "arithmetic" of magnitudes.[36] Second, and this is a very positive point, it is perhaps the abstraction of this same theory of proportion which leads Viète to construct a theoretical calculus on letters. As with ratios, one doesn't have to know what the letters and their operations mean, but only how the letters operate one on the other.

Is Viète's Zetetics an algebra in the sense of Mahoney? I would say yes for the first two characteristics, having an operational symbolism and giving prime importance to relations.[37] As for the third, I consider that Viète's approach is abstract more than intuitive, but it is clearly not completely free from ontological questions and commitments. It has the aim of being structured in such a way as to make possible the translation of its results in purely geometrical terms.

Viète is ahead of his time. His *Algebrâ Novâ*, as he calls his analysis, has very few followers before the 1630s. Soon after the renewal of interest in Viète's works, Descartes publishes *La Géométrie* (1637). Viète is then permanently displaced by this new and more convivial approach.

5.2. *Descartes*

We are back to Descartes. As with Viète, symbolic algebra is for him a legitimate tool of solving problems. But Descartes does not impose a homogeneous mode of solution. The algebraic solution of a geometrical problem does not have to be rewritten in purely geometrical terms. Algebra is not only different from geometry or arithmetic; it is a superior tool with which to solve problems. Therefore, Descartes does not need the cumbersome theory of proportions to be embedded in his algebra. As we have seen in the introduction, he goes further than an "arithmetic" of magnitudes by a conscious use of a unity of measure.

So what is the difference between Descartes and the algebraists before Viète, Chuquet for example? There are two.

Descartes' introduction of a unity of measure is not for finding a numerical representative of a magnitude (a segment, for example), it is for allowing a simplification in the manipulation of symbolic expressions. He writes:[38]

> It should also be noted that all parts of a single line should always be expressed by the same number of dimensions, provided unity is not determined by the conditions of the problem. ... It is not, however, the same thing when unity is determined, because unity can always be understood, even where there are too many or too few dimensions; thus, if it be required to extract the cube root of $a^2b^2 - b$, we must consider the quantity a^2b^2 divided once by unity, and the quantity b multiplied twice by unity.

Equations then do not have to be homogeneous.

Descartes has a symbolic notation which is simple and which corresponds essentially to ours. This is an advantage compared to Viète's and his predecessors' primitive symbolism. Descartes' symbolic algebra has a status similar to Viète's *Algebrâ Novâ*. It is self sufficient. But by letting go of the homogeneity, Descartes' algebra is even more abstract than Viète's. His passage from an "arithmetic" of geometrical magnitudes to an "arithmetic" of letters has left behind more of the properties of geometric magnitudes than did Viète's Zetetics. In terms of ontological commitment, Descartes has gone a step further than Viète.

6. ALGEBRAIC WAY OF THINKING!

With the historical sketches presented in this chapter, we attempted to give some indication of the nature of the links between algebra and geometry. To summarize those links, the focus in this last section will be on four points.

6.1. *Algebra is not only an Extension of the Numerical Domain*

One often hears in school that x, z, a, b, or c are like numbers. They behave like numbers. This emphasis on numbers hides the fact that many other mathematical objects also behave like those algebraic symbols, or should I say, the other way around, that algebraic symbols behave like other mathematical objects. Segments, areas, and volumes are in some ways even closer to algebraic symbols than numbers. When two numbers are added one to the other, the result is a new number symbolized by a completely new symbol. The two original numbers have vanished in the process. With geometrical magnitudes, it is different. Putting two segments together, one after the other, gives a new segment. If one uses a real line as the representation of this new segment, it has within itself the representation of the two original segments. If one uses a letter, then the letter representing the new segment does not have a meaning by itself and therefore cannot be interpreted without specific references to the original segment.

History provides a warning about viewing algebra simply as an extension of arithmetic. From the middle of the 14th century to the middle of the 16th century, algebra is seen mostly in numerical terms. Algebra is found in books for merchants. Those algebraists are practitioners of commercial arithmetic. New rules are found by them, but, as a mathematical domain, algebra does not progress fundamentally during this period. As stated in Section 4.1 on Chuquet, those algebraists also deal with geometrical problems--but those problems are seen as numerical problems. They have great difficulty with general reasoning.

Descartes seems to put forward numbers. But, looking at Descartes' algebra with a critical eye, one sees that his symbols do not represent segments with all their properties, but only segments with certain properties--thus belonging to an "arithmetic" of rarefied segments.

6.2. *Algebra is not only a Question of Symbolism*

On the one hand, symbolism is central to algebra. One may see symbolism as a way to condense the presentation of an argument. Symbolism is then a language. But first of all, symbolism allows the naming of something that has no name *a priori*. We have shown how giving a name is important in the analytical way of tackling a problem.

On the other hand, symbolism is not the whole of algebra. François Viète has a special name for his corpus of symbolic calculations: *logistica speciosa*. For him, *logistica speciosa* is a tool for the Zetetics. It does not have an aim within itself. It is a means to solve problems. It is a means to accomplish the analysis of a problem. It is not for the efficiency of his symbolism that Viète is called the "father of algebra." His symbolism is quite rudimentary. The power of Viète's algebra comes from the fact that operations on letters are defined operationally but not semantically. To be efficient, the symbolism must be put in a broad context with theoretical aims wider than just symbolical calculations.

6.3. *Algebra is a Way to Manipulate Relations*

Algebra is foremost a means to manipulate relations. Dealing with relations implies dealing with objects related one to the other. To which point shall one take into account all the different characteristics of those objets, and all their relations?

Making such a choice is to construct an "arithmetic."[39] The analogies between such an "arithmetic" and the arithmetic of whole numbers may lead to a certain identification of those two arithmetics. This identification, if formalized, becomes a theory of measure. But, to deal with those analogies in an intuitive and informal way produces difficulties. The multifaceted multiplication used by Cardano (area and repeated addition) is an example of this kind of difficulty. The question of the homogeneity of symbolic expression is another one.

The analogy between an arithmetic of geometrical magnitudes, such as the arithmetic of segments, and the arithmetic of whole numbers is strained by the fact that some magnitudes are incommensurable. This asymmetry was certainly an incentive to designing a theory of proportions in order to take care of this last problem, without referring to numbers.

At every step of its evolution toward a "science of relations," algebra, whatever it was, had to develop ways to represent relations between numbers or between magnitudes, or between the arithmetic of numbers and the arithmetic of magnitudes. In doing so, it develops little by little the modern notion of number. A number is a representation of a relation between two magnitudes. In that sense, letters in algebra and numbers behave similarly, but one may see that only if one knows what numbers are. It is not surprising that algebra emerges as a field of knowledge at a time when decimal numbers begin to be used in calculation and when measurements become central to physical sciences.

High school students have a poor understanding of what numbers are. They know natural numbers. But what about rational numbers? ... and irrational numbers? They have almost no experience of number as a measure of magnitude. They get used to numbers before having the chance to grasp how marvelous numbers are because they never use magnitudes in a way that does not involve numbers as their measures. They know the arithmetic of numbers without knowing the arithmetic of magnitudes underneath. How then can they understand the beauty of algebra?

6.4. *Analysis is at the Heart of Algebra*

Analysis is a process central to solving problems algebraically. Arithmetical processes are not usually analytical. In geometry, analysis revolves around the search for what is known among what seems to be unknown. The core of analysis is the hypothesis, that is, the assumption that the problem is solved. As said before, it imposes the development of a certain way of representing the unknown magnitudes that are considered given by hypothesis. In that process, all lines or parts of a figure are dealt with in the same way. Relations between those lines are studied, whether the lines are given or not.

This is similar to what Descartes said in our quotation at the beginning of this chapter: "Then, making no distinction between known and unknown lines, we must unravel the difficulty." In other words, in an algebraic expression, the unknown and the coefficients are manipulated in the same way. This is one main improvement of Viète's *Algebrâ Novâ*. It is not then surprising that Viète uses almost solely the term "analysis" for his algebra and that thereafter, for the next century, the terms algebra and analysis were synonymous.

NOTES

1 Quoted in Unguru (1975, p. 77). Quotation translated from Mahoney (1971, pp. 16-17). See also Mahoney (1972, p. 372) and Mahoney (1980, p. 142).
2 Descartes (1637/1954, pp. 2, 9).
3 Euclid (1925/1956, vol. 1, Book I, p. 155). The fourth and fifth Common Notions are: Things which coincide with one another are equal to one another; The whole is greater than the part. The fourth one expresses clearly the geometrical nature of equalities in the *Elements*. See Section 2.1.2 for a brief analysis of this formulation of equality.
4 Notice that Al-Khwarizmi follows Euclid's order, defining al-jabr before al-muqabala.
5 Euclid (1925/1956, vol. 2, Book III, pp. 71-72).
6 Ptolemy (1984, pp. 48-56).
7 This has been challenged in Unguru (1975). A strong reaction to this article has been written by van der Waerden (1976).
8 Euclid (1925/1956, vol. 1, Book II, p. 372).
9 Euclid (19251956, vol. 1, Book II, p. 382).
10 van der Waerden (1961).
11 Euclid (1925/1956, vol. 1, Book II, pp. 402-403, 409-410).
12 Euclid (1925/1956, vol. 2, Book V, p. 114).
13 *Idem.*
14 Euclid (1925/1956, vol. 1, Book I, p. 155).

15 Euclid (1925/1956, vol. 2, pp. 112-113).
16 The discovery of incommensurable magnitudes has an important role in the edification of a general theory of proportions. Ratios of commensurable magnitudes may be represented by two natural numbers, but with incommensurable magnitudes such a representation is impossible. See Heath's Introductory note to Book V, Euclid (1925/1956, vol. 2, pp. 224-225).
17 On the difficulties related with the finding of a fourth proportional, see Heath's note to proposition V-18 in Euclid (1925/1956, vol. 2, pp. 170-174). When a and b are commensurable magnitudes, the ratio $\frac{a}{b}$ may be represented by the ratio $\frac{n}{m}$, where n and m are the measure of a and b by a unity of measure u which measures exactly a and b. This particular case of finding a fourth proportional is dealt with in proposition IX-19. See Euclid (19251956, vol. 2, Book IX, pp. 409-412).
18 The information contained in this section comes from Boyer (1956, pp. 24-31).
19 Of course, Apollonius always gives these relations in rhetorical language.
20 Boyer (1956, p. 28-29).
21 Mahoney (1968, p. 322).
22 It is an example made famous by Hankel at the end of 19th century. It has been reproduced by Heath in his introduction to *The Elements*, in Euclid (1925/1956, vol. 1, pp. 141-142). Euclid has only figure 4a. Figures 4b and 4c have been added here to facilitate the understanding of the reasoning. Euclid does not specify the name of the point H (Figure 4c).
23 Cardano (1968, p. 82). The problem has been translated into a modern form by Witmer.
24 In French algebra books, such a number was called *nombre nombrant* (numbering number).
25 Cardano (1968, pp. 96-99). This is in chapter XI.
26 It would have been very interesting to study Fibonacci. Luis Radford pointed out to me that in the *Liber Abbaci* (1202), the passage from geometry to algebraic reasoning is not as straightforward as in Chuquet.
27 The rule of "first terms" refers to rules of manipulation of the first (prime) power of the unknowns. See Chuquet (1880, part III) and Flegg, Hay, & Moss (1985, pp. 144-191). *La Géométrie* was first published in 1979, with an extensive introduction by Hervé l'Huillier in Chuquet (1979). A partial English translation of the whole Chuquet manuscript has been published by Flegg, Hay, & Moss (1985).
28 I use the translation by Flegg, Hay, & Moss (1985, p. 266).
29 Regiomontanus (1533/1967). The 1967 book contains a reproduction of the original publication and an English translation.
30 *Ibid.,* p. 31.
31 *Ibid,* p. 31
32 *Ibid.,* p. 35.
33 This filiation has been revitalized by Mahoney (1973, p. 26; 1980, pp. 147-150). See also Van Egmond (1988, pp. 142-3).
34 See also Charbonneau and Lefebvre, Section 2.3 (this volume).
35 On the difficult task of correctly defining the Poristics, see Viète (1983, pp. 12-13, note 6) and Ferrier (1980, pp. 134-158).
36 This lack of numerical concern allowed Viète to develop the powerful and inventive *genesis triangulorum,* a calculus on triangles with its two operations of addition of triangles. This calculus will be revived by the manipulation of complex numbers in the 18th and beginning of the 19th centuries. See Bashmakova & Slavutin (1977).
37 This question of Zetetics as giving prime importance to relations has not been discussed thoroughly here. It would have needed a more complete description of Viète's theory of equations. See Charbonneau and Lefebvre, Section 2.3 (this volume). The *genesis triangulorum* is another example of this relational point of view.
38 Descartes (1637/1954, p 6). This quotation is immediately followed by the second paragraph of the previous quotation from *La Géométrie* given at the beginning of my chapter.
39 Here, "arithmetic" stands for a structure that mimics the structure of the natural numbers, with an "addition" and a "multiplication."

CHAPTER 3

THE ROLES OF GEOMETRY AND ARITHMETIC IN THE DEVELOPMENT OF ALGEBRA: HISTORICAL REMARKS FROM A DIDACTIC PERSPECTIVE

LUIS RADFORD

In order to provide a brief overview of some of the historical affiliations between geometry and arithmetic in the emergence of algebra, we discuss some hypotheses on the origins of Diophantus' algebraic ideas, based on recent historical data. The first part deals with the concept of unknown and its links to two different currents of Babylonian mathematics (one arithmetical and the other geometric). The second part deals with the concepts of formula and variable. Our study suggests that the historical conceptual structure of our main modern elementary algebraic concepts, that of unknown and that of variable, are quite different. The historical discussion allows us to raise some questions concerning the role geometry and arithmetic could play in the teaching of basic concepts of algebra in junior high school.

1. THE ROOTS OF ALGEBRA: ARITHMETIC OR GEOMETRY?

1.1. *The Geometric Current*

The translation and interpretation of ancient Babylonian tablets, during the first half of this century, by Neugebauer (1935-1937), Neugebauer and Sachs (1945), and Thureau-Dangin (1938a) provide us with a wealth of knowledge on one of the earliest forms of mathematics practiced. Many of these tablets deal with numerical problems. In most cases, the problem-solving procedure is not completely explained and appears as a sequence of calculations. This makes it difficult to understand the way of thinking followed by the scribe in solving the problem. However, when interpreted in the wake of present day algebraic concepts and symbols, the calculations acquire some sense. This interpretation led the translators mentioned above, as well as some historians of mathematics (e.g., Boyer & Merzbach, 1991; Kline, 1972; van der Waerden, 1961, 1983), to claim that the Babylonians had developed a "Babylonian algebra." This algebra has been seen to be different from our modern elementary algebra, principally in its lack of symbolic representations.

The first problem of tablet BM 13901[1] is an example of this Babylonian algebra. Classically translated, the problem can be formulated as follows (cf., Thureau-Dangin, 1938a, p. 1, or van der Waerden, 1983, pp. 60-61) : "I have added the surface and the side of my square, and it is 3/4."

The classical interpretation of the solution is as follows:

Take 1 to be the coefficient [of the side of the square]. Divide 1 into two parts. 1/2 x 1/2 = 1/4 you add to 3/4. 1 is the square of 1. You subtract 1/2, which you have multiplied by itself, and 1/2 is [the side of the square].

N. Bednarz et al. (eds.), Approaches to Algebra, 39-53.

The classical interpretation "sees" the equation $x^2 + x = 3/4$ in the statement of the problem and "sees" the sequence of numerical operations leading to the solution as:

$$x = \sqrt{\left(\frac{1}{2}\right)^2 + \frac{3}{4}} - \frac{1}{2}$$

These calculations correspond to our own general formula for such a problem:

$$x = \sqrt{\left(\frac{b}{2}\right)^2 + c} - \frac{b}{2}$$

which yields the positive solution for the equation of type $x^2 + bx = c$.

Thus, according to the classical interpretation, Babylonian mathematicians would have known the general formula without being able to express it as such, because they lacked the symbols to do so.

The main argument supporting the idea of a Babylonian algebra is, thus, the possibility of translating the Babylonians' problems and calculations into modern arithmetic-algebraic symbolism. However it is an argument whose validity is not supported by historical evidence (see Unguru, 1975).

There is, however, a completely different interpretation of this type of problem. In fact, during the past years, Høyrup has studied the problems found in the Babylonian tablets as well as the terms used in them (Høyrup, 1987). Høyrup (1990) claims that:

> Old Babylonian "algebra" cannot have been arithmetical, that is, conceptualized as dealing with unknown numbers as organized by means of numerical operations. Instead it appears to have been organized on the basis of "naive," non-deductive geometry. (p. 211)

This non-deductive geometry, developed extensively in Høyrup (1985, 1986), consists of a "cut-and-paste geometry" in which the complicated arithmetical calculations resulting from the classical interpretation correspond to simple naive geometric transformations.

For example, Høyrup translates the problem shown above (BM 13901) as, "The surface and the square-line I have accumulated: 3/4."

His translation of the solution is as follows:

> 1 the projection you put down. The half of 1 you break, 1/2 and 1/2 you make span [a rectangle, here a square], 1/4 to 3/4 you append: 1, makes 1 equilateral. 1/2 which you made span you tear out inside 1: 1/2 the square line. (Høyrup, 1986, p. 450)

As one can see from the diagram which accompanies Høyrup's explanation,[2] the procedure consists of *projecting* a rectangle of base equal to 1 on the line-side of the square (Figure 1). The projected rectangle is cut into two rectangles, the base of each one being equal to 1/2. The rectangle on the right is then transferred to the bottom (Figure 2). Then a small square is added (Figure 3) in order to obtain a complete square from which the sought square-line can be found (Figure 4).

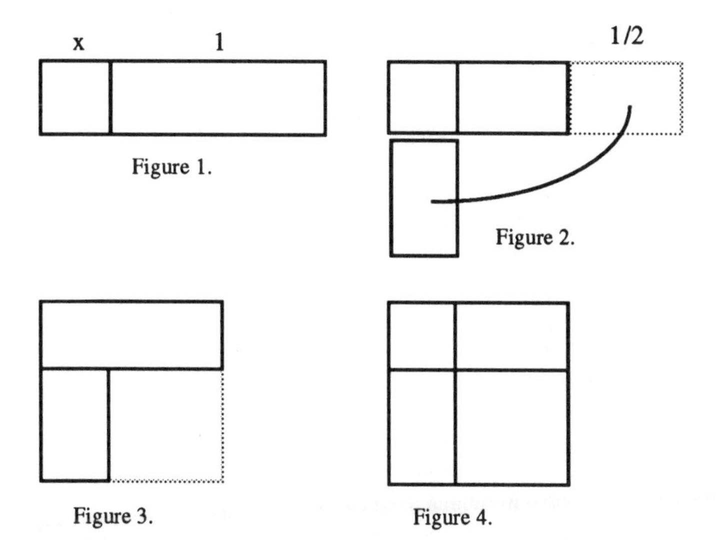

Figure 1.

Figure 2.

Figure 3.

Figure 4.

This geometric interpretation is quite different from the classical interpretation which views such problems as problems dealing with numbers whose solutions are based on arithmetical (or algebraic) reasoning. Thus, while Neugebauer claims that "geometric concepts play only a secondary role" (Neugebauer, 1957, p. 65) in Babylonian algebra, Høyrup argues that *naïve geometry* is the basis on which such problems were posed and solved.

The new interpretation of the Babylonian algebra leads us to a re-evaluation of the role of arithmetic and geometry in the emergence of algebra.

We will discuss in Section 1.3 the influence of the "cut-and-paste geometry" in the emergence of algebra. Now we have to turn briefly to a Babylonian numeric current.

1.2. The *Numeric Current*

This new interpretation of Babylonian algebra, which concerns especially those problems whose translation into modern symbolism can be seen as second degree equations, does not preclude the existence of a numeric current in Babylonian algebra (see Høyrup, 1985, pp. 98-100). In fact, there are many problems, especially those concerning Babylonian commerce, which do not lend themselves to a geometric interpretation. This is also the case for certain first-degree problems (i.e., VAT 8520, No. 1) and the problems found on tablets VAT 8389 and VAT8391--both of which date back to the first Babylonian dynasty (circa 1900 B.C.).

The problems on tablets VAT 8389 and VAT 8391 concern grain production in two fields. If we designate α as the production per unit of area of the first field, X as the area of the first field, β as the production per unit of area of the second field, and Y as the area of the second field, most of the problems on the tablets can be formulated, using modern notation, as follows:

$$\alpha X \pm \beta Y = c; \qquad X \pm Y = \delta$$

(α, β, c, and δ having numerical values particular to each problem).

In modern notation, Problem 1 of tablet VAT 8389 corresponds to the following linear system of equations:[3]

$$\alpha X - \beta Y = c$$
$$X + Y = \delta$$

The calculations suggest that to find X and Y, the scribe first takes a *false solution*: in this case, $X_o = Y_o = \delta/2$ (satisfying the condition $X + Y = \delta$). He then calculates the production of each field (which he names "false grain"), which is, in modern notation, αX_o and βY_o.

Next, the scribe calculates the excess production of the first field over the second: $\alpha X_o - \beta Y_o$. Let c_o be this excess. He then calculates the production which is missing to satisfy the conditions of the problem. In our notation the production missing is $c - c_o$.

After that, the scribe calculates $\alpha + \beta$. This quantity is precisely the excess production of the first field over the second when we add one unit of area to the first field and we subtract one unit of area from the second field.

To compensate for the missing production $c - c_o$, we must make the excess $\alpha + \beta$ equal to $c - c_o$, a problem which can be solved through the tools of proportional quantities, a field of study in which the ancient pre-Greek mathematicians excelled. In fact, making the excess $\alpha + \beta$ equal to $c - c_o$ is achieved by multiplying $(\alpha + \beta)$ by the number of units of area to be added to X_o. The resulting quantity must then be equal to $c - c_o$.

The scribe knows that the number of units of area to be added to X, which we can designate as z, is obtained by multiplying the inverse of $(\alpha + \beta)$ by $(c - c_o)$. The number z thus obtained by the scribe is added to the first field and subtracted from the second field, thereby providing the real areas of the fields which are consequently $X_o + z$ and $Y_o - z$ respectively.

The procedure to solve this problem is clearly arithmetical and not geometric. The tablet allows us to see that it consists of an arithmetical method of *false position:* the scribe begins by assigning a numeric value, which is recognized as being false *a priori*, to the sought quantities (i.e., the areas of the fields). Using the false values and the data given in the problem's statement, he obtains new data. The new data (here the "false grain") can then be compensated for, to ultimately yield a correct solution. This method of *false position* is used to solve many of the problems on the Babylonian tablets (see, e.g., Thureau-Dangin, 1938b--tablets Str. 368, VAT 7535, and VAT 7532).

From a historical perspective, it is difficult to establish a link between the geometric and numeric currents in Babylonian algebra (see, however, Høyrup, 1990). It is also difficult to ascertain exactly what influence either of the currents may have had on the initial development of algebra. In fact, many of the most important early works containing basic algebraic concepts, such as Diophantus' *Arithmetica*, contain no explicit references to antecedent sources of inspiration. Nevertheless, we can trace certain elements in Diophantus' *Arithmetica* to the numeric and geometric currents of Babylonian and Egyptian mathematics.

1.3. *Diophantus' Arithmetica*

Diophantus' *Arithmetica* (circa 250 A.D.) is a collection of problems divided into 13 books, 3 of which remain lost. To clearly trace the links between Diophantus' algebra and some antecedent mathematical traditions, we must first recall that Diophantus, like Aristotle, conceived of number as being composed of discrete units. Moreover, Diophantus believed that numbers can be divided into "categories" or "classes," each category containing the numbers that share the same exponent: the first category being the squared numbers (designated as Δ^{Υ}), the second being the cubed numbers (designated as K^{Υ}), the third being the squared squares (designated as $\Delta^{\Upsilon}\Delta$) and the cubo-cubes (designated by $K^{\Upsilon}K$). The problem statements are written as *relations* between these categories. Hence, there are no particular numbers given in the problem statements. Following are two examples.

Book IV, Problem 6:

> We wish to find two numbers, one square the other cubic, which comprise [i.e., that their product is] a square number. (Sesiano, 1982, p. 90)

Book I, Problem 27:

> Find two numbers such that their sum and their product equal the given numbers. (Ver Eecke, 1959, p. 36--our translation).

After having classified numbers on the basis of their exponents at the beginning of Book I, Diophantus introduces one of the most important concepts to this discussion: the *arithme* (the "*number*") which is "an undetermined quantity of units" (Ver Eecke, 1959, p. 2). The arithme is introduced for the purpose of representing the unknown in a problem; it is a heuristic tool in the context of problem solving.[4]

Problem 27, Book I of the *Arithmetica*, referred to above, establishes a link between Diophantus' algebra and antecedent mathematical traditions. The solution begins as follows:

> The square of half of the sum of the numbers we are seeking must exceed by one square the product of these numbers, which is figurative. (Ver Eecke, 1959, p. 36--our translation)

Thus, Diophantus begins by giving a condition which the numbers must fulfill, in order that the problem could be solved. The fact that this condition is "figurative" suggests that Diophantus is referring to a condition which can be visualized through a geometric representation (Ver Eecke, 1959, p. 36-37). Furthermore, Diophantus does not make his thought explicit, which suggests that he is referring to something well known to the reader (cf., Høyrup, 1985, p. 103). It is possible that Diophantus is actually referring to a cut-and-paste procedure (see Figure 5):

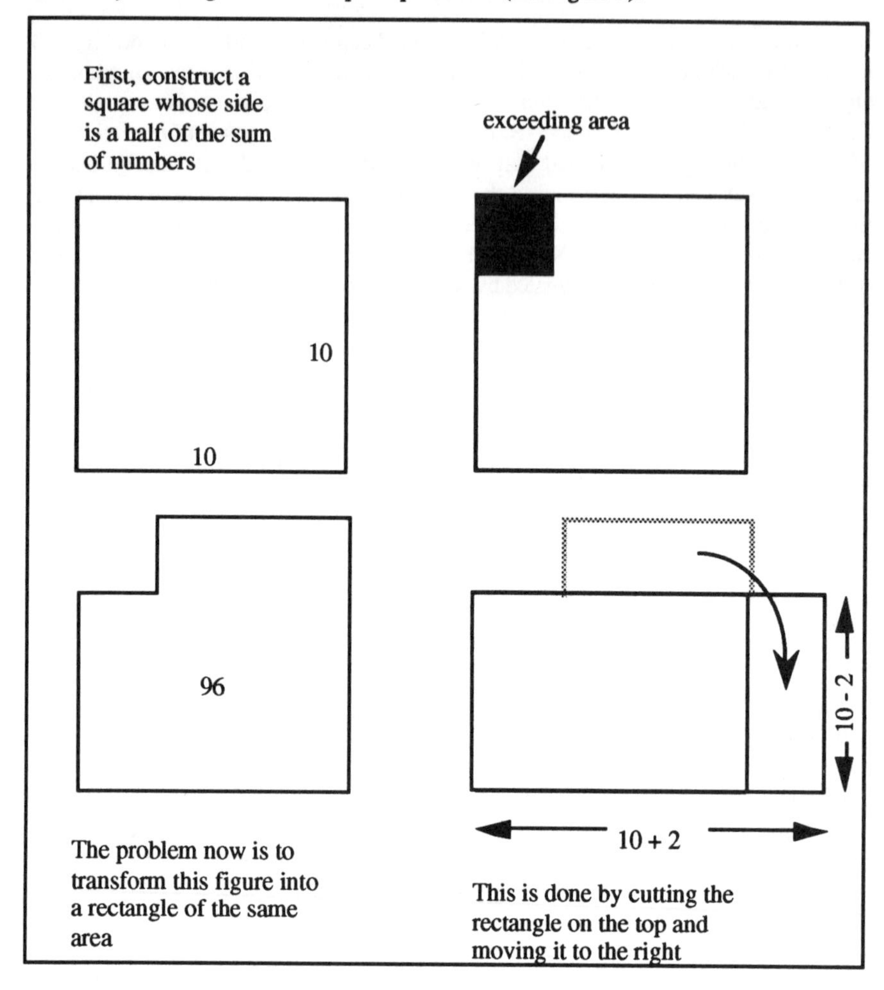

Figure 5.

For the problem to be solved numerically (using positive rational numbers, the only numbers considered by Diophantus), the exceeding area must be a square. As we see, the "figurative" condition mentioned by Diophantus seems to refer strongly to the above cut-and-paste procedure and suggests a link between Diophantus' methods and those of the Babylonian mathematicians belonging to the geometric current. To continue, Diophantus' solution to this problem is as follows: "Let the sum of the numbers equal 20 units, and their product equal 96 units." Diophantus then specifies the parameters of the problem:

Let the quantity in excess of the two numbers be 2 arithmes. Thus, because the sum of the two numbers is 20 units, which when divided by two yields two equal parts, each part will be half of the sum, or 10 units. Therefore, if we add to one part, and subtract from the other, half of the quantity in excess of the two numbers, that is 1 arithme, it follows that the sum will still equal 20 units and that the quantity in excess of the numbers will remain 2 arithmes. Thus, let the greater number be 1 arithme increased by 10 units, which is half of the sum of numbers, and consequently, the lower number is 10 units minus 1 arithme; it follows that the sum of these numbers is 20 units and that the quantity in excess of the numbers is 2 arithmes.

The product of the numbers must equal 96 units. This product is equal to 100 units minus 1 squared arithme; which we equate to 96 units, thereby causing the arithme to be 2 units. Consequently, the greater number will be 12 units and the lesser number will be 8 units. These numbers satisfy the initial conditions. (Ver Eecke, 1959, pp. 37-38--our translation)

To find the greater and lesser numbers, Diophantus used a method very similar to the one used in tablet VAT 8389 (a method that van der Waerden, 1983, pp. 62-63, called the "sum and difference"). In fact, both procedures take half of the sum of the numbers into account. They find a number that is added to and subtracted from half of the sum of the numbers. This resemblance of methods strongly suggests a link between Diophantus' methods and those of the Babylonian mathematicians belonging to the numeric current.

The similarity between Diophantus' method and the "false position" Babylonian method has been discussed by Gandz (1938) and by Thureau-Dangin (1938a, 1938b). Thureau-Dangin claims that the *false position* is in fact an *algebraic procedure,* where the unknown is *represented* by the number "1," for the Babylonians did not develop any symbolism allowing them to represent the unknown by a letter. However, this argument rather forces the old Babylonian thinking to fit into the modern algebraic thinking. On the other hand, Gandz states that Diophantus' algebraic method, shown in Book I, Problems 27-30, is a method which was known by the Babylonian mathematicians and which allowed them to find the solutions for the mixed second-degree equations. However, it supposes that ancient Babylonian mathematicians had at their disposal a complex symbolic language (see Gandz' calculations, pp. 416 ss.), but there is no historical evidence for this.

On the other hand, we can appreciate that Thureau-Dangin's and Gandz' main aim is to try to understand Babylonian mathematics. They go from Diophantus to Babylonian mathematics. We try to go in the other direction: from the Babylonians to Diophantus. We think that ancient mathematicians developed arithmetical methods, such as that of *false position,* in order to solve problems. Such a method is not based on algebraic ideas but on an idea of *proportionality.*

Yet another link between the arithmetical current and Diophantus' *Arithmetica* can be found in an Egyptian papyrus (known as Michigan papyrus 620) written before Diophantus (circa 100 A.D.) (Robbins, 1929). This papyrus contains a series of numerical problems that closely resemble many of the problems in Diophantus' Book I. Here is an example:

There are four numbers, the sum of which is 9900; let the second exceed the first by one-seventh of the first; let the third exceed the sum of the first two by 300, and let the fourth exceed the sum of the first three by 300; find the numbers.

This papyrus contains the symbol ς which Diophantus used in *Arithmetica.* Furthermore, the choice of the unknown in the papyrus is 1/7 of the first number--

not unlike Diophantus' choice of the unknown in his problems (a choice which is surely made to avoid calculations involving fractions: e.g., Book I, Problem 6).

To sum up our discussion, we can say that, from a historical point of view, the roots of Diophantus' algebra seem to be found in both: (a) a geometric Babylonian tradition (related to a cut-and-paste geometric current) and (b) a numeric Babylonian-Egyptian tradition (related to the false position method).[5] These mathematical traditions led to two kinds of different ancient "algebras":

1) On one hand, the "algebra" related to geometry, cultivated, in all likelihood, within the communities of surveyors who dealt with problems about geometrical figures. One of the paradigmatic problems of this current is to find the length or width of a rectangle satisfying certain conditions (e.g., the surface and the side equal to 3/4, as in BM 13901 seen above). In this algebra, the side of a figure can be seen as a side as well as a rectangle--the rectangle whose height (or *projection*) is equal to 1. Thus, the length of a side can also represent the area of the projected rectangle. This kind of algebra is underlined by the visual conservation of areas of figures submitted to the cut-and-paste procedures. Algebraic equality refers here to equalities between areas. Algebraic methods are essentially based on a sequence of geometric transformations, T_i, starting with the given figure, F_1 and ending with a square F_n of a known area:

$$F_1 \xrightarrow{T_1} F_2 \xrightarrow{T_2} \ldots \xrightarrow{T_{n-1}} F_n$$

The unknowns of the problem (e.g., the lengths of the sides of a rectangle) are taken into account through the problem-solving procedure. However, the unknowns are not the object of calculation.[6]

2) On the other hand, the numeric tradition led to a "numeric algebra" that dealt with theoretical problems about numbers (such as those found in Diophantus' *Arithmetica*, but with some *riddles* also; e.g., to find the amount of apples divided between a certain number of people according to specific conditions). This kind of algebra is based on a different conceptualization from the geometric one. Here one does not have segments to represent the unknowns. Furthermore, there is not a "natural" name (like "side" or "height") to be used in speaking about the unknowns. A closer look at the problem-solving methods included in the *Arithmetica* shows that all the numeric algebra deals not with several unknowns but with a single unknown. Diophantus just called the unknown, the arithme, that is, the number (which should be understood as *the* number) for which we are looking. As in the case of the algebra of the geometric tradition, the numeric algebra supposes a hypothetical thinking: one reasons as if the number sought was already known. However, in contrast to the algebra of the geometric current, in the numeric algebra one calculates with and on the unknown. The calculation with/on the unknown makes it possible for the emergence of a new kind of calculation that is independent of the context and of the problem, a *formal calculus* (in the sense that it takes into consideration only the form--the ειδοσ, *eidos*--of the mathematical expressions). Within this new calculus, algebraic equality refers to equality between *species* (Radford, 1992),

something that we can translate into modern terms by *monoms*. The algebraic methods used to solve problems are based here on transformations of monoms. The goal of these transformations is to arrive at a monom equal to another monom (i.e., in modern terms, to an equation of the form $a \cdot x^n = b \cdot x^m$), and then to arrive at an equation of the form, a monom equal to a number (i.e., $c \cdot x^k = d$). As we can see, the heuristic behind the resolution of a problem in each algebra is different.

The mutual influence of these two kinds of algebras is difficult to detect in Babylonian mathematical thought. However, this influence is perceptible in Diophantus' *Arithmetica*, as we saw before in our example of Problem 27, Book I. Later on, these algebras seem to have followed separate paths. We can see cut-and-paste geometry reappearing in Arabic soil some centuries later, in the work of Al-Khwarizmi and in that of Abû Bakr. The algebras of numeric and geometric origin will meet together during the awakening of sciences in the late Latin Mediaeval Age, as a result of an intensive translation of Arabic and Greek mathematical works into Latin. Without a doubt, both algebras merged in the *Liber Abbaci* of Leonardo Pisano, in the beginning of the 13th century. Even though our modern elementary algebra looks more like the numerical algebra than the cut-and-paste one, we should be aware that the development of mediaeval algebra (and to some extent the algebra of the Renaissance) was organized according to some "types of equations" whose distinction was completely guided by cut-and-paste geometry.[7]

Concerning the ancient algebra of the numeric current, the historical records available today make it possible to reconstruct a scenario of the conceptual relationships between the false position methods and the Diophantine algebraic ones. We cannot discuss here such a scenario. Let us simply mention that the "jump" from arithmetic thinking to algebraic thinking seems to be located in a reinterpretation of the false position method based on the search for a shorter and direct method for solving problems according to which one no longer thinks in terms of false values but in terms of the unknown itself.

2. UNKNOWNS AND VARIABLES: TWO DIFFERENT CONCEPTUALIZATIONS

Our discussion thus far has centered on the concept of unknown. There is, however, an equally important concept which we have yet to consider: the variable. While the unknown is a number which does not vary, the variable designates a quantity whose value can change. A variable *varies* (see Schoenfeld & Arcavi, 1988, p. 421).

Where can we find the origin of variables in the history of algebra? As in the case of the unknown and in light of our current historical knowledge, it is difficult to accurately pinpoint the variable's "big-bang." In a certain sense, we find some traces of the concept of variable in some ancient Babylonian tablets. For instance, there are tablets for reciprocal numbers. But these tablets seem to stress a *relationship* between numbers rather than a variational property of a mathematical object. We can, however, trace certain elements of a more elaborate concept of variable in Diophantus' book entitled, *On Polygonal Numbers.*

In order to better understand the scope of Diophantus' conceptualization of variables, it is important to note that polygonal numbers emerge in a philosophical context of classification of numbers, which dates back to the era of the first

Pythagoreans.[8] An important book which includes a detailed treatment of polygonal numbers is Nicomachus' *Introduction to Arithmetic* (2nd century A.D.). In this work, Nicomachus tries to discover patterns between numbers, for example, that every square number is the sum of two consecutive triangular numbers (D'Ooge, 1926, p. 247).

However, the observed patterns are not proved (at least, not in the Euclidean sense; this is an altogether different type of mathematics in which philosophical considerations require no *deductive* proof).

A result, concerning triangular numbers, probably obtained from a few concrete numerical examples, is stated by Plutarch (who lived in Nicomachus' time) in the following terms: "Every triangular number taken eight times and then increased by 1 gives a square" (Heath, 1910/1964, p. 127). It is the generalization of this result to other polygonal numbers that interested Diophantus in his book, *On Polygonal Numbers.*

The book, of which the last part is lost, consists of four *deductively* connected propositions, concerning arithmetical progressions. The third, for instance, states that in an arithmetical progression "the sum of the largest and smallest [terms], multiplied by the quantity of numbers, form a number [equal to] twice the sum of the given numbers"[9] (Ver Eecke, 1959, p. 280--our translation). Using the first three propositions, Diophantus proves the fourth one. Using modern symbols, this proposition can be stated as:

$$S_n \times 8d+(d-2)^2=\left[(2n-1)\,d+2\right]^2$$

where n designates the side of the polygonal number, d designates the difference, and S_n designates the polygonal number.

Using the fact that $a = d + 2$ (cf., Footnote 8), the above proposition can also be written as:

$$S_n \times 8\,(a-2)+(a-4)^2=\left[(2n-1)\;(a-2)+2\right]^2$$

Diophantus uses this last proposition to obtain an *explicit formula* to calculate the polygonal number S_n when the "side," n, is known. This proposition also allows Diophantus to give a formula to calculate the "side," n, when the number S_n is known.

Translated into modern notation, the first of the formulas can be written as:

$$S_n=\frac{\left[(2n-1)\;(a-2)+2\right]^2-(a-4)^2}{8\;(a-2)}$$

(Take twice the side of the polygonal number; from this subtract one unit; multiply the result by the number of angles minus 2; then add 2 units. Take the square of the resulting number. From this, subtract the square of the number of angles minus 4. Divide the result by 8 times the number of angles minus 2 units. This gives us the polygonal number we are looking for.) (based on the translation of Ver Eecke, 1959, pp. 290-291)

It is important that we stress the conceptual nature of the numbers S_n, n, d, and a in the preceding propositions (or rather, their link to the concept of variable). To do this, we can refer to certain passages in the proof of proposition 3 mentioned above, that is: "the sum of the largest and smallest [terms], multiplied by the quantity of numbers, form a number [equal to] twice the sum of the given numbers."

The proof is divided into two cases: the first case is related to an even quantity of numbers and the second, an odd quantity of numbers. In the first case, Diophantus develops his reasoning taking into account only six numbers, represented by: α, β ,γ, δ, ε, ζ. (In the second case he represents five numbers.)[10] The fact that he limits himself to six or five numbers is a result of the symbolic system of representation he uses (a system which is in fact borrowed from Euclid; for an example, see *Elements*, Book IX, prop. 20); nevertheless, the generality of the proof (in the Greek sense) is not lessened by its limitation to six (or five) numbers.

The quantity of numbers (which we would call "n" in our modern symbolic system) is represented by a *segment*.[11] This is the segment $\eta\upsilon$ in Figure 6. In accord with Diophantus' concept of number, the segment $\eta\upsilon$ can be divided into its units: λ, μ, χ...

The "numbers in equal difference," that is, the numbers α, β, γ, δ, ε, ζ, are placed on the *segment*.

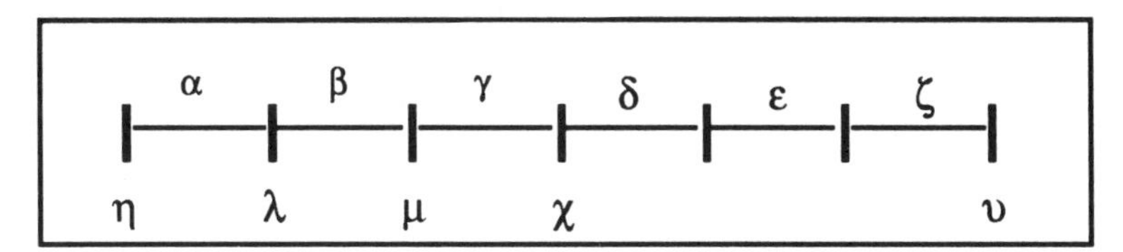

Figure 6. (From Tannery, 1893, Vol. II, p. 457)

This geometric representation allows Diophantus to organize the proof, which begins by noting (translated into modern notation) that:

$$\gamma - \alpha = \zeta - \delta$$

because of the equal difference between numbers. Then, Diophantus transforms this equality into:

$$\gamma + \delta = \zeta + \alpha$$

which he also writes as:

$$\gamma + \delta = (\zeta + \alpha) \times \eta\lambda$$

because $\eta\lambda$ is equal to the unit.

In the same way, Diophantus obtains:

$$\varepsilon + \beta = (\zeta + \alpha) \times \lambda\mu.$$

We do not need to go beyond this point in the proof. For our purposes, we only need to see that the quantity "n" of numbers and the number S_n, which is equal to $\alpha + \beta + \gamma + \delta + \varepsilon + \zeta$, are still not seen as *variable* numbers. In fact, n and S_n do not vary during the proof; "n," for instance, is a given number, chosen from the beginning.

Neither are the numbers n and S_n *empirical set values* in Diophantus' work, as is the case in Nicomachus' development of Polygonal Numbers Theory. Let us explain this idea. Patterns in Nicomachus' work state relations between numbers, but what supports the validity of patterns (in addition to philosophical premises) is the fact that when the numbers to which he refers in the patterns are replaced by concrete numbers, the resulting calculation is true. Thus, the numbers involved in Nicomachus' propositions are set values, or in order to stress their relationship with the concrete numbers, we can call them *empirical set values.* However, Diophantus never gives a numerical example. The propositions in Diophantus' theory of polygonal numbers are supported by a *deductive organization,* which breaks with Nicomachus' concrete-arithmetical treatment of numbers, and confers on numbers a different status. We can say that numbers (such as the numbers n, d, and S_n) become *abstract set values.*

We have said before that the numbers n, d, and S_n are not variables in the proof of a proposition given by Diophantus. Significantly, from the moment Diophantus states explicitly that one must find the polygonal number when the *side is given* (which arrives after all the deductive treatment of Proposition 4 is done), the numbers n and S_n are no longer *abstract set values;* they become *dynamic quantities* that can have different values depending on the values taken by the other quantity. They become variable mathematical objects.

This is, of course, a long way from the concept of variable that we find in the 18th century or by Oresme in the 14th century where the variable is especially linked to continuous quantities and seen as "intensities" associated to "qualities" (see Clagett, 1968). From the medieval context of the study of motion of bodies, variables will appear as flowing quantities, and the main problem will be that of describing the effect of the variation of one variable over another. This description will lead to the concept of *function.*

Diophantus is concerned about variables, not through the concept of function, but through the concept of formula. His concept of formula is not based on flowing continuous quantities, but on: (a) an explicit relationship between numbers that are seen as monads (i.e., unities) or fractional parts of monads and (b) an explicit sequence of calculations allowing one to determine a number given the identity of another number.

What is it, in Diophantus' work, that makes the emergence of the concept of variable possible? One reason may be that the problem is transformed from a *relational problem* to one involving *calculations.* This step can be achieved only in a "calculator's mind" (see Rey, 1948). The presence of the calculator's mind is obvious in the passages that we showed in the proof of Diophantus' Proposition 3: He uses the unity as a "real" number (and not only as the generating monad of the early Greek mathematicians). He performs calculations with the unity: this is why he can transform $\gamma + \delta = \zeta + \alpha$ into $\gamma + \delta = (\zeta + \alpha) \times \eta\lambda$.

This step requires--to pose the problem in the Greek context--that *Logistic* (which concerns the calculations of numbers, originally calculations with the material aspect of numbers, and later the way in which calculations are performed) be considered a theory of the same intellectual value as *Arithmetic* (which refers originally to the study of numbers in its deeper significance, in their essence, without any relation to the concrete world).[12] Arithmetic is concerned with propositions or theorems that state relations between numbers and/or properties of numbers. The step that joins *Logistic* and *Arithmetic* is not taken by Euclid, but is taken by Diophantus, in combining two intellectual inheritances: the Babylonian and Egyptian traditions of calculator mathematicians (as we have seen previously in Section 1) and the theoretical tradition of his predecessor in Alexandria, Euclid himself.

We now turn to the differences between unknowns and variables. The first difference is found in the context (more specifically, in the goal or intentionality) in which they appear. In fact, we can note that the main topic of *On Polygonal Numbers* is quite different from that of *Arithmetica.* In the former the goal is to establish *relationships* between numbers, while in the latter it is to solve word problems (i.e., to find the value of one or more unknowns). In *On Polygonal Numbers*, the relations between numbers are stated in terms of *propositions*; furthermore, they are organized in a deductive structure. Variables are derived from the passage from a *relational problem* to one dealing with *abstract set values calculations*. This is not the case in *Arithmetica*, where the problems refer to relations between "classes" or "categories" of numbers (squares, cubes, etc.) that constitute the Theory of Arithmetic.

The second difference between unknowns and variables is found in their representations. In the case of *On Polygonal Numbers*, the key concept is the *abstract set value* (which leads to that of variable). Abstract set value is represented geometrically or by letters, which are included in a geometric representation. In the case of *Arithmetica*, the key concept is the *unknown* (the arithme) which is not represented geometrically.

While both of these concepts deal with numbers, their conceptualizations seem to be entirely different.[13]

3. IMPLICATIONS FOR THE TEACHING OF ALGEBRA

What role could the preceding historical analysis play in the modern day teaching of mathematics? It seems to me that the historical construction of mathematical concepts can supply us with a better understanding of the ways in which our students construct their knowledge of mathematics. But aside from its contribution to our comprehension of certain phenomena arising in the psychology of mathematics, the epistemology of mathematics can also provide us with information that allows us to improve our teaching methods. Thus, we can ask ourselves certain questions:

1) Is it to our advantage to develop certain elements of "cut-and-paste geometry," as developed by Høyrup, for use in our classrooms in order to facilitate the acquisition of basic algebraic concepts? Can the teaching of "cut-and-paste procedures" awaken in students the analytical thinking that is required in learning algebra?
2) Is it to our advantage to introduce certain elements of the "false position" method prior to introducing the concept of the unknown in the solution of word problems?
3) Could we use proportional thinking as a useful link to algebraic thinking?
4) Can current teaching methods introduce some appropriate distinction between the concepts of unknown and variable, using the historical ideas seen here?

5) Is it reasonable to believe that the development in a deductive context of the concept of *abstract set value*, as developed by Diophantus, is a prerequisite to a deep learning of the concept of variable?

These questions deserve a great deal of reflection and discussion, which are beyond the scope of this chapter. Question 5, for instance, suggests a way of thinking about variables that contrasts with the usual way taken to introduce variables in junior high school, where the inductive context (based on the empirical set value concept of number) is preferred to a deductive one. Of course, from a behavioral point of view, the inductive tasks can be accomplished more quickly in the classroom than the deductive ones. But in terms of real understanding of what a variable is, our analysis can at least indicate some new didactic dimensions to be explored, in order to reach a deeper knowledge of the concept of variable.

Our analysis also suggests that symbolism does not capture completely the deep meaning of the concept of variable, as the teaching of mathematics seems to imply.

However, the results of this epistemological analysis should not be construed as normative for teaching. Rather, such an analysis is meant to give us a better understanding of the significance of mathematical knowledge and to provide us with a path, one of many, for its construction.

NOTES

1 The numbers in the following two translations are in base 10 (and not in base 60); as well, all fractions are represented using the modern symbol "/".
2 This diagram is implicitly referred to in the problem-solving procedure through the *meaning* of geometrical operations (see Høyrup, 1985, pp. 28-29).
3 The statement of this problem (based on Thureau-Dangin's French translation, 1938a, p. 103-105) is as follows: "By *bur*, I perceived 4 *kur*. By second *bur*, I perceived 3 *kur* of grain. One grain exceeds the other by 8`20 <*sila*>. I added my fields: 30` <SAR> What are my fields?" (NB: *bur* is a measure of surface; *kur* and *sila* are measures of capacity. A complete commented solution of this problem can be found in Radford, 1995a).
4 The arithme is designated as the letter ς, probably, as Heath (1910/1964) suggested, because it is the last letter in the Greek word arithme (arithmos, αριθμος). It is important to note that the symbolism used to designate the arithme and its powers (square, cube, etc.) leads us to the first symbolic algebraic language ever known (see Radford, 1992).
5 There are other problems in Diophantus' *Arithmetica* whose problem-solving procedures recall the false-position method: for example, Book "IV" problem 8 (Ver Eecke, 1959, pp. 119-120); Book "IV" problem 31 (Ver Eecke, 1959, pp. 155-157). These problems are discussed in Katz (1993, pp. 170-172).
6 For instance, if we refer to the first problem of tablet BM 13901, shown earlier in this chapter, we can see that the unknown is found by displacements of figures. The unknown is not *really* involved in calculations, which are performed or executed with known quantities (Radford, 1995a). Although sometimes the scribe takes a fraction of the unknown or the unknowns in order to cut a figure (for some examples, see Høyrup 1986, pp. 449-455), it does not constitute a truly extended or generalized calculation *with* or *between* unknowns. The limits of such a geometric calculus can be better circumscribed if it is compared to the algebraic calculus which Diophantus develops at the beginning of his *Arithmetica* (see Ver Eecke, 1959, pp. 3-9).
7 See Radford (1995b).
8 Remember that polygonal numbers are obtained as the sum of the first numbers in an arithmetical progression with difference d and first term $a_1 = 1$. If we designate, in modern notation, the arithmetical progression as $1, 1 + d, 1 + 2d,...$, then the polygonal numbers are:

$$S_1 = 1,\ S_2 = 1+(1+d),\ S_3 = 1+(1+d)+(1+2d),\dots,\text{etc.}$$

If $d = 1$, we obtain the triangular numbers, if $d = 2$, we obtain the square numbers, if $d = 3$, we obtain the pentagonal numbers, etc. The first triangular numbers are: 1, 3, 6, 10, ... Note that the angles a of a polygonal number are obtained by adding 2 to its generating difference d, that is $a = d + 2$. Thus, the *angles* of any triangular number are $a = 1 + 2 = 3$. The *side* of a specific polygonal number $S_n = a_1 + a_2 + ... + a_n$ is n. Thus the side of the triangular number 6 is 3.

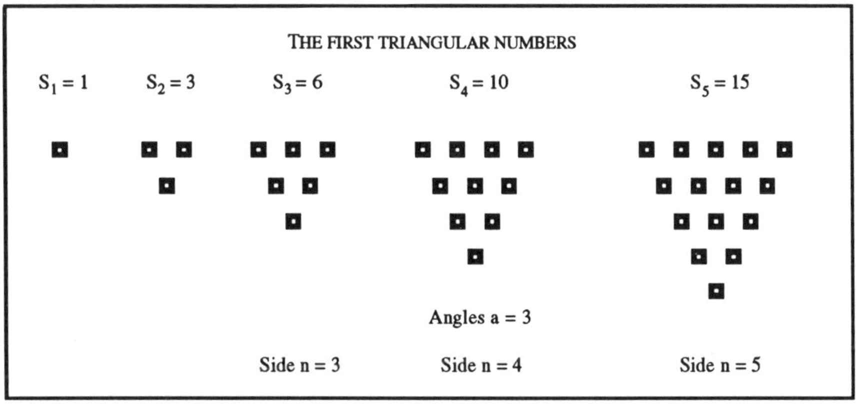

THE FIRST TRIANGULAR NUMBERS

$S_1 = 1$ $S_2 = 3$ $S_3 = 6$ $S_4 = 10$ $S_5 = 15$

Angles a = 3

Side n = 3 Side n = 4 Side n = 5

9 In modern notation, this proposition can be stated as: $(a_n + a_1)n = 2S_n$.
10 A discrete representation of the numbers (i.e., by small circles or other objects, as used by Nicomachus), would not have suited this proof. In fact, in order to keep the numbers unspecified, new "abstract" representations are required. The choice of letters as representations satisfies this requirement.
11 It should be noted that geometrical segments are used here in a completely different way from that of Babylonian mathematics! (see Section 1.1).
12 The next passage from Olympiodorus' *Scholia* to Plato's *Gorgias* explains the initial difference between *logistic* and *arithmetic*: "It must be understood that the following difference exists: arithmetic concerns itself with the kinds of numbers; logistic, on the other hand, with their material" (cited by Klein, 1968, p. 13).
13 The distinction between variables and unknowns is not stated explicitly by Diophantus. This distinction is explicitly made in the 18th century by Leonard Euler, who sees unknowns as objects belonging to "ordinary analysis" (i.e., algebra) while variables are seen as objects belonging to "new analysis" (i.e., infinitesimal analysis) (see Sierpinska, 1992, p. 37).

CHAPTER 4

THE ROLE OF PROBLEMS AND PROBLEM SOLVING IN THE DEVELOPMENT OF ALGEBRA

TERESA ROJANO

The well known separation that students tend to make between algebraic manipulation and its use in modeling and solving problems may have its origins in an educational approach based on an oversimplified vision of algebra, which hides the semantic background of its grammar. In this chapter, some decisive factors in the evolution of the constitution of algebraic language are discussed in order to extract lessons from history that have a bearing on the teaching of this language today.

1. INTRODUCTION

Probably one of the oldest errors committed in algebra teaching is that of trying to communicate to students, from their very first contact with the subject, the qualities and virtues of mastering its syntax in relation to its usefulness in modeling and solving word problems. It is well known that this usefulness is the result of the refinement of the so-called Cartesian method,[1] whose application, widely encountered in the modern era, has led to overrating the value of the instrumental aspect of algebra.

While it is important to preserve an appreciation for the use of algebra as an indispensable linguistic support in the development of mathematics, the conception of symbolic algebra as a linguistic vehicle for "detaching" the semantics of a word problem, and thereby enabling an "automatic" solving of the problem, tends to lose sight of the principal facts in the constitution of the algebraic language. For example, this conception overlooks the following:

1) In spite of the fact that the algebraists of the mid 16th century were well aware that algebra was applicable to a wide range of problems, it was the formulation of problems that were not simple exercises of application which guided their research in the second half of that century. This research led to the generation of a more general theory, with a scope beyond the limits originally defined by the problems themselves. It is worth mentioning, as relevant examples, the geometric problems taken from the classic texts of Greek geometry, the irreducible nature of cubic equations (or the problem of the trisection of an angle), the problem of negative and imaginary roots, and that of the relationship between the coefficients of an equation and its roots.
2) The birth of literal calculus, usually associated with the figure of Viète in his *Introduction to the Analytic Art* (Viète, 1983), possesses an intimate link with the world of geometry. This is seen both at a conceptual level, in the conforming of algebraic expressions to the law of homogeneity (or equidimensionality), and at a notational level, in the designation of powers in geometric figures with nominal symbols (number (N), square (Q), cube (C), squared square (QQ)).

N. Bednarz et al. (eds.), Approaches to Algebra, 55-62.

3) Among the most significant antecedents of the birth of symbolic algebra is the development of the so-called Practical Mathematics of the Renaissance, of which the most representative printed works are the "Abbacus Books" (Van Egmond, 1980). An intensive use of techniques and strategies for practical problem solving in an entirely rhetorical language can be observed in these documents. Their importance as a stage preceding the appearance of symbolic algebraic language derives from the fact that these texts constitute the most feasible way of assimilating Indo-Arabic mathematics into the civilization of Western Europe.

The culmination of the evolution of symbolic algebra and its everyday use in the modeling of problems from other sciences has led to a situation where, according to Klein (1968), it is impossible to separate content from form in physical-mathematical science. This finished version of instrumental algebra, with all its potential as a synthetic and formal language, is what teaching sometimes attempts to communicate prematurely to the student. The well known separation that students tend to make between algebraic manipulation and its use in problem solving may originate in an educational approach based on this oversimplified vision of algebra, which hides the significance of its origins and the semantic background of its grammar.

In the following sections, Points 1, 2 and 3, which are indicated above as decisive factors in the constitution of the algebraic language, are discussed in order to extract lessons from history that have a bearing on the learning and teaching of this language in the present day.

2. IN THE BEGINNING WERE THE PROBLEMS

From its beginnings in Babylon to its culmination in the *cossists* tradition of the Renaissance, algebra has constituted a sophisticated form of solving arithmetic problems (Treutlein, 1879; Tropfke, 1933). Up until that time, with some exceptions, such as that of Diophantus and Jordanus de Nemore (Hugues, 1981) among others, it can be said that both the statement of the problem and the steps towards its solution were expressed entirely in natural language. The technical baggage that such solutions presupposed were the six arithmetic operations of addition, subtraction, multiplication, division, raising to a power, and the extraction of roots, and also the manipulation of unknowns as if they were known numbers.

During this extensive period, "the problem" and "the equation to solve it" are indistinguishable. Thus, any reflection on the role of the solution of problems in the development of algebra will necessarily have to be located in the area of the problems themselves and in the techniques and strategies for solving them.

We will start by looking at one of the ancient Greek sources, the "*Palatino*" or Greek Anthology, which contains a group of 46 numerical problems, stated in an epigrammatic form and compiled in 500 A.D. These problems, alluded to by Plato, are similar to some of those found on the Rhind Papyrus (1650 B.C.) and correspond to the categories which traditionally appear in algebra school textbooks, that is, problems of "distribution," "work," "cisterns," "mixtures," and "ages." While there are common traits in the processes of solution of similar problems, the procedures as such develop around the specific numerical characteristics of each problem.

This development of multiple strategies corresponding to the multiplicity of problems, which to the modern eye (e.g., with the perspective provided by the Cartesian model) could be identified as belonging to the same family, can be clearly appreciated in the abbacus books.

As mentioned in an earlier paragraph, the abbacus texts are the most important testimony of Renaissance Practical Mathematics. The first of these texts was written in 1202 by Leonardo Pisano (better known as Leonardo Fibonacci) and was entitled *Liber Abbaci*. It is a compendium of the practical-commercial mathematics known up to that time, based on the Indo-Arabic system (Van Egmond, 1980). Regarding their symbolic characteristics, the abbacus books involve rhetorical algebra, and in each of them (about 400 are known to date) more than 400 problems are solved in natural language. Although there is the use of rules and some general methods (such as the Babylonian method for solving the quadratic equation), the diversity of strategies applied to problems which, as we have said before, are "identical" from the point of view of the symbolic algebra currently in use, is notorious. An example of this is seen in the problems which are solved using systems of equations with the same operational structure and which differ only regarding coefficients (see the examples from the *Trattato di Fioretti* analyzed in Filloy & Rojano, 1984b).

At this pre-symbolic level, the lack of a language with which to express the procedures used from problem to problem does not permit the application of these procedures in a generalized fashion to problems pertaining to the same class or family. On the contrary, it is the specific numerical characteristics that guide the strategy for attacking each problem (Colín & Rojano, 1991). Today, if looked at from the point of view of the languages structured by adults, the language of the abbacus books constitutes a dead language. If translated to current symbolic language, sequences of actions appear that lead to the same results as ours, but which follow paths we have not thought of. Their presence, from problem to problem, book to book, shows us abilities that do not fit with those which we have developed when constructing and using our algebraic language and which we have never been driven to develop when we confront problems with our arithmetic abilities and knowledge (Filloy & Rojano, 1984b).

Perhaps, without the historical experience of Renaissance Practical Mathematics, we would not be able today to contrast so precisely pre- and symbolic algebraic thinking, where problem solving is concerned.

While medieval works, such as that of Fibonacci, represent an important antecedent in the development of Renaissance mathematics, the principal contribution of the texts and schools of abbacus is of a social and economic nature. On the other hand, the evidence of their intensive practice of solving thousands of problems and of their ways of doing so show us the limited generalization of methods and communication of solution strategies which the symbolic insufficiency of that practical mathematics implied.

3. SYNCOPATION OF ALGEBRA: PROBLEMS AND EQUATIONS

Syncopation of algebra is indicated by Eves (1983) as one of the great moments for algebra before 1650. Diophantus is considered to be the principal protagonist of the syncopation of Greek algebra for having introduced the shorthand abbreviations to denote the unknown and its powers, as well as subtraction, equality, and the reciprocals.

With numerals and stenographic abbreviations, the equation

$$x^3 + 13x^2 + 8x$$

would be written in this way: $K^{\Upsilon}\alpha\ \Delta^{\Upsilon}\iota\ \xi\ \eta$.

In his *Arithmetica*, Diophantus solves around 130 varied problems, most of which lead to indeterminate equations of the first degree in one unknown, and of the second degree in two or three unknowns. Here it is also surprising to find the absence of general methods for problem solving on the one hand, and the development of ingenious mathematical tools especially designed for the needs of each problem, on the other (Eves, 1983). Some of Diophantus' problems have become central problems in the Theory of Numbers and current challenges for professional mathematicians (e.g., Fermat's Last Theorem). But apart from this type of contribution to mathematical knowledge in general, Diophantus' *Arithmetica* presents us with the possibility of distinguishing the statement of a problem from the equation which solves it, thanks to the potential presence of two representations: the rhetorical and the syncopated.

In contrast with the work of Diophantus, where general methods are absent, *De Numeris Datis* is a syncopated work from the 13th century, written by Jordanus de Nemore, that contains a general strategy. This strategy, which could well be attributed to the symbolic level of the language he uses, puts the work at a more advanced stage than that of the abbacus texts with regard to the evolution of symbols and methods.

The language of this book is characterized by the use of letters to denote both the unknown and known quantities (unknowns and coefficients respectively) of an equation or system of equations, and the attempt is made to discover numbers of which some relationships given by constants are known. Thus, for example, in one of these relationships, the author talks about finding three numbers whose sum is known ($x+y+z=a$), instead of the case in which the given sum is a specific number ($x+y+z=228$) as appears in *Abbaco* (Mazzinghi, 1967). This not only puts *De Numeris Datis* at the level of syncopated work, but also from the outset in each case, because of its form of expression, it is solving a whole family of problems. Furthermore, the sequence in which the propositions are solved explicitly shows the reduction of each new problem to a canonic form corresponding to some problem that has been solved previously. The feasibility of recognizing analogous problems, corresponding to systems of equations with an identical operational structure, is based on the fact that the coefficients and the constant terms are generalized numbers.

De Numeris Datis is considered by some historians to be the first textbook of advanced algebra because it was written with the intention of presenting the university students of the time (presumably the University of Toulouse) with non-routine, practical problems.

To briefly recapitulate, it should be pointed out that it is in the area of problem solving, more specifically in the case of certain rhetorical algebraists of the 13th and 14th centuries, where the limits of the scope of this level of algebra can be seen. In the two cases of syncopation mentioned here, that of *Arithmetic* and that of

De Numeris Datis, the engine of symbolic evolution is located in the same terrain. Turning once again to the classic mathematical problems of Greek geometry, it is also in the area of problem solving that qualitative changes in approach occur and where a new mathematics is developed, that of the symbolic algebra of the second half of the 16th century.

4. CLASSICAL PROBLEMS AND THE VIETAN PROJECT

The shift in cultural perspective in Europe in the mid-16th century gave rise to one of the most spectacular and dizzying changes in mathematics, that of the construction of symbolic algebra. Never before had humanity been capable of creating an autonomous, specifically mathematical language, in which it was not only possible to state problems and theorems, but also to express the steps of solution and proof. According to Klein (1968), the Zetetics rules or syntactic rules of the *Introduction to the Analytic Art* by Viète in 1593 represent the first modern, axiomatic system because they create the systematic context that "defines" the object to which these rules apply.

The birth of this new system and its potential autonomy with respect to other languages (such as the geometric and the natural) mark the beginning of the dichotomy between two types of representations of a problem: the word statement and the equation or equations that solve it. However, the starting point is once again the intention of solving problems (on this occasion, they are not routine problems, but problems that the mathematics developed up to that time had left without solution). That is, the fundamental motivation of the Vietan project, according to the author in the epilogue of the *In Artem Analyticem Isagoge*, is that of *not leaving any problem unsolved*. This declaration obviously includes the Greek problems unsolved with a ruler and compass. It is thus that, at its origin, the algebraic syntax inaugurated by Viète is intimately related with elements of Greek geometry.

The problem of the trisection of the angle is an example whose history illustrates the determining role played by the revival of the Greek classics in the evolution of algebra. This problem reappears in the 16th century with very different conditioning factors from those of the constructions with ruler and compass (characteristic of the Greek approach) and unconnected with the calculation of an approximate solution (i.e., by using fractions, as found in the Arabic approach). The problem appears to be linked to the algebra for solving equations. At that time, cubic equations had been solved, but the equation solving the trisection of the angle is irreducible and the formulas of Cardano are not applicable. Thus, the problem would be reformulated as follows: Is there such a close link between the irreducible cubic equations and the problem of the trisection of the angle that one problem determines the other? (Paradís, Miralles, & Malet, 1989).

Viète and Bombelli approach the problem from different perspectives. The latter applies formal calculus and perceives by intuition the role that complex numbers would come to play, but he does not fully solve the problem. The former approaches it from a geometric perspective and, returning to the classic construction of the trisection of an angle, examines the relationship between the bases of two isosceles triangles (which appear in this construction) with equal acute angles and finds a third-degree polynomial relationship between the bases of these triangles. This relationship generates all the irreducible cubic equations when the angles are varied. This allows him to identify the two problems and to solve the irreducible cubic with trigonometric methods. All this was the result of bringing together a classic perspective and the use of the new algebraic tools of the period.

This combination of geometrical practice in problem solving and the aim of reformulating the art of analysis in order to achieve a new *Logistics*, the *Logistics of the Species*, is what leads Viète to take the great step from the language of proportions to the language of algebra. Thus it can be said that the language of literal calculus, while it appears to be autonomous to us today, is clearly derived from the geometric conceptions of its founder. An adequate reading of the following expression (a product of the application of the rules of Zetetics) allows us to understand that while the choice of literal symbols denoting both unknowns and constants is arbitrary, the way in which these symbols are "arranged" to form the expressions or the equations is not.

The laws of homogeneous terms, the writing of the powers of the unknown, and the affection of a power and the function of the supplementary terms or coefficients (i.e., a term that is added to or subtracted from a power of the unknown) permit Viète to write the expression,

$$\frac{A\ CUBUS\ -\ B\ SOLIDO\ 3}{C\ in\ E\ CUADRATIVA}$$

which in modern notation would be:

$$\frac{X^3 - 3b}{cY^2}$$

This significant progress towards the production of what would correspond to current algebraic expressions is based on a clear change in the conception of the mathematical entities involved in such expressions but, at the same time, implies a conciliation of the new concepts with a basic requirement of classical mathematics, that of the first and eternal law of equations and proportions. In modern terminology this is that *all the numbers in an equation should have the same dimension* (Klein, 1968). Thus, Parshall (1988) writes that a glance at the two versions of the above expression shows the vestiges of geometric-algebraic lineage (dimensionality) in Viète's work. But it is precisely there, where there was a need for purely syntactic progress when confronted with the difficulty of adding (or subtracting) terms with potentially different dimensions, that the importance of having conferred on the symbols or *species* the geometric meanings, which allowed the terms in question to be "homogenized," is appreciated (Rojano & Sutherland, 1991a).

5. WORDS AND THINGS

In Foucault's analysis of the experience of language in *Words and Things* (Foucault, 1966/1981), he mentions the moment at the end of the Renaissance when language ceases to be the material written expression of things and becomes part of the general regime of representative signs. In this new order, says Foucault, the question of *how to recognize that a sign designates what it signifies* is transformed and, from the 17th century onwards, becomes *how a sign can be linked to what its signifies*. The answer to this question in modern thought is given by the analysis of meaning and signification and, from this moment onwards, "the deep mutual possession of language and the world is undone, words and things separate and discourse will have the function of saying what is, but will be no more than what it says" (Foucault, 1966/1981, p. 50; our translation of the Spanish edition).

In the case of mathematical language, the birth of symbolic algebra clearly marks this moment of separation between signs and what they represent. Algebraic syntax gains a life of its own and becomes disconnected from the semantics of the problems from which it originated and, in a dramatic change of roles, becomes the problem-solving tool *par excellence*.

6. LESSONS FROM HISTORY

In this brief review of the great moments in the history of algebra, demonstrating the determining role of problem solving in the generation of new knowledge in this discipline up until the culminating moment when literal calculus intervenes, we can see a constant (which consists of the determination of a sort of situation-at-the-limits to which mathematics is carried at each one of these moments) as a function of the families of problems that can and cannot be solved with existing knowledge.

The latter can be taken as a lesson from history as to the meaning conferred by problem solving on new knowledge. This lesson, taken up at an ontological level and in the field of teaching and learning algebra, warns us of the risk of placing symbolic manipulation as an object of knowledge in advance of the situations which can give it meaning.

Secondly, the history of the birth of *Species Logistics* or Vietan algebra calls our attention to the original meanings of the symbols involved in algebraic expressions and to the laws governing syntax or grammar which make writing them possible (the terms "supplementary" or "coefficients," for example, have the function of "compensating" the dimensions of the terms containing the unknown or one of its powers, in order to conserve the homogeneity of an equation). This leads to a reflection on the effects of starting algebra at school with a syntax but without references immersed in contexts that are familiar to the students.

A paradoxical aspect of the figure of Viète as the father of symbolic algebra is that which shows, on the one hand, his brilliant invention of the Logistics of the Species as the point of rupture with previous algebraic thinking, which subordinated the development of methods and strategies for solving problems to the specificities and semantics of the particular problems and, on the other, his ability to use the knowledge and skills developed by classical Greek mathematics (specifically, geometry and arithmetic) to achieve his aims. This apparent contradiction, found in one of the central protagonists of these profoundly significant changes in the history of algebraic thought, points towards the importance from the outset of not denying the value of the knowledge, methods, and skills that the students already have (however primitive or informal they may be) when they are to be initiated into a new area of knowledge, specifically, symbolic algebra and its use.

Finally, the birth of Cartesian or analytic geometry, identified with the moment in which geometry ceases to play a fundamental role in the justification of algebraic theorems and, on the contrary, starts to use the instruments of symbolic algebra for its own ends, indicates an exchange of roles between these two mathematical disciplines. Furthermore, it shows a recognition of a situation-at-the-limit in the evolution of one of them (geometry) and the potential of the other (algebra) to intervene in overcoming the stagnation of the former. This lesson from history has implications for teaching in the sense that the potential of dominating algebraic syntax will not be appreciated by students until they have experienced the limits of the scope of their previous knowledge and skills and start using the basic elements of algebraic syntax.

What we have called here "lessons from history" could be used as the basis for formulating a theoretical framework for the elaboration of proposals for approaches to the teaching of school algebra. The development of such a project goes beyond the scope of this article. However, it is worth mentioning that results obtained in recent empirical studies on the acquisition of algebraic language point in the same direction as some of the lessons mentioned above and can be considered the empirical counterpart of the theme with which we have dealt.

NOTES

1 Polya (1966a) adapts the Cartesian rules for problem solving as follows:

1) In the first place, get a good grasp of the problem and then convert it into the determination of a certain number of unknown quantities (Rules XIII to XVI).

2) Examine the problem in a natural fashion, thinking of it as already solved and putting in order all the relationships between the unknowns and the data that should be verified, according to the conditions stated (Rule XVII).

3) Separate out a part of the conditions which allows the expression of the same quantity in two different ways, thus obtaining an equation between the unknowns. Eventually, separate out the conditions into various parts. In this way a system with as many equations as unknowns will be obtained (Rule XIX).

4) Transform the system of equations into a single equation (Rule XXI).

PART II

A GENERALIZATION PERSPECTIVE ON THE INTRODUCTION OF ALGEBRA

CHAPTER 5

EXPRESSING GENERALITY AND ROOTS OF ALGEBRA

JOHN MASON

This chapter makes the case that the heart of teaching mathematics is the awakening of pupil sensitivity to the nature of mathematical generalization and, dually, to specialization; that children who can walk and talk have shown plenty of evidence of the requisite thinking; that algebra as it is understood in school is the language for expression and manipulation of generalities; and that the successful teaching of algebra requires attention to the evocation and expression of that natural algebraic thinking. There is no single program for learning algebra through the expression of generality. It is a matter of awakening and sharpening sensitivity to the presence and potential for algebraic thinking. Some examples of tasks which have been exploited in this way are presented with comments about some of the difficulties encountered.

1. PREFACE

My long-standing conjecture, has been, and still is, that when awareness of generality permeates the classroom, algebra will cease to be a watershed for most people; that algebra as it is usually interpreted at school is a dead subject, akin to conjugating Latin verbs and memorizing parts of flowers--that is, unless and until, expressing generality becomes natural and spontaneous in the conduct of mathematics, imbuing every mathematical encounter and topic. Generalization is the heartbeat of mathematics, and appears in many forms. If teachers are unaware of its presence, and are not in the habit of getting students to work at expressing their own generalizations, then mathematical thinking is not taking place.

School algebra tends to be associated with numbers, and latterly, with functions on numbers. There are other domains in which symbolic expression of generality can be studied and developed (see, e.g., Bell, this volume). The reason for emphasizing expression of generality in number patterns is only to provide experiences which highlight the process.

One way to work at developing awareness of generality is to be sensitized by the distinction between *looking through* and *looking at*, which leads to the primal abstraction and concretization experiences, namely *seeing a generality through the particular* and *seeing the particular in the general.* These are enhanced by distinguishing between working through a sequence of exercises, and working on those exercises as a whole. The difference between looking-through and looking-at, and working-on and working-through applies to the use/misuse of manipulables (whether concrete apparatus or abstract symbols) and is described in a spiral of development from confident manipulation, to getting-a-sense-of, to articulating that sense, to that articulation becoming itself in turn confidently manipulable. Others refer to this in different language, and with different emphasis, for example, as

N. Bednarz et al. (eds.), Approaches to Algebra, 65-86.

reification (Sfard, 1991, 1992), as reflective abstraction (Dubinsky & Levin, 1986), as concept image (Tall & Vinner, 1981), and so on.

2. INTRODUCTION

The theme of generality has been implicitly employed and explicitly described many times by many people, from Pappus to Wallis to Polya and Krutetskii, from Viète to Jordan to MacLane, and generalization lies behind a variety of algebra curricula (e.g., the Nottingham Mathematics Project, the New South Wales Curriculum). I mention three resources that I have been involved with, not because they are especially important, but because I am most familiar with them:

Routes to / Roots of Algebra (Mason, Graham, Pimm, & Gowar, 1985)
Expressing Generality (Mason, 1988)
Supporting Primary Mathematics: Algebra (Mason, 1991b)

The inspirational sources for these works were many, and often ancient (e.g., Babylonian Tablets, the figurate numbers of Nicomachus, Egyptian puzzles, medieval puzzles, etc.). But they are based on the premise that, of the four principal roots of algebra we identified:

Expressing Generality
Possibilities and Constraints (supporting awareness of variable)
Rearranging and Manipulating (seeing why apparently different expressions for the same thing do in fact give the same answers)
Generalized Arithmetic (traditional letters in place of numbers to express the rules of arithmetic),

expression of generality is of paramount importance precisely because it is so often overlooked and under-played. Facility in manipulation of generality follows as confidence in expression develops and as multiple expressions for the same thing arise. The use of algebra to solve problems depends on confident expression of generality using the as-yet-unknown (Tahta, 1972), supported by awareness of the role of constraints on variables. That is not to prescribe an order in particular cases, but rather in the overall development of individuals' appreciation of generality. Some manipulative facility can be gained prior to appreciation, comprehension, or understanding, but it is maintained only if explicit outer success is accompanied by implicit inner success at making sense.

Generality is so central to all of mathematics that many professionals no longer notice its presence in what is, for them, elementary. But it is precisely the shifts of attention that experts have integrated into their thinking which are problematic for novices. Generalization is not just the culmination of mathematical investigations, as many seem to think. It is natural, endemic, and ubiquitous.

How might my conjectures and assertions be tested? Certainly not by describing yet another curriculum project with materials that teachers can administer to pupils. I am referring to a manner of thinking and acting, a cultural shift, in which teachers feel comfortable in acting mathematically with and in front of their pupils; in which pupils are enculturated into mathematical thinking and expression just as naturally as they are into listening to and speaking their native tongue. Hence the title of Pimm's (1987) book, *Speaking Mathematically*, which focused attention on the variety of ways in which we unwittingly, socially and culturally, misdirect pupil attention in the mathematics classroom. Seeger (1989) offers similar views:

> Algebraic thinking has to be cultivated ... to the same extent that nowadays the practical, hands-on approach, and the use of manipulative aids is cultivated.

He built on Davydov (1990) who stresses the need to achieve a balance between the theoretical and the empirical, even at the youngest age:

> Emphasis on "empirical thinking" ... explains "what went wrong" in the primary grades.

I find that the most potent force for developing my own sensitivity to the role of generality is focused by questions such as:

> What are students attending to? What am I stressing? What am I seeing and saying, and where are they in the spectrum from particular to general?

3. PARTICULARITY AND GENERALITY

When teachers or authors do an "example" for students, their experience is often completely different from that of their audience. For the teacher, the example is an example *of* something; it is a particular case of a more general notion. As the teacher goes through the details, the specific numbers or items are experienced as place-holders, as slots in which different particulars could appear. For the student, the example is a totality. It is not seen as illustrating a generality, but as complete in itself. Items, which to the teacher are particular instances, are for many students indistinguishable from the other elements of the example. The students' task is to re-construct generality from the particular cases offered. Often students do this brilliantly, but inappropriately, because they unwittingly stress aspects that the teacher does not, and vice versa.

3.1. *Example: Angle Sums*

The sum of the angles in a triangle is 180 degrees. What is the most important word, mathematically, in that assertion? I suggest that it is the most innocuous, the smallest, least noticed word, and that the true meaning of that word is construable only when you appreciate the nature of mathematical assertions. The word I have in mind is *a*. That tiny indefinite article signals the adjectival pronoun *any*, which in turn signals the adjective *every*, which refers to the scope of variability being countenanced. I suggest that the second most important word is the definite article *the* modifying *sum*, for it asserts something particular which complements the generality, namely that in the domain of all possible triangles, as you change the triangle you look at, the angle-sum remains constant. It is invariant. The fact that this constant is 180 degrees is of relatively little importance. Yet in many classrooms, it is the 180 that is stressed, presumably so that students will remember it. But failure to use this fact is rarely due to forgetting whether it is 180 or some other number, but rather, due to lack of appreciation of the generality, the invariance, being expressed.

The essence of the angle-sum assertion, and indeed, I conjecture, of most mathematical assertions, lies in the generality which can be read in it. There is some attribute which is invariant, while something else roams around a specified or implied domain of generality. Do pupils appreciate this? Are they aware of the stressing that a mathematician makes, at least implicitly if not explicitly, when reading mathematical assertions?

Notice that I have begun my investigation into generality with a particularity, a single mathematical assertion taken from the secondary curriculum, but offered as a generic example. It is essential, therefore, that you pause and consider the generality concerning definite and indefinite articles, and the need to read implicit generality in what often appears to be very particular assertions that are being put forward. To do that effectively, you need to specialize, testing it in your own experience.

3.2. *Example: The Square-Enigma*

Consider the diagrams shown in Figure 1.

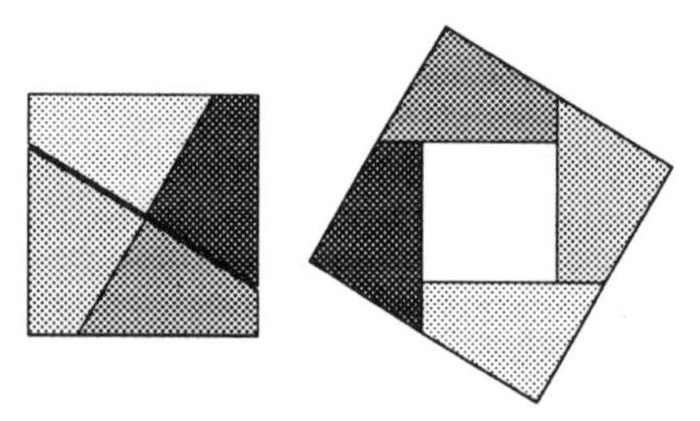

Figure 1.

Normally these two figures appear as a puzzle: You are given the four pieces of wood or plastic, and asked to "make a square," then to "make another square." Once the two squares are found, moving from one to the other is not difficult, but attention *on* that movement reveals that the pieces slide but do not rotate. It is possible to formulate a story[1] that connects the two, drawing on jigsaw-awareness that the area of the pieces is invariant under movement of the pieces: "The difference in area of the outer and inner squares on the right is itself the area of a square."

Implicit in the particular puzzle is the generality that the difference of *any* two squares can be depicted similarly. And hiding in the indefinite *similarly* is another aspect of generality.

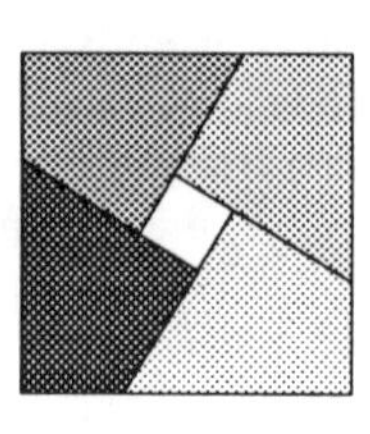

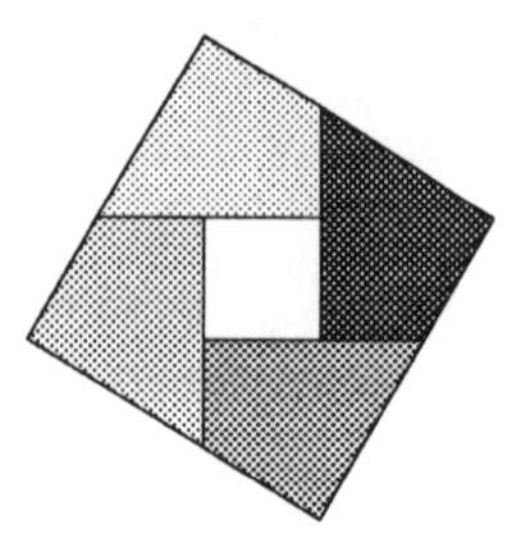

Figure 2.

What is particular, and what general, about the particular pieces of the puzzle? Can the pieces be seen not as given, but as indicating a construction that can be performed? What aspects (angles, lengths, ratios) are essential, and what variable? What aspects can be changed while preserving similarity with the particular case given? For example, the arrangement of Figure 2 have some similarities with those of Figure 1, and the following images in Figure 3 might even help as intermediate frames in imagined transformation from the first pair to the second. Such transformations could be done with physical shapes, animated on a computer, or achieved in a geometrical drawing package.

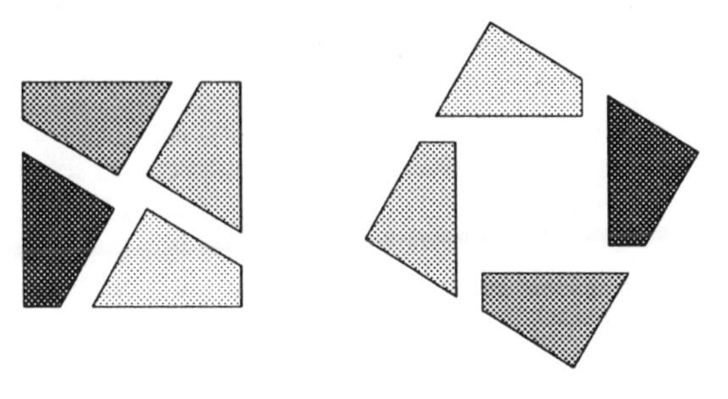

Figure 3.

Finally, there is the connecting question about how this puzzle is related to Pythagoras' theorem, and the extending question as to whether similar dissections are possible with other shapes, such as equilateral triangles, pentagons, and so on.

Students are imbued with the cultural myth that mathematics is the one subject where you know when you have got the correct answer. But this is only part of the story, for it is also the case that mathematical investigations are never finished, merely suspended. A satisfactory place to leave an exploration is with a conjecture, together with whatever supporting evidence and reasoning have been accumulated so far.

Generalization, in which I include variation and extension, as well as pure generalization, is one means for broadening the scope of reference and application of a result, thus placing it in ever broader contexts by removing particular restrictions. Another is connection-seeking, staying with the particular, but trying to see what other aspects of mathematical thinking, both topics and processes, are related or drawn upon.

In the case of the puzzle just presented, there are connections to area and jigsaws, to Pythagoras and to number theory, drawing upon *seeing generality through particularity*, and transition from geometry to algebra and back again. The fact that a teacher has these awarenesses says nothing about what pupils experience, but if teachers are unaware of these dimensions, most pupils are unlikely to appreciate them. That does not mean that the teacher has to be explicit. When actions are informed by awareness, there is a better chance of pupils' awareness being activated than if the teacher acts mechanically.

The square-enigma puzzle is offered as a generic example of the use of manipulative apparatus. Apparatus, and indeed any confidently manipulable entity whether physical, diagrammatic, or symbolic (Mason, 1980) can act as a means for individuals to relate to a question, and to resort to, when they want something particular to manipulate. But its very particularity has to be used as a window to be looked through, rather than as a wall to be looked at (Griffin & Mason, 1990). The *generalized arithmetic* root of algebra offers but one instance of an opportunity to draw attention away from the particular and onto the processes which "make those objects what they are."

3.3. *Example:* $3 + 2 = 2 + 3$

The structure of arithmetic, when expressed, produces algebra as generalized arithmetic--one of several roots of algebra.

After a succession of three paired questions, of the form 3 + 2 is ... 2 + 3 is ..., a five-year-old suddenly made the spontaneous utterance: "Five plus something ... something plus something is ... the same as something plus... something."

And there is a further potential shift in perspective to be able to say: "Order doesn't matter."

It is unclear whether the pupil appreciates that $3 + 2 = 2 + 3$ because both are equal to 5, or because of deeper, structural reasons (order does not matter). Awareness can vary from the tacit and action based, to the explicit and articulate. Children can act as if they are aware that all the chairs in the classroom have four legs without saying anything about it; a child can say "a chair in the classroom has four legs" without necessarily recognizing implicit generalities (what is a chair, *all* the chairs, etc.). Neither action nor utterances alone are guarantors of perceived generality, merely indicators. Contemplating the properties of numbers is one way of standing back from engagement in the particular, and becoming aware of processes. Gattegno (1990) proposed that algebra emerges when people become aware of the awarenesses that permit them to operate on objects (in this case number). Generalized arithmetic is but one context for this process of standing back from the particular.

3.4. *Example: Materiality and Particularity*

Children often evince great pleasure and effectiveness in generalization, in asking what-if questions, pondering alternative realities, and playing with language. At school, they show a predilection for the concrete, and this has led to apparatus-based practice being promulgated throughout the educational community. Teaching theorists, building on misinterpretations of Piaget's epistemological insights concerning the role of concrete objects, compound this apparent preference for the material world, to produce programs of instruction that stress the particular. The effect is to draw attention away from the general, and sometimes to emphasize the particular in ways that make it harder rather than easier to appreciate the general. Davydov (quoted earlier) agrees that this is where major pedagogical errors have been made.

Is using a particular compass at a particular setting to draw a particular circle any indication of awareness of the circle-drawing potential of compass-like tools? Does manipulation of Dienes' blocks or Cuisenaire rods necessarily lead to appreciation of what the educationalist sees as being instantiated? Does software that provides particular instances necessarily lead pupils to awareness of generality?

Clearly all the evidence is against such conjectures. More is needed. The presence of someone whose attention is differently structured, whose awareness is broader and multiply-leveled, who can direct or attract pupil attention appropriately to important features, is essential. It is customary to cite Vygotsky (1978) as justification for proposing that students get access to higher psychological functioning by first being in the presence of such functioning and being embedded in practices employing these. However I believe that this notion is at least implicit in the teaching of mystics throughout the ages, not to say educators in every age. Vygotsky did put it particularly succinctly however.

3.5. *Example: Remainder Arithmetic*

Remainders, and modular arithmetic, provide a rich context in which to practice expressing generality, while illustrating some of the many elliptical expressions and implicit referents in mathematical discourse.

Notice that 1, 4, and 7 leave a remainder of 1 when divided by 3. Any number which has that property can be written in the form $1 + 3n$ where n is any integer. The form, $1 + 3n$, is succinct, and peculiar. In one sense, it denotes a single number, depending on the value of n; in another sense it describes the structure of a number, or indeed a set of numbers, namely those that leave a remainder of 1 on dividing by 3; in another sense it *is* a number, just as 3 denotes or names a number but is often taken as being a number; in another sense it is a rule for calculating a number, an expression; in another sense it is the result of carrying out that calculation.

And notice the implicit shifts in the following:

- "leaves a remainder of 1" in three instances,
- "leaves a remainder of 1 in three instances,"
- "leaves a remainder" as a property of numbers in general,
- "leaves a remainder as a property of numbers in general,"
- "leaves a remainder of one on dividing by three" as a potential property for numbers in general.

More generally, $r + 3n$ describes all numbers that ... and $r + kn$ denotes .... There are usually many levels of generality signifiable by even the simplest particularity.

3.6. *Example: Exercises*

Students are often set the task of completing a collection of exercises in order to achieve facility with a technique. The very term *exercises* suggests a metaphor of physical training in which muscles are developed and reactions quickened through repetition. But doing mathematical exercises is about recognizing "types" of questions and about being able to re-construct a general technique for oneself. Automating a technique is only one aspect of "practice to make perfect." The trouble with mindless practice is that it takes place in a rarefied context, and often mindlessly. Practice tends to focus attention on precisely those aspects of a technique which have to be done without attention when the technique is mastered.

3.7. *Example: Language*

There is nothing odd or peculiar about seeing through a particular instance to a generality. It is part of everyday activity and the competent use of language in particular. In speaking, children show the power to generalize because words are general not particular, and speaking represents a movement away from dwelling with the particular (through pointing), to dwelling with the general that is signified. To speak is to signal recognition of a situation as similar to others, to participate in classification and generalization. It is to see the general through the particular and the particular through the general (Mason & Pimm, 1984). Seeing the particular in the general is specializing, which makes application of theory, of accumulated experience, of perspective, possible. It is a non-trivial act, requiring a letting go or de-stressing of the particular and a seeing through to the essence, the form of the situation.

3.8. *Reflection*

I have offered things that I have labeled *examples*. You must work at what it is you think is being exemplified. That is how examplehood works. Until you can see my "examples" as *examples of* something, they remain isolated and of little meaning.

In the 1960s, there was a growing awareness that graduate students in mathematics were induced to dwell in the general to such an extent that they lost contact with the particular. The slogan, "Be wise, generalize," took hold of graduate schools. Halmos (1975) tried to regain a balance by praising the particular. But always there is that human need to generalize, and indeed pure mathematics provides a home for people seeking to avoid the nasty complexities of the particular.

Man muss immer generalisiern.
[One should always generalize.]
(Jacobi, in the 1840s, cited in Davis & Hersh, 1981, p. 134)

I find it sensible and consistent to consider Plato's world of forms as a world that can be experienced, and manipulated, in concert with the material world. The material world has no generality. It is entirely particular. Measured angles of physical triangles do not add up to exactly 180 degrees; jigsaws do not exactly partition. The world of electronic screens is similarly remarkably particular. As soon as expression in language takes place, the world of forms is invoked and broached. Islamic educationalists use a similar justification for compulsory mathematics at school: It provides access to and experience of the kind of abstract thought that is essential for spiritual awareness. I am using *language* in a very general sense here, following Maturana (1978): "Language is the consensual coordination of the consensual coordination of action."

Despite apparent abstractness and generality, it is in fact through the world of forms that coordination in the material world takes place. Penrose (1991) takes a very firm if extreme stance along these lines:

"Seeing" ... is the essence of mathematical understanding. When mathematicians communicate, this is made possible by ... the consciousness of each being in a position to perceive mathematical truths directly. Since each can make contact with Plato's world directly, they can more readily communicate with each other than one might have expected. (p. 428)

In the midst of seeing, it is rare to be aware of the stressing and ignoring (another Gattegno phrase), which is making that seeing possible. And it is also the case that the general can usually only be approached through the particular, which is why, for example, I began this section with a particular assertion rather than a general thesis, and why most arithmetic textbooks (I include Babylonian tablets, Egyptian papyri, Chinese manuscripts, commercial arithmetics, and the like) have traditionally instructed in the performance of technique by providing *worked examples*, from which the student is to infer the general.

The whole notion of *example* depends upon and draws out the notion of generality. My opening assertion about the sum of the angles of a triangle, or any other assertion, cannot be an example *of* anything until someone comes along and sees it as such. If a teacher puts "examples" on the board, or an author puts "examples" in a text, they remain undifferentiated text until a student construes them as examples *of* something, and that can be done only by seeing through the particular to a generality. Then there is the question of whether the student's generality is the same as the author's. Many pupil confusions are literally that--fusions of unintended (by teacher or author) student stressing, leading to a wide range of errors and bugs, the study of which constitutes a significant proportion of mathematics education research and publications (Brown & Burton, 1978; van Lehn, 1982).

4. ALGEBRA ITSELF

Algebra is not a single thing. The word *algebra* resonates with (brings into awareness) a variety of experiences, some culturally shared, some idiosyncratic. It is derived from problems of *al-jabr* (literally, adding or multiplying both sides of an equation by the same thing in order to eliminate negative/fractional terms), which were paralleled by problems of *al-muqabala* (subtracting the same thing from or dividing the same thing into both sides). The root meaning is to equate, to compare, to pose opposite. It is purely chance that we are discussing *Algebra* rather than Almukabala, or even Mukjabra! Van der Waerden (1980) suggests that for a while, historically, the combination of the two words, *al-jabr wal-muqabala* was sometimes used in the more general sense of performing algebraic operations.

Brahmagupta (7th century) is translated as saying:

Since some questions can scarcely be known without algebra, I shall therefore speak of algebra with examples. By knowing the pulverizers, zero, negative and positive quantities, unknowns, elimination of middle terms, equations with one unknown, factum and square-nature, one becomes a learned professor amongst the learned. (Quoted in MA290, Open University Course)

Algebra is usually what people have in mind when they think of mathematics as a language: strings of algebraic symbols. But strings of symbols are not in themselves algebra. The meaning of algebra has developed and broadened, passing from process (algebra) to object (an algebra), in time-honored mathematical fashion. At school, algebra has come to mean "using symbols to express and manipulate generalities in number contexts." Yet every discipline is concerned with expressing generality, with differences only in what the generality is about, and how generalities are justified (Mason, 1984).

Generalization is the life-blood, the heart of mathematics. Certainly there can be very difficult particular questions, such as trying to find the sum of the squares of the reciprocals of the positive integers, the 397th digit in the decimal expansion of $\sqrt{2}$, the quadrature of a particular region, or finding the solution set of a particular polynomial, and these questions have indeed occupied many mathematician-years of effort. But in almost every case, there is an implied generality, either because it is hoped that the techniques will apply more generally, or, as in the case of the squared reciprocals, because it can be recast as a statement about an infinite set of numbers approaching a particular limit.

Bednarz and Janvier (this volume) demonstrate the contrasts between arithmetic thinking (direct, known to unknown) and algebraic thinking (indirect, unknown via generality to known) in the context of word problems. In an atmosphere of generality, the as-yet-unknown is as familiar and confidently manipulable as the known, and often better than most!

Gattegno (1990) coined the assertion: "Something is mathematical, only if it is shot through with infinity." I take this to mean that to be fully mathematical, there has to be a generality present. His assertion can be quite useful as a reminder, in the midst of a lesson, to attend to the infinity oneself. He suggested that algebra as a disciplined form of thought emerged when people became aware of the fact that they could operate on objects (numbers, shapes, expressions), and could operate on those operations. Thus, when you are able to think about combining arithmetical operations, you have begun to do algebra, since you are operating on operations of objects. The algebras of tertiary mathematics (groups, rings, fields, semigroups) all arise as the structure of operations on operations on objects; so this view of algebra provides a connection between school algebra (as it could be) and higher algebra. Word problems are a context in which to display one's ability to operate on operations by denoting and manipulating them to express relationships.

Generality is not a single notion, but rather is relative to an individual's domain of confidence and facility. What is symbolic or abstract to one may be concrete to another (Mason, 1980). Put another way, attention is attracted by the particular, by what is confidently manipulable. It may require conscious effort of letting go to "perform a Mary Boole" and acknowledge ignorance, and then to use the expression of that ignorance to express relationships and resolve problems.

Algebraic awareness requires, perhaps even consists of, necessary *shifts of attention*, which make it possible to be flexible in seeing written symbols

- as expression and as value,
- as object and as process.

Transitions from one way of seeing to another invoke experiences which echo similar struggles of our intellectual ancestors. For example, number names used as adjectives to describe order in a sequence, become nouns. Attention shifts, and at the same time, the shift applies not just to number names actually used so far, but to all potential number names. Three million, four hundred and fifty-six thousand, five hundred and forty-three has probably rarely been used by children in counting a set, yet they reach a perception in which such a symbol sequence is seen as a potential number, not just a number of things.

5. DIFFICULTIES IN THE CLASSROOM

The following examples of difficulties with generality encountered by pupils are not intended to be comprehensive. All arise from exposure to generalization without it pervading the classroom atmosphere sufficiently.

5.1. *Example: Rush to Symbols*

Rushing from words to single letter symbols has marked school algebra instruction for over a hundred years. Formal treatments, which have been very successful for the quick-thinkers and the non-questioners, have dealt with symbols as unknown numbers, and embarked on a series of games with rules for manipulating these. But meaning-seekers and those less able to achieve quick success by recognizing simplification patterns spontaneously, have deserted mathematics, maintaining algebra as the principal mathematical watershed for society as a whole.

I set a group of students the classic medieval eggs problem, in which the eggs carried to market leave remainders of 1, 2, 3, 4, 5, and none, when put in groups of 2, 3, 4, 5, 6, and 7 respectively. What is the least number of eggs brought to market?

The students rushed in to write down equations involving unknowns, showing a fine sense of expressing generality. However they were unable to do anything with all their equations. This is an algorithm-seeking question, not a simple algebra question.

5.2. *Example: Pattern Spotting Can Remain Trivial*

Tony came up and said he'd found a pattern (in looking at numbers that have an integer as the result of subtracting 1, then dividing by 4, then multiplying by 3) "All the second numbers are in the three times table." I asked him if that was always true and he went off to check. Several children noticed that all the successful starting numbers were odd.

How was he to check? Try lots more examples? How can a teacher assist the pupil's attention from a pattern observed in answers, to the method of generating those numbers? Is it effective use of Tony's time to spend much of a period until he sees that since he has multiplied by 3, the answers will be in the three-times table?

5.3. *Example: Formulating Formulas*

In seeking formulas for counting figurate number patterns, such as triangular numbers, there are three major perceptions and approaches:

- Manipulating the figure itself to make counting easier (a standard mathematical device, manifested in the famous story about Gauss and the triangular numbers, in summing of an arithmetic progression more generally, and varied for use in summing of a geometric progression);

- Finding a local rule (recursion) which reflects one way to build the next term from previous ones (a standard mathematical device which developed into the study of finite differences and generating polynomials);
- Spotting a pattern which leads to a direct formula (a preoccupation of pre-computer mathematics influenced by the Cartesian program to exploit the mechanical lawfulness of nature by finding explicit formulas).

All three of these approaches are for generating conjectures. Wallis, in his 1685 book *A Treatise of Algebra*, seems to have been the first to call this idea *his method of investigation*, and seems to be the source of the now popular term *investigational work* in mathematics education in the UK and Australia. Wallis' approach was castigated by Fermat as unmathematical behavior, because he wanted generalizations to be deduced not induced. Once students have a conjecture, they then have to verify that it always works by recourse to the source of the sequence. There is a world of awareness of generality in the *always works,* which seems to escape most students. Unfortunately there is already an established practice of making a table, guessing a formula, checking that it works on one or two more examples, and then moving on to the next question. I suggest that students remain unaware of the generality in a formula they conjecture because, in most mathematical topics, teachers collude with them to keep their attention focused on the technique in particular cases, and not on the technique qua technique, together with questions about the domain of applicability of the technique. With the rapid growth of software that will "do" most mathematical questions that are traditionally set at school, it would be possible for students and teachers to stress generality and the domain of applicability and reduce time spent on mastering the application of specific techniques.

Arzarello (1991b), studied the responses of some 11- and 16-year-olds who were presented with both a closed formula and a recursive formula for the sequence of figurate square numbers and were asked to do the same for the triangular numbers (presented as the familiar staircase). The difficulties experienced by the students of Arzarello's study reflect some of the classic struggles found in the research literature regarding:

- handling subscripts (e.g., using $T_{n_1} - T_{n_2} = n$ to express the difference n between consecutive triangular numbers; being uncertain as to whether $T_{n+1} = T_n + n + 1$ and $T_n = T_{n-1} + n$ say the same thing) (see Mason, 1989);
- confusing the particular with the general (e.g., moving from the case of $n = 5$, in which $5^2 - (1 + 2 + 3 + 4)$ gives the right answer, to: $T_n = (n+5)^2 - (n+1 + n+2 + n+3 + n+4)$);
- moving from a similarly incorrect generalization to a form that can be made to work (e.g., finding that $T_3 = 3^2 - 3$, proposing that $T_n = n^2 - n$, finding it gives wrong answers, so adjusting it to $T_n = n^2 - (T_n - x)$ where x has to be determined);
- using a version of the area of a triangle to find a pattern in the numbers that have to be multiplied to give the answer in each case and generalizing to a specific large case ($n = 100$) and thence to n;
- looking for a local rule (i.e., a recursive formula) rather than a global rule (i.e., a closed formula).

Many teachers at all levels[2] have found that student attention is often caught by parallel rules of growth in two columns of a table (x grows by 1 and y by 10). Pushed to express this as a formula, students quite naturally try things like $x + 1 = y + 10$, while the relation $y = 10x + 3$ remains hidden from view.

Part of the enculturation process into the ways of mathematical thinking includes exposure to fruitful ideas and ways of thinking: manipulating diagrams (so as to display the squares as $T_n + T_{n-1}$), use of subscripts as a convenient (functional) notation, techniques for passing from a recursive to a closed formula, and so on.

Furthermore, there is a world of difference between being at the mercy of yet another task from the teacher, and generating your own tasks yourself. The thrust of expressing generality is that students take over more and more responsibility for recognizing and expressing generality, and verifying the associated conjectures.

Gardiner, quoted in Tall and Thomas (1991), set the following task in a school problem solving contest:

> Find a prime number which is one less than a cube. Find another prime number which is one less than a cube. Explain! (p. 127)

Many students were reported as finding that $2^3 - 1$ is prime, but without the transition to form rather than particular, to $n^3 - 1 = (n - 1)(n^2 + n + 1)$, their searches would prove fruitless. As Tall and Thomas say:

> There is a stage in the curriculum when the introduction of algebra may make simple things hard, but not teaching algebra will soon render it impossible to make hard things simple. (p. 128)

They go on to report students' persistent behavior, on encountering the task, "Factor $(2x + 1)^2 - 3x(2x + 1)$," of expanding the whole, regrouping, and then trying to spot factors, rather than directly seeing the common factor in the form presented. They account for this behavior in terms of an addiction to serialist processing based on instrumental understanding, responding to the first approach that comes to them without standing back and contemplating the task and the given.

Generalization is usually taken to be an inductively empirical activity in which one accumulates many examples and detects pattern. But the most powerful generalization is usually quite different. Hilbert (Courant, 1981) and Davydov (1990) both refer to mastery of a single example that, with appropriate stressing and consequent ignoring of special features, serves as a generic example from which the general can be read. This is not a distant feature of advanced mathematicians, but can be experienced at all levels. For example, simple matchstick patterns can often be split up and "seen" in a variety of ways (see *Rows and Columns* later in this section; Mason, 1988; Mason, 1991b; Mason et al., 1985). It takes only the splitting of one example to be able to read the general, once you have seen it donc and participated yourself. Providing a generic example is really what teachers are doing when they do an example publicly: They are offering a generic example. But it is usually assumed that the students see what is particular and what is general in the same way as the teacher, despite evidence to the contrary every time the example-exercise pedagogy is executed.

Lee and Wheeler (1989) studied the responses of 15-year-olds to questions on the boundary between algebra and arithmetic, in order to gain access to their conceptions of algebraic generalization and justification. Significant discordance was found between arithmetic and algebra: Different rules applied, and answers obtained in one domain did not have to agree with answers obtained in the other. They show that the superficial reliance on algebra as generalized arithmetic, as letter-arithmetic is in itself insufficient to obtain mastery and smooth transition.

5.4. *Example: Arithmagons*

As part of the assessment for an in-service course on mathematics education, taken predominantly but not exclusively by teachers teaching from reception through to age 13 or 14, students were asked to explore the reasonably well known setting in which three numbers are given at the vertices of triangle, and the task is to find numbers that will go on the edges so that the edge numbers are the sums of the vertex numbers. That task is of course unproblematic. Then they are asked to consider the associated undoing task: Given numbers on the edges, are there corresponding numbers on the vertices. Extensions and generalizations in many directions are both possible and fruitful: Changing the operation, and extending to other polygons and networks, are the two usual directions. There was not time for all of that in the examination, but what did emerge was a wide variation in manifested generality and particularity.

The question offered a particular set of three numbers on the edges, as illustrated in Figure 4,

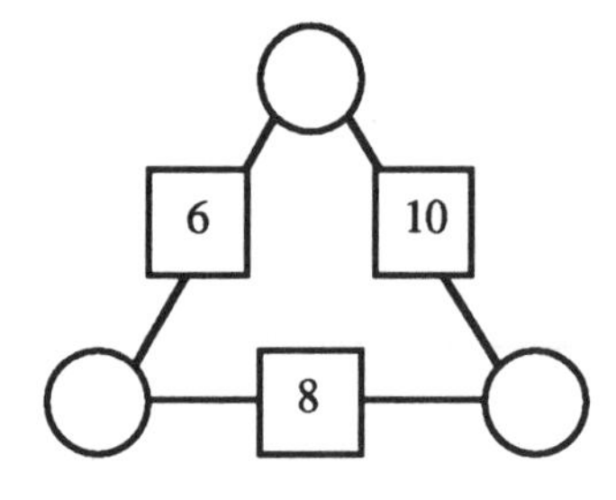

Figure 4.

and asked if there is more than one solution, and then asked for investigation of those sets of three numbers which could be put on the edges so that a solution could be found for the vertices. It was left entirely open as to what sorts of numbers would be permitted.

Most candidates struggled merely to cope with the particular, and barely approached the more general investigation. Activities noted included:

- trying numbers apparently at random, and finding (or not finding) a solution in the particular case;
- writing out all the edge numbers as a sum of two positive whole numbers, then (mostly) spotting a solution;
- starting with one entry and working out the other two (usually getting a conflict), then changing the starting number by one and recalculating

(an approach which Indian mathematicians formalized into the technique of false position).

Where candidates got beyond the particular, activities included:

- trying other sets of three particular numbers, and seeking solutions, then trying to find a pattern in those solutions, some generalizable, some very particular (e.g., relationship between differences);
- using letters to express the unknown vertex numbers, but then resorting to specializing in order to solve those equations;
- trying cases that are multiples of the original edge numbers, but not seeing that the solutions are likewise multiples;
- using algebra to resolve particular equations, but not general equations (with unknowns for the edges);
- encountering fractions and negatives, but accepting these as solutions in particular cases (while others clearly worked only with positive whole numbers);
- using non-examples (solutions not whole numbers) to assist in seeking pattern.

The course had spent some time on the notion of expressing generality, but no mention was made of techniques for solving simultaneous equations.

Conjectures involved:

- believing that only one solution is possible, based on an inability to find any others in the particular case through inspection and adjustment;
- seeing that the three vertex numbers add to half the sum of the edge numbers, but not combining that with a particular edge sum to deduce immediately the opposite edge number;
- seeing from two cases that, if the edge numbers increase by 2 in a clockwise direction, the vertex numbers do the same in an anti-clockwise direction, but with no time to explore this and no hint of the potential generality of 2;
- thinking that perhaps the middle number has to be even with an equal amount added or subtracted to make the other two, but that doesn't meet all the cases found.

Candidates who made explicit use of the course rubric (framework of what to do when you are stuck: Acknowledge, locate what you want and what you know, and try specializing in order to re-generalize)[3] were able to make more substantial progress than those who did not. They were frequently able to say what they would do if there were more time.

In one or two responses, there was just a hint of appreciation of what it would mean to characterize the triples of edge numbers for which corresponding vertex numbers would be integral.

What did not emerge from the two hundred students was the observation that, to get whole numbers on the vertices, the sum of the edges must be even; or put another way, either exactly one of, or all of the edge numbers, has to be even.

An algebraic expression in and of itself is not liberating if it does not release or trigger techniques for reaching a solution. It is unlikely that you would be able to treat the general triangle-arithmagon unless you already had manipulative confidence with particulars, or be able to treat general-arithmagons on polygons unless you already had manipulative confidence with quadrilaterals as well. But the point is not to resolve arithmagons, rather to use it as experience to catch yourself in various useful and not-so-useful thinking postures, as preparation for the future.

5.5. *Example: Rows and Columns*

In-service teachers doing a course, *Developing Mathematical Thinking,* faced the following question in the examination:

The picture below (see Figure 5) shows a rectangle made up of two rows of four columns and of squares outlined by matches. How many matches would be needed to make a rectangle with R rows and C columns?

Figure 5.

Students had been exposed to the notion of expressing generality in simple matchstick and other patterns, and to various techniques, such as specializing, specializing systematically, and attending to how you draw chosen examples, but only as a small part of the whole course of 400 hours.

Many of the mathematically less experienced candidates found great difficulty with this question. They were all able to specialize in one direction, and then to make progress in specializing in the other direction. Some were even systematic in taking 1, then 2, then 3 rows, and developing formulas for the number of matches required to make C columns in each case. But only a few were able to address the double generality, finding a formula involving both R and C. Many made comments along the lines of: "I have a sense of what is wanted, but I can't quite see it".

There is much more to be gained from simple pattern generating and generality expressing than may appear at first sight. Variations, extensions and connections are manifold:

- Multiple seeings: finding several ways to *see* how to count the number of matchsticks and expressing these as general formulas; for example, which of the following seeings (see Figure 6) for the case $R = C$ will generalize[4]?

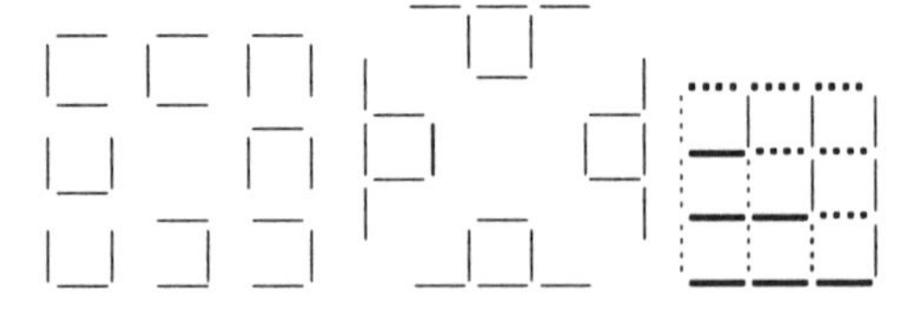

Figure 6.

- Reversing seeings: taking an equivalent arrangement of the expression and trying to arrange the counting process to produce this as a seeing;
- Relating different forms of the expression to other contexts in which the same or similar formulas arise, such as triangular and square numbers, and using these to see the connection directly;

- Doing and Undoing: taking doing as finding the number of matchsticks for a given number of rows and columns, and undoing is deciding if a given number could arise as such an answer, and trying to characterize the form of such numbers;
- The configuration can move into higher dimensions; diagonal matches can be inserted; the number of squares and rectangles of different sizes can be counted; the minimum number of matches that can be removed so that there are no complete squares (rectangles) can be sought, and even turned into a game, in which players alternately remove matchsticks.

6. THEORETICAL PERSPECTIVES: FROM ACTION TO EXPRESSION

Birds have wings, pencils make marks, paper is flat, ... are examples of *generality in action* (extending a notion of Vergnaud, 1981) in the sense that very young children act as if they perceive these generalities. Bringing them to expression is another matter. It is so very tempting to ask students questions in an attempt to bring their tacit awareness to expression. "What do birds have that other creatures do not?" so very easily leads to "Guess what is in my mind," what Bauersfeld called funneling, and Holt (1964) exemplified beautifully with Clare--all because the teacher has something specific in mind (Mason, 1992). "What do we do with pencils, Johnny?" can be seen as socialization or as genuine inquiry (Ainley, 1987), as well as an attempt to attract Johnny's attention. Only teachers can know for sure, and often even they may not be aware of how pupils interpret their questions.

How teachers can attract student attention, evoke awareness, and assist them to experience requisite shifts of attention is a huge topic of its own, and cannot be pursued in detail here. MacLane (1986) draws attention to various forms of generalization, such as from cases, from analogy, and by modification. He observes that abstraction is basically essence seeking, or in other words, de-contextualization. This clearly is based on stressing and ignoring as well, and so is akin to generalization, though the relationship is a matter of definition and has been much debated (Harel & Tall, 1991). MacLane suggests that abstraction takes place through "deletion, analogy, and shift of attention" (pp. 434-436).

Harel and Tall (1991) draw a distinction between *expansive, reconstructive,* and *disjunctive* generalization. An expansive generalization extends the range of an existing schema, assimilating the new particular in an old generality; reconstructive generalization involves rebuilding, accommodating the old generality to subsume the new particular; disjunctive generalization adjoins the new particular as an extra case.

A framework I have found particularly informative is derived from Bruner's notion of *enactive*, *iconic*, and *symbolic* representations. These were recast (Open University, 1982) as *manipulating, getting-a-sense-of,* and *articulating* (see also Mason, 1989). As mere words, they are no different from any other sets of words, but if introduced as descriptions of recent experiences shared by teachers, they can develop into potent sensitizers to future classroom opportunities.

We find it informative to think of them as phases in a developing spiral (see Figure 7), in which:

- manipulation (whether of physical, mental, or symbolic objects) provides the basis for getting a sense of patterns, relationships, generalities, and so on;

- the struggle to bring these to articulation is an on-going one, and that as articulation develops, sense-of also changes;
- as you become articulate, your relationship with the ideas changes; you experience an actual shift in the way you see things, that is, a shift in the form and structure of your attention; what was previously abstract becomes increasingly, confidently manipulable.

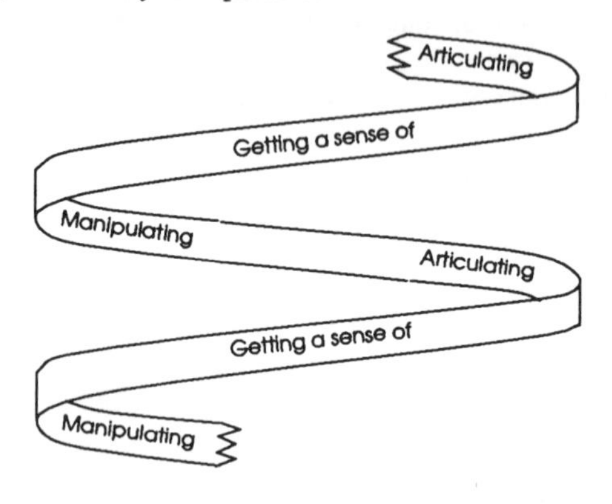

Figure 7.

Once you can add three-digit numbers together, there is no need to use pictures of bundles of ten, or physical bundles either. If a difficulty is encountered, you can track-back down the spiral, or fold-back to an earlier level of facility in order to rebuild confidence and understanding. If students remain stuck in the material, stuck even in the particular, their mathematical being is truncated and may atrophy. It is essential to draw attention out of the manipulated and particular and into the general if mathematical thinking is to take place.

The helix image and its triad of labels are redolent of many authors, from Dewey to Kieren and Pirie (1992). Hiebert (1988) suggested a five-phase theory in which each phase was not only essential, but necessarily ordered: *connecting* individual symbols with referents; *developing* symbol manipulation procedures; *elaborating* procedures; *routinizing* procedures; and *building* more abstract symbol systems from these.

Whereas Hiebert stresses the importance of meaningful referents for symbols at the beginning, followed by de- and then re-contextualization, I take the view that abstraction and generality involve letting go, even while doing things with referents. That is, they involve stressing and consequent ignoring, which Gattegno (1990) also identified as the essence of generalization.

Gattegno's approach is supported by a technique of Brown and Walter (1983) in which you take any mathematical assertion whatsoever (my favorite is "Three points determine a circle"), and you get each participant to say it out loud but each stressing a different word. It is amazing how the stressing not only draws attention to that word, but also invites consideration of "what if ... it were changed," leading to further exploration, variation, extension, and generalization.

The spiraling helix offers a less prescriptive description than does Hiebert, but with a similar flavor. It tries to take into account the movement from the confidently manipulable to the growth of a sense of relationship or pattern, and the expression of that pattern with increasing facility, which eventually inspires enough confidence that these articulations can themselves become sources of confidence and hence components to yet more patterns and relationships. The helix notion attempts to connect similar yet different states, while suggesting that manipulation changes as pattern is sensed, that attempts at articulation may cause re-thinking and re-manipulating, but that through a fluid almost symbiotic process, increasing facility and confidence develop.

The proof of a framework of labels is in future experience. Does it inform practice by sharpening sensitivity and providing opportunity to choose to respond rather than to react mechanically?

Teaching algebra the way we currently do is like teaching people to speak by making them move their mouths into certain positions, over and over--ridiculous! We don't teach people to walk by moving their legs for them, or drawing feet on the floor for them to step on; walking is an expression of the urge to be elsewhere. We do teach people to read by rehearsing combinations of letters (and look at the success rate); but reading is an expression of the urge to participate in asynchronous communication. Detecting sameness and difference, making distinctions, repeating and ordering, classifying and labeling are expressions of the urge to prepare for the future, to minimize demands on attention. They are the basis of what I call *algebraic thinking* and the root of what we have come to call algebra.

School algebra is currently a manifestation of generalization about number. Excessive concentration on manifestation rather than on awareness of the natural experience of algebraic thinking, on behavior rather than process, has, through the worst effects of what Chevallard (1985) called the didactic transposition, produced the unbalanced emphasis on behavior over awareness and emotion (cognition and affect, if you must) which characterizes our pragmatic society at the end of the 20th century (see, e.g., Kang & Kilpatrick, 1992). Behavior is important, but it must be integrated with awareness and emotion to be of any use. All this is summarized for me in three assertions (derived from ancient literature and stimulated by Gattegno) and elaborated elsewhere (Mason, 1991a):

Only awareness is educable.
Only emotion is harnessable.
Only behavior is trainable.

How does all this relate to the perennial failure of most pupils to be able to solve the word problems so beloved of mathematics educators and puzzle addicts since the beginning of records? I conjecture that, as Bill Brookes once pointed out, a problem is only problematic when it is a problem for me. I must feel it as problematic. I must feel comfortable in expressing generality myself, before I am likely to make sense of manipulating other people's generalities in traditional generalized-number exercises. Then I may be attracted to resolving contrived situations, perhaps even constructing some of my own.

By drawing attention away from generality, in the hope that it will make learning easier, we are in fact removing the birthright of every pupil to experience and work confidently with generality as well as particularity, to see the general through the particular and the particular in the general in mathematics. Lessons that are not imbued with generalization and conjecturing are not mathematics lessons, whatever the title claims them to be.

7. APPENDIX: SAY WHAT YOU SEE AND MULTIPLE EXPRESSIONS

The following is an extract from *Supporting Primary Mathematics: Algebra* (Mason, 1991b).

SEQUENCE: Say to yourself (or to a colleague) what you see in the picture sequence. Then state a rule in words for extending the sequence of pictures indefinitely. Work on it for yourself before reading on.

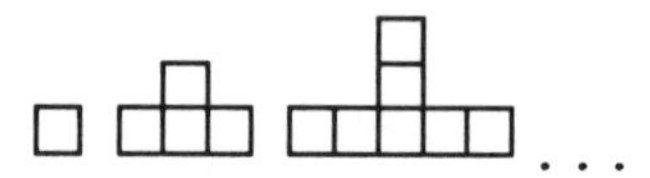

A useful strategy is to make several models, or to draw several pictures in the sequence, and to attend to how you do the making/drawing.

Some people see stands, others upside-down Ts or crosses with an arm missing. There are several ways of stating a rule. For example:

"At each stage, you add one more square to each arm."

"At each stage, there is one central square with three arms of equal length increasing by one square at a time."

"At each stage, there are three equal arms which overlap (indicated by shading) in a common central square."

"At each stage, there are two identical Ls overlapping in a central column."

Notice that it may take a few seconds to make sense of someone else's picture or words, but your own seeing comes in a flash. Sometimes it is then difficult to see in a fresh way.

Notice also the difference between the first way, which grows the pictures one by one, and the other three, which show how to build a picture directly.

Taking time to state your rule in simple language helps you to find a formula. By inviting children to express their rules out loud, you create an opportunity for children to exercise their powers of observation and description to an audience, to show tolerance and appreciation of other people's struggles to express themselves, and to encounter other ways of seeing and thinking which might be useful in the future.

Each way of seeing gives rise to a way of counting, but before launching into counting, how do you go about detecting a pattern in a sequence of drawings?

There is no mechanical method, no rule, but on the other hand pattern spotting is what human beings are good at doing. You do need to become immersed in the pattern, to participate by drawing or counting things yourself, changing the way that you do it to find an efficient and generalizably systematic approach. It may be worth developing facility by working on several examples before continuing.

HOW MANY? How many squares will be used to make the 7th, the 37th, the 137th picture? How many in general? Try using several ways of seeing in order to do the counting.

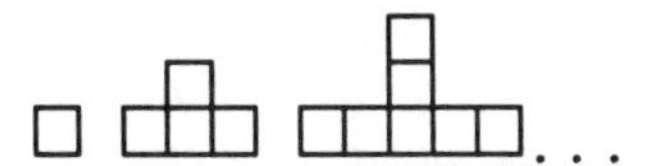

Direct counting can be used for the 7th, and perhaps, though tediously, for the 37th. The invitation is to see how in general the number of squares needed is related to the picture number for the sequence being considered. The first seeing may not be the most efficient for counting, nor the most easily expressed.

The four ways of seeing shown earlier give rise to different ways of counting.

At each stage, you add one more square to each arm. So the number of squares required is 1, 1 + 3, 1 + 3 + 3, ... The 7th will require 1 + 3 + 3 + 3 + 3 + 3 + 3 squares.

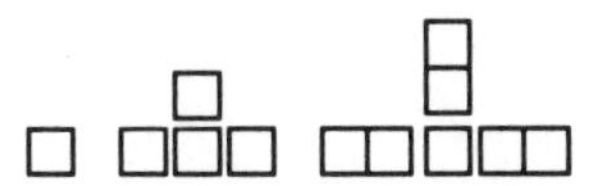

At each stage, there is one central square plus three arms of equal length. The number of squares required is 1, 1 + 3 x 1, 1 + 3 x 2, 1 + 3 x 3, ..., so the 7th will require 1 + 3 x 6.

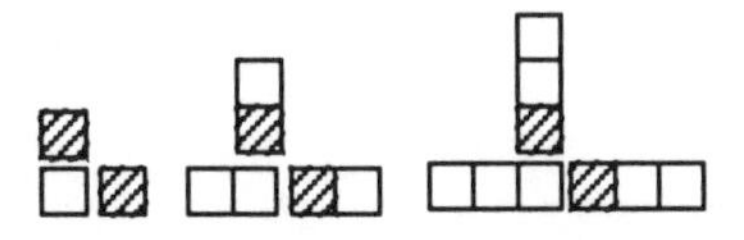

At each stage, there are three equal arms overlapping in a common square. The number of squares required is 3 x 1 – 2, 3 x 2 – 2, 3 x 3 – 2, ..., so the 7th will require 3 x 7 – 2.

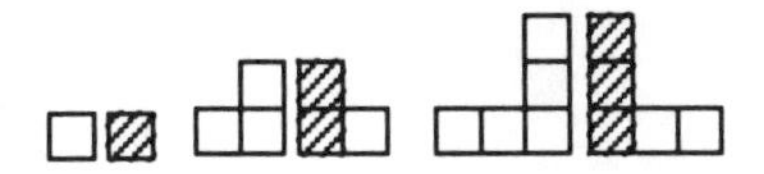

At each stage, there are two identical Ls overlapping in a central column. The number of squares required is 2 x 1 – 1, 2 x 3 – 2, 2 x 5 – 3, ..., so the 7th will require 2 x 13 – 7.

Before going further with this particular sequence, try expressing how to count the number of squares needed to make the 7th and 37th pictures in other picture sequences like the ones below, or of your own devising. Remember to begin by saying what you see, because this will form the pattern rule on which your counting will be based. Look for different ways of seeing as well.

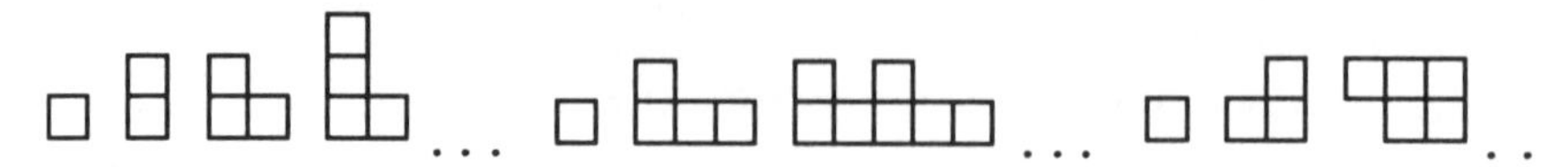

It is important to see and state a rule, for all sequences can be extended in more than one way, and children are especially good at seeing novel ways. The important thing about a rule is that it should account for all the terms shown, and be capable of being extended indefinitely. This is one major aspect of mathematical sequences.

Once you have a way to extend the sequence, try to count the number of squares (or edges, or vertices, . . . challenge yourself!) in more than one way.

The purpose of counting in particular cases such as 7th, 37th, and so on is to prompt you to see what is happening not just in particular cases, but in all pictures simultaneously without having to see or draw them all.

NOTES

1 Precursor to getting a formula.
2 For example, it has been the subject of some 30 or more electronic communications in a discussion chaired by Jim Kaput, during 1992.
3 See, for example, Mason, Burton, and Stacey, 1984.
4 My attention was directed to these by Eric Love's account of a workshop run by him with Dave Hewitt and Barrie Galpin.

CHAPTER 6

AN INITIATION INTO ALGEBRAIC CULTURE THROUGH GENERALIZATION ACTIVITIES

LESLEY LEE

Considering algebra as a culture, this chapter looks at the introduction of algebra as an initiation process where generalization activities can be extremely effective. After a reflection on my own immersion into algebra and the evolution of attitudes toward the teaching of algebra, a teaching experiment using generalization activities is presented. Two generalizing activities are described in some detail, looking at the behavior of adults in the experimental group in the light of research results of high school students on tests and interviews involving the same activities. The paper concludes with a "cultural" reflection on the teaching experiment and a more general consideration of the role of generalization in the introduction of algebra.

1. WHAT IS ALGEBRA?

I find it helpful to think of algebra as a mini-culture within the wider culture of mathematics. This allows me to integrate algebra as a set of activities, algebra as a language, and all the other metaphorical and metonymic ways of defining algebra. It is also helpful in thinking about the interaction of language and knowledge in the gradual process of algebraic acculturation that takes place in the classroom, as well as the interaction between algebra and other mathematical cultures such as arithmetic.

The term "cultural shock" can then be useful in thinking about students' entry into the algebraic culture.[1] And the whole question of introducing students to algebra becomes one of initiating students into a somewhat curious culture with its own "customary beliefs, social forms,"[2] way of communicating (relying heavily on written and visual form), selection of topics, structures, code of conduct, and so on. Compared to other cultures, algebra has developed remarkable universality which allows us to talk at international meetings and in publications of "algebra," "Vietan algebra," "school algebra," hardly acknowledging the impact of other cultures on our mathematical one.

I think of algebra as a culture in the same way I think of American football as a culture, though it is easier to coldly examine the latter since I am not directly involved. In football too, one finds a community of people who share the same language (also relying a lot on visual codes, particularly in drawings of play strategies), rules, mode of relating to each other, belief in their own superiority and the ultimate importance of the game (which to an outsider can appear totally futile[3]), heroes, and so on. When I look at the success school football has enjoyed, compared to the impact of school algebra, I think we might learn some lessons from studying the acculturation strategies of the former.[4]

N. Bednarz et al. (eds.), Approaches to Algebra, 87-106.

2. INTRODUCING ALGEBRA

If algebra is a culture, then its introduction is the first stage in a long process of acculturation, an initiation into this new culture.

I had the fortune of going through school behind an older sister who took great care to warn me of the pitfalls--advising me, for instance, that I should know how to read before I started kindergarten. She too had friends a year ahead of her who had friends a year ahead of them ... And so the word was passed on. By grade five or six, when all was going well in arithmetic, I had got the clear warning about algebra: "You may be doing fine now. That's the easy stuff. But wait until you get algebra!"

Algebra happened in high school. You knew you'd "got" it because the textbook cover displayed *ALGEBRA* in inch-high letters.[5] It was, as I fully anticipated, a move into the fourth dimension, another planet. In Chapter I, we got the "fundamental operations of algebra" out of the way and tackled such practical problems as:

A courtyard is in the form of a rectangle x feet y inches by y feet x inches. Part of it is paved with 48 square tiles each measuring (x-y) /2 inches on the side, and the remainder is covered with grass. Find in square inches the area of the grass. (p. 18)

Chapter II dealt with "Equations" beginning with the "Solution of Equations of the First Degree in One Unknown." Two examples were given and we were launched into the "exercises" before we moved on (two pages later) to "The General Equation of the First Degree in One Unknown" followed immediately by the "Solution of Systems of Two Equations of the First Degree in Two Unknowns." But back to the solution of first-degree equations in one unknown. Here is example 1:

$$\text{Solve } 8x - 11 = 5x + 4.$$

Example 2 involved rational expressions with the x appearing twice to the left of the equal sign and once to the right. No shying away from the "didactic cut" (Filloy & Rojano, 1984b) in those days!

As for the "word problems," how about:

Find a fraction equal to 3/4 such that 1/3 of the denominator exceeds 2/9 of the numerator by 8. (p. 49)

We were truly on another planet. And I loved it! For me, algebra was a place where anything was possible. Nothing was bogged down with the gray and heavy reality of the world around me (a military establishment in post-war Europe). Play and imagining were possible. No one, that I remember, tried to link algebra to arithmetic, much less to the world around us. And I thank them for it.

There were no attempts at "real life" problems of the missiles and money types although the authors do say in the preface that "an honest attempt has been made to introduce more practical problems where possible."

They offer as an example of this attempt their section on the "use of indices to represent very large or very small numbers." But for the most part we were asked to reflect on problems like:

The difference between the digits of a number less than 100 is 6. Show that the difference between the number and the number formed by reversing the digits is always 54. (p. 45)

A number consists of a units digit and a tens digit, the units digit being the greater by 1. The sum of the digits is less than twice the number by 2. Find the number. (p. 45)

My introduction to algebra took the form of a total cultural immersion on a par with moving from a small university town in Nova Scotia to a Quebec industrial city. Learning the language was a small though integral part of the total acculturation process in both the move from arithmetic to algebra and the move from Nova Scotia to Quebec. Acculturation involved learning what is "sayable," what we talk about and how we talk about it; what we do not talk about; what level of formality is used in various writing situations; what experiences and words are untranslatable; what are the gestures and symbols, the worlds of sense around objects, dates, rites; what are the sacred cows; what is the shared history of institutions, families, communities; what are the objects of thought; what is funny and what isn't.

3. TEACHING ALGEBRA

As a teacher of algebra I gradually learned that the way I had been taught algebra was all wrong. It was apparently in the old "drill and practice" school with no emphasis on understanding and no connection with "real life." "Students want to know what algebra is for," we were told. "Problem solving is where it's at." Function, which I didn't hear about until many years after high school algebra, "is the primitive algebraic object," "what algebra is all about."

I still remember my dismay when the word came down in the 1970s that we must re-orient our Cegep (post secondary) mathematics courses to deal with real-life problems. There we were, sixteen mathematics teachers, and we couldn't come up with even three real-life problems for our introductory calculus course. At that point we were teaching functions as ordered triplets with a half-page long definition to be memorized. I believed that the only area of mathematics where real-life problems were appropriate was in statistics--which I really didn't feel was a full-fledged member of the mathematics community.

Then I was puzzled. Why, if we were working so hard to make algebra "meaningful," did the students seem to find it meaningless? Why were students not enjoying this "new improved algebra"? Indeed they appeared to hate it. And why didn't they succeed?

In about 1985, David Wheeler and I began looking at students' high school algebra. After an exploratory year we got funding to test and interview all grade 10 students in three Montreal high schools. Three years later, we began a second study involving adults in the prerequisite elementary algebra course offered at Concordia University. Our aim was to find similarities and differences in the two populations. As a final stage in the research on adults, we planned a teaching experiment based on some of the generalization and justification elements from our tests and interviews. I conducted that teaching experiment in 1991 with a group of six volunteers enrolled in the elementary algebra course. It is within the context of that teaching experiment that I would like to situate my discussion of generalization.

4. TEACHING GENERALIZATION

The experimental group consisted of four women and two men ranging in ages from the late twenties to the mid-forties. The last math courses taken were: O-level/Cambridge in 1960, high school functions in 1982 and 1985, grade 9 algebra in 1964, "algebra" in 1981, grade 8 math in 1973 with a "failed intro. to algebra" the previous term. The latter student wrote on the back of her pretest:

> Can't do word problems. Was told I was not good at factoring. Failed [math] 200. Spent money on tutoring for fractions. Figured them out (sort of) and still failed. Have problems with remembering formulas. Need to pass 201 or I dead [sic].

Another unsolicited paragraph on the back of the pretest said:

> I have not done any formal courses in algebra since high school. I am planning to do 200 level algebra 1992. Math was always a problem for me. Fractions, ratios. I have had to struggle through a statistics course some 11 years ago. This was mostly programmed learning. I was frustrated with the course.

Another said she would like to know about "rooting."

The four women had families and three were working in full time jobs and studying evenings. We met on Friday evenings in my office on the remote (and dark) Loyola campus for six weeks in February and March.

The discussion below will focus on two of the six sessions. Both sessions were structured around questions that had been used in the original work with high school students, as well as in the testing and interviewing of adults. A brief review of the source questions will give some idea of how students at both levels had responded to them and provide some appreciation of the context in which the teaching experiment took place.

4.1. *Results on the First Source Problem*

The first source problem was one we called the *Consecutive Numbers* problem:

> Show, using algebra, that the sum of two consecutive numbers (i.e., numbers that follow each other) is always an odd number.

In analyzing the generalization (and justification) elements in this question in our 1987 report (Lee & Wheeler, 1987), we said:

> The consecutive number question does not require students to express generality in the same sense as that required in the fourth question [dot rectangles which we will look at next], where students must perceive a pattern and then express and test it. ... Generalised statements about odd and even numbers must be made though. For instance, students must realize that every pair of consecutive numbers contains one even and one odd number. The fact that even numbers are divisible by two and vice-versa is a helpful generalisation as well. In other words, the generalisation that is required here is at the level of what might

be called "number facts." Pre-algebra students would probably be as aware of these generalisations as the students interviewed here. The algebraic generalisation of any pair of consecutive numbers as x and $x + 1$, while helpful in the addition question, is certainly not necessary. (pp. 63-64)

We pointed out that:

When two consecutive numbers are symbolised by x and $x + 1$, there is nothing in the definition or the representation which draws attention to the "oddness" or "evenness" of the numbers. Only their "consecutiveness" is at issue. But when the sum of these numbers is conceived and represented as $x + (x + 1)$, a new element enters, brought by the dynamic of simplification ("collecting like terms," in this case). Simplifying the representation to $2x + 1$ may immediately suggest "odd number": the $2x + 1$ may be perceived as constructed by adding 1 to $2x$, an "even number," and so obtaining an odd number. The matter of parity may now be read back into the starting situation, drawing attention to the fact that x and $x+1$ are an odd and even number *though we don t know which is which.* (p. 64)

We recognized that using the algebraic language to "think with" in this way is full of subtleties and it is not surprising that novice algebraists do not yet have a grasp of these possibilities.

This question had been given to 113 high school students on a test and 4 of these had been interviewed about it. Only 8 of the 113 students worked the problem through correctly. Nevertheless, 49 did express two consecutive numbers as x and $x+1$ and most of these saw the sum as $x + (x + 1)$, if not $2x + 1$. Thirteen students assigned a different letter to each of the consecutive numbers (i.e., x and y) and 15 made errors in expressing consecutive numbers with a single variable (i.e., $1x$ and $2x$). Of those who expressed the sum of two consecutive numbers in the form $x + (x + 1)$ or $2x + 1$, nine substituted a few values for x to show its "oddness," eight set the expression for the sum equal to a fixed (or several fixed) odd number(s) and solved for x, four set the sum equal to 0 and solved for x, and one set it equal to itself. Several insisted that x was the even number and $x + 1$ the odd, and several more confused even numbers with positive numbers and odd numbers with negatives.

The same problem, but without the request to use algebra, was given to another 118 students and, for the 27 students who attempted to use some algebraic symbolism, behavior was very similar to the above. Most of the above behavior was exhibited as well by the four high school students who were interviewed on this question.

The responses of the adult students were almost as varied as those of the high school students, and proportionately speaking, almost identical. It was the interview behavior of the adult students that revealed slight differences with their high school counterparts. Adults showed less concern about putting some algebra down on their papers and did not go off on meaningless algebraic manipulations to the same extent as high school students. They demonstrated a similar, if not greater, inability to use algebra. Generalization was a little less successful for the adults interviewed. None were able to let x be any number. For those who used x and $x + 1$, the x was very definitely an even number and considerable time was devoted to getting them to see that it could also be odd. It was pointed out in the report on high school students that the generalization here involved mainly "number facts." Because the adult students experienced some problems with these (i.e., that even numbers are divisible by two, that consecutive numbers refer to whole numbers) and were distracted by strong word associations or misunderstandings, it is understandable that they appeared to experience more difficulty in these generalization tasks.

The problem was given in a pretest to all six students in the experimental group; their now familiar responses are displayed in Figure 1.

$x + (x + 1) = y$ $y \pm 1 =$ divisible by 2 I don't think using algebra. This is like a puzzle, using different language. I can turn parts of it into a formula but I can't figure out how to put "even" or "odd" into an equation	$1 + 2 = 3$ $2 + 3 = 5$ $3 - 2 = 1$ $5 - 2 = 3$
$N + N + 1 =$ odd number 1 + 2 | 3 2 + 3 | 5 3 + 4 | 7 Well, it has to be so because with any two consecutive numbers, one of them is going to be odd. N + N + 1 = odd number Two odd added together makes an even number → $2N = -1$ $N = -1/2$ adding or subtracting 1 from an even number will make it odd	$x + (x + 1) =$ $2x + 1 =$ It is divisible by 2, therefore it is even and then by adding 1 it becomes odd.
is odd number 5 + 6 = 11 it can be divided by 2 45 + 46 = 71 it is odd number because it can be divided by two	no can do

Figure 1. Student responses to the consecutive number problem

4.2. *The Experimental Class Based on the First Source Problem*

This was the second meeting with the group. The previous week we had worked on problems involving "hidden" consecutive numbers such as: "The sum of two consecutive numbers is 111. Find the numbers." By the end of that evening we had worked our way up to verbal and algebraic expressions of general solution methods for problems involving the sum of n consecutive numbers.

In the introduction to this class, students were reminded how they had represented consecutive numbers algebraically and how we had initially used x to represent a "hidden number." Some examples of the use of letters in the role of "any number" were presented and students were told that this week we would let x and $x + 1$ represent any two consecutive numbers and prove theorems about them.

Starting with showing that the sum of two consecutive numbers is always odd, students moved on to considerations of the sum of 3, 4, 5, ... , 10 consecutive numbers and formulated (by the end of Task V, the fifth task of the evening) a general theorem on sums of consecutive numbers:

The sum of n consecutive numbers is odd if n equals 2, 6, 10, 14, ...
The sum of n consecutive numbers is even if n equals 4, 8, 12, 16, ...
If n is an odd number, then the sum can be even or odd.

In Task VI they applied this theorem to predict the evenness of the sum of 20 consecutive numbers. The subsequent sequence of tasks concentrated on exploring the divisibility of the sum of n consecutive numbers by n, starting with $n = 3$. By Task XII, students were able to complete the theorem, "The sum of n consecutive numbers is divisible by n if ... ," with the words, "n is odd."

Next ,we started with three consecutive numbers and the rule: Square the middle one and subtract the product of the other two. In each of these series of tasks our first attempt involved each student picking consecutive numbers, performing the operations, and comparing the results with others (knowing none had picked the same numbers). Students considered whether this theorem was generalizable to more than three consecutive numbers and decided it was not.

The next tasks involved series of consecutive multiples of four, then five, and so on, with the formulation of the following theorem by students:

For three consecutive multiples of n, the square of the middle one minus the product of the other two is n^2.

And the final problem "for the road" was to consider any arithmetic sequence of three numbers going up by 4 and to find a general result after "squaring the middle one and subtracting the product of the other two." Subsequently they might try an arithmetic sequence going up by 5 and so on.

4.3. *The "Rules of Algebra"*

Looking at students' weakness in both arithmetic and algebraic syntax as evidenced in their pretests on this question, it was to be expected that these somewhat elaborate generalizations would lead to some syntactic discussions. But rather than focusing on syntax errors or inabilities, we kept our attention on the generalizing. Student motivation to produce ever more elaborate generalizations was so high that it carried the day. They tended to help each other out with syntax. For instance, although not everyone knew how to square $x + 1$, a couple of students managed to figure it out and share this.

One student, who had never had any schooling beyond grade 8, took a while to be convinced that $x + (x + 1)$ was $2x + 1$ and not $2x + x$ as she thought she had just learned in her prerequisite algebra course. Once that had been settled, she was quite happy with adding three, four, five, and more, consecutive numbers. Moving on to multiplication, she had difficulties as well and had to learn how to multiply $x(x + 2)$ for instance, as well as how to square $(x + 1)$.

Quite a discussion arose when one student declared he had switched to $x - 1$, x, and $x + 1$ as his three consecutive numbers and found things worked out just as well and were, in effect, simpler. One or two students were convinced by his demonstration and began experimenting with these as well.

There was increased excitement during this course. Students seemed to feel the power of their new found ability to make up theorems about numbers themselves and then prove them. It was hard to end the class and a couple of students stayed afterwards to chat. They seemed to want to tackle the entire universe of mathematics--wanting to know how to do a cube root "by hand," what were "discrete structures," the use of "successive approximations," and the prices of various computers. The student who had used no algebra in the pretest and who had got 71 for the sum of 45 + 46 announced the following week that she thought she would major in mathematics!

4.4. *Results on the Second Source Problem*

The second problem that had been used in testing high school and adult students and that gave rise to an evening of generalizing activity with the experimental group was the "dot rectangle" problem shown in Figure 2.

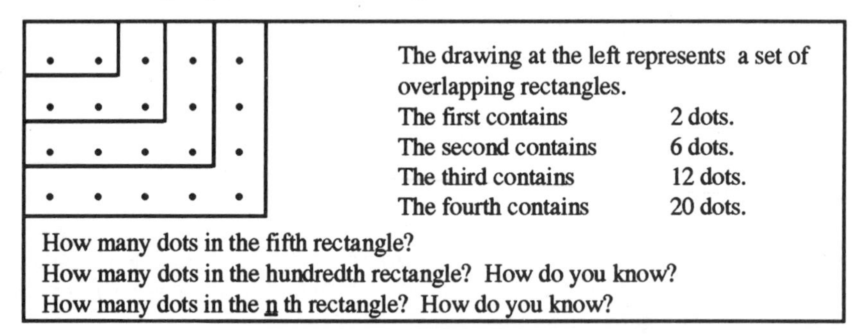

Figure 2. The dot rectangle problem

In analyzing the demands of this generalization problem we found the categories used by the Open University team on very similar type problems useful. As the authors of *Routes to / Roots of Algebra* (Mason, Graham, Pimm, & Gowar, 1985) suggest, this problem requires students to "see," "say," "record," and "test" a pattern. The key to success seemed to be at the first stage of pattern perception, where a certain flexibility was necessary to hit on a mathematically recordable pattern.

Of the 176 high school students given this question, 26 had correct answers and explanations for all three questions. Only 13 of those who attempted the first question concerning the 5th rectangle did not get a correct answer. This indicates that the problem did not lie in students inability to "see a (correct) pattern." In fact, 10 of the 13 gave the answer 10 for the fifth rectangle which suggests that their initial perception of pattern involved only the borders. Seventy-one students gave correct answers to the second question as well, though a number of these did not give their reasoning ("How do you know?").

From the interviews, three perceptions of pattern were identified and student movement through these various perceptions could thus be viewed. The three patterns were:

- "Borders" or increasing numbers of dots in the border along the bottom and up the right hand side (2, 4, 6, 8, ...).
- Boxes or rectangles increasing in size (there were both static and fluid perceptions of these increases--static perceptions involving a series of rectangle states and fluid perceptions, the transformations from one state to another).
- A number pattern: 2, 6, 12, 20, which had two forms: "add on" and multiplicative.

One student's evolution of pattern perception was, for example, analyzed as follows: borders -> static boxes -> boxes without pattern -> fluid boxes -> static boxes.

The major problem was not in "seeing a pattern," it was in perceiving an algebraically useful pattern. For instance, a borders perception does not lead to a general solution; nor does an "add on" number pattern perception. Yet once students had fixed on an initial perception of pattern, it was very hard to get them to abandon it. A "borders perception" led to a conflict with the numbers given to the right. And yet students would often return several times to a borders perception in the course of their work.

In examining the transcriptions for this and other problems in the series, we realized how the interviewer's perception of pattern can interfere with the students' progress and the comprehension of some protocols.

> Some comment must be made here about interviewer pattern perception because it plays such a major role in these problems. It is evident that the interviewer had his own pattern perception in each of these questions. This in itself need not be a problem. However, when there is a quantity of interviewer input, as evidenced here, it does become important. It can be noted that the interviewer consistently talked to his own pattern and at times seemed unaware of the student's perception. And so, in the confused interchange, it was as if student and interviewer were talking about two different problems. In most cases interviewer intervention led only to more confusion and in some cases to the abandonment of the problem by the student. The dominant role played by one's own pattern perception is also evident in reading over these protocols. It is much harder to analyse a protocol when the student's perception is foreign to the reader. (Lee & Wheeler, 1987, pp. 109-110)

As mentioned above, seeing a useful pattern (and hopefully one that was shared by the interviewer) seemed to be more of a problem than simply "seeing a pattern." Perceptual agility seemed to be a key: being able to see several patterns and willing to abandon those that do not prove useful. We concluded that "pedagogically, the question of 'seeing a pattern' might be supplanted by 'seeing patterns,' teaching perceptual agility" (p. 109). As for "recording a pattern," the majority of the high school students seemed to have the algebraic tools necessary to record the patterns they saw here. But they did not often "say a pattern" nor did they "test their patterns," as suggested in *Routes to / Roots of Algebra*.

The adult students out-performed high school students on this question. Whereas only 19% of high school students got the correct answer for the *n*th rectangle, 42% of adults did so. Fifteen percent of high school students got correct answers and explanations for all three questions; whereas 32% of adults did so. Only 2% of high school students produced a general formula first, as compared to 29% of adults.

In their interviews, adults used different strategies than high school students to get the fifth rectangle. Whereas high school students focused on the dot patterns, adults seemed to center on the number sequence 2, 6, 12, 20, written to the right of the problem. However, one-third of the adults interviewed did demonstrate a "borders" perception at some point, and in general they did display, though to a lesser degree, the static and fluid boxes perceptions as well. Two number pattern perceptions appeared that had not surfaced with the high school students. One involved exponentials (recently studied in their class work), 2^x, and one involved a doubling pattern that may have originated in a borders perception. Concerning the former, we (Lee & Wheeler, 1990) noted in our report on adults that:

> The inferred search for an exponential pattern reminds one that perceiving a pattern isn't just a question of noticing "what is there." There is an element of *familiarity*--what patterns has one had experience of--and of *construction*--what patterns can one *make* out of the data. (p. 46)

In interviews, adults demonstrated much greater pattern flexibility than high school students on this question. The problem was given to the experimental group in their pretest, and once again the results (see Figure 3) reflected the test work of other adults with about the same overall success rate and the tendency of going for the general formula first.

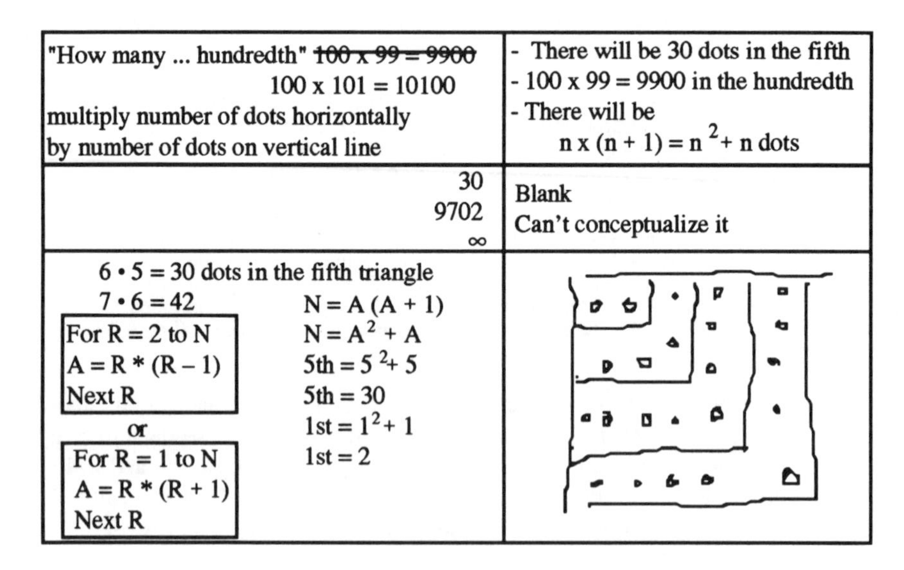

Figure 3. Student responses to the dot rectangle problem

4.5. *The Experimental Class Based on the Second Source Problem*

Given the tendency among adults to focus on number patterns, the tasks for the fourth evening were based uniquely on dot patterns. The first three tasks involved the rectangular dot patterns shown in Figure 4.

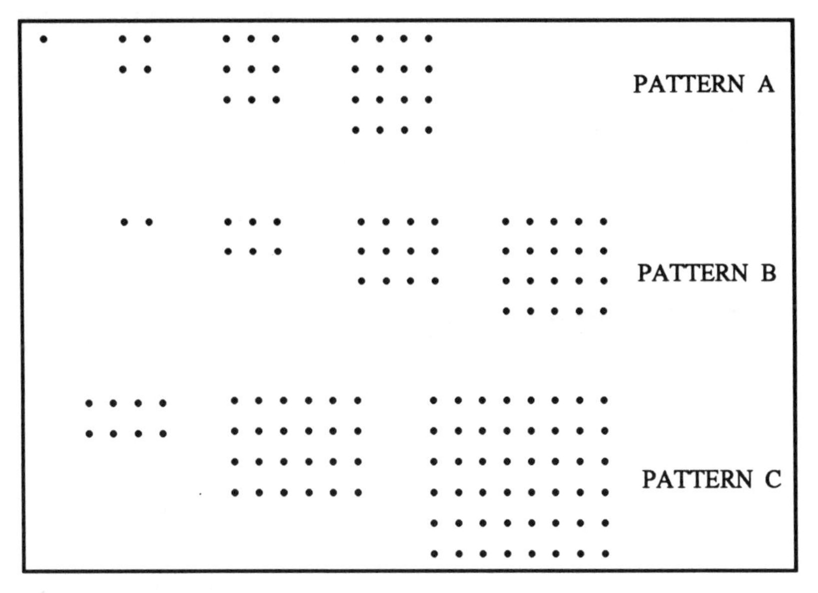

Figure 4. Rectangular dot patterns--the first three tasks

Students were asked for the number of dots in the next, tenth and *n*th rectangles. They seemed to find these very difficult and progress was slow. Two students were unable to see any pattern and one of these could be seen to be counting dots one at a time. At this point I introduced sub-questions in order to focus attention on rows and columns. The position of each rectangle in the sequence was also emphasized. (i.e., "In the third rectangle there are how many rows? how many columns?")

Still one student's difficulties persisted. Her behavior is worth looking at because it illustrates some problems with the business of pattern perception that I certainly did not expect. This was the student who, on her exam, was unable to recopy the given dot rectangle. That she saw neither rows nor columns is hinted at by her somewhat haphazard and skewed placing of the dots (particularly the lack of rows). One wonders if she perceived any pattern at all.

When we began the dot pattern exercise, she persisted in counting dots one at a time. My sub-questions obliged her to identify rows and columns, but she did not seem to know or believe that one could multiply the number of rows by the number of columns to find the total number of dots.

I knew that this student had never learned her multiplication tables (each multiplication had to be done on her calculator), but what I now suspect is that multiplication had no meaning for her either. But she quickly learned to play the game: To answer to "How many rows? How many columns?" she would punch these in on her calculator and chant out the answer.[6]

When we moved on to the second pattern she declared it "stupid"--since it was a repeat of the first. My first reaction was: She can't see the difference! It took us all a few seconds to see what she saw when she declared: "It's the first pattern--just missing a row." No one else had noticed this. And so the number of dots in the nth could be found by subtracting $n + 1$ from her previous $(n + 1)^2$ response.

Another interesting perception of hers emerged sometime during the discussion of the first two patterns. There seemed to be some problem in her numbering of the rectangles in the sequence. When the rest of us were talking about the third rectangle, she appeared to be talking about the fourth. The confusion was cleared up when she declared that surely you can't consider that the single dot in the first pattern and the double dots in the second constitute rectangles. She had quite simply rejected them and begun numbering her rectangles starting with what was for us the second.

By the time we had moved on to the third pattern I thought we had pretty well cleared things up for her. But it quickly became obvious that she had gone back to counting dots one at a time.[7] Forcing the issue, we all counted the number of rows and columns in each rectangle and it was at this point that I realized that identifying rows and columns was not an easy task for her. The word, the concept, of "rows," could not have been new. But perhaps the context, rows of dots, was new enough to cause problems. And what about the straightness of rows?[8] And columns? Where in her post-war experience would she have encountered columns?

The business of columns came up a little later with another student who had kept very quiet and seemed to be having difficulty. She could not predict the number of columns in a dot pattern produced by one of the other students. From her somewhat reluctant responses, it looks as if she was not seeing columns or was confusing rows and columns.

Once students had agreed on a pattern, the algebraic expression for the nth rectangle did not seem to be a major hurdle. However, manipulation of that general expression was shaky. The first example came up in pattern B when, after agreeing on $n\ (n + 1)$ as the general formula, we were forced to accept $(n + 1)^2 - (n + 1)$ coming from the "missing row perception." These two expressions certainly did not *look* the same. Then in pattern C, when we had agreed on $2n\ (2n + 2)$, I wrote beside it: or $4n^2 + 4n$. This led to a long discussion on multiplying with brackets which some remembered how to do, but I'm not sure any believed or knew why.

Here another point arose that took me a bit by surprise. Having developed the $4n^2 + 4n$ expression for pattern C and discussed it at length, I asked, "How many dots in the third rectangle according to this?" No response. So I reformulated the question asking first what n was in the third rectangle. When we agreed n was 3, I worked slowly through substituting with them $n = 3$ into $4n^2 + 4n$ to get 48. We then checked this against the third rectangle. This was repeated with several other rectangles checking calculations against predictions. The fact that these students had developed the formula $2n\ (2n + 2)$ did not mean they really believed $4n^2 + 4n$ was equivalent and even if it was, that it could be used to predict the number of dots in any rectangle.

At this point I asked students to make up their own dot patterns by inventing a multiplicative formula of the above type where factors represented the number of rows and columns respectively, then substituting in $n = 1, 2, 3, \ldots$ to draw the first few rectangles.

The first pattern produced is shown in Figure 5.

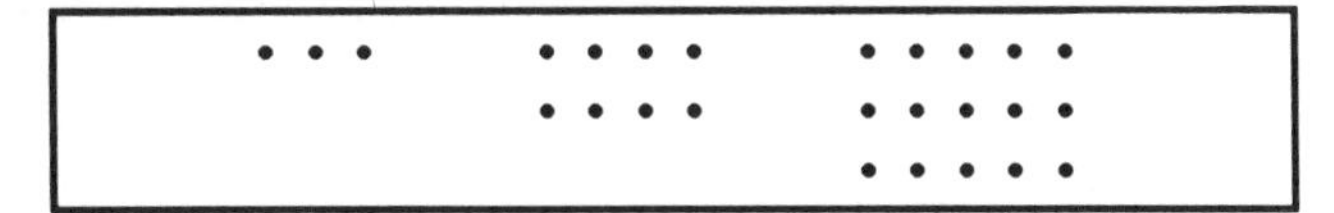

Figure 5. The first student dot pattern

We worked through it together quite quickly. The inventor confessed he had originally wanted to try $(n - 1)(n + 1)$, but realized he would have trouble with his first rectangle.

The next pattern (Figure 6) was more elaborate and the originator admitted, "It doesn't quite make sense to me yet." In the ensuing discussion, the student revealed that she did not start with a formula but with dot arrangements. Yet students did find a pattern of sorts. For instance, one student said there will be 10 rows in the fourth rectangle: "The way I figured out the fourth one was that--she's skipping like that--she's missing one diagram between the first two and between the second and third she's missing two, and so she is going to miss three, and so the fourth one will be like 10 by 12."

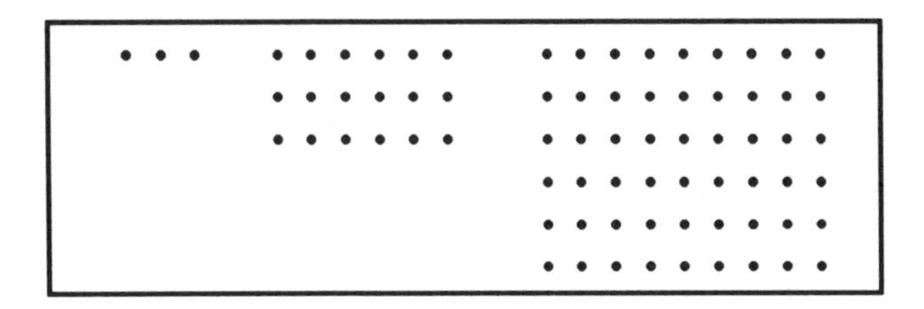

Figure 6. The second student dot pattern

Eventually we agreed to remove the last row in the third rectangle and came up with the formula $(2n - 1)3n$. This provided some scope for expression manipulation and checking against given rectangles.

We moved on to the border problems E and F (shown in Figure 7) during the last 10 or 15 minutes of the session. The task was to find how many dots in the first, next, tenth, and nth borders.

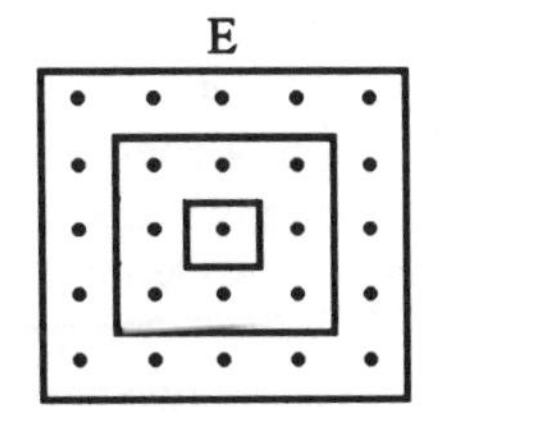

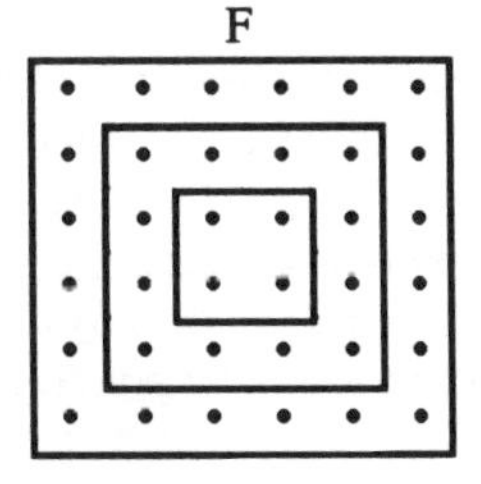

Figure 7. Border problems E and F

Here two different pattern perceptions led to algebraically equivalent expressions and gave rise to some interesting discussions (explanations of the perceptions that gave rise to them as well as their algebraic equivalence). One student confessed she had counted dots and dealt entirely with number patterns, which was how she had dealt with the dot rectangle on the test. So after two hours of concentration on dot patterns, she had decided to go back to her number pattern strategy. Students were asked to consider a dot triangle pattern for the following week and some did spend considerable time on it.

4.6. *Reflections on these two Generalization Sessions*

Considering algebra as a culture, one might say that the student who had the greatest difficulties with these problems could be viewed as a "foreign visitor." Cultural knowledge and practice that I had taken for granted, such as identifying rows and columns, knowing that multiplying the number of dots in each would give the total number in the array, having some meaning for multiplication and the ability to multiply single digit numbers, and being able to add x and $x + 1$, had to be explained to her. Like a foreigner, this student helped us recognize and make explicit our shared perceptions and skills. In asking the "stupid questions," she also helped others for whom some of the same issues were fuzzy. And because she was seeing things afresh, she allowed us to have new and enriching perceptions (as when she saw pattern B as A with a row missing). Her refusal to play the "as if" game, which had the rest of us seeing a single dot as a one-by-one rectangle, really underlined an almost invisible cultural practice which only one other student in the hundreds we tested had ever remarked on. Looking through her eyes at the dot rectangle test question, one had to recognize that it could be seen, in a less rectangular culture, as a bunch of dots in some stacked ┘ shaped pieces or even a bunch of lines with some dots scattered between them. She made us realize just how much our perceptions in algebra had already been trained and our potential for "seeing" had essentially been whittled down over the years.[9] Like any foreign visitor, she forced us to see ourselves.

It was obvious that our foreign visitor was not the only one to have difficulties. Having accepted x and $x + 1$ as any two consecutive numbers did not mean that these students did not put some restrictions on them. The "+ 1" in the second seemed to endow it with an uneven quality whereas the x was definitely even. This was quite useful when we wanted them to look at $2x + 1$ and "see" its oddness,[10] and it did not interfere with most of their work, but at some stage it might have caused some insurmountable problems.

On many occasions in the dot problems, it was obvious that students slipped back into their own ways, after sometimes as much as a couple of hours of our trying to initiate them into "our ways" of seeing, saying, writing, or doing. Going back to counting dots one by one and shifting back to considering number patterns after an evening of considerably directed work with dot patterns are good examples of student resistance to new practices. Foreign visitors do not always see our ways of doing things as "right" or "better." Those who had already been initiated into an acceptable practice were every bit as reluctant to change theirs. For example, having been shown by one student that $x-1$, x, and $x+1$ were more efficient representations of three consecutive numbers for certain problems, the others did not rush in to try these. It was almost as if students felt threatened and did not want to particularly hear about other pattern perceptions or ways of doing things--afraid it might "mix me up" was heard on several occasions.

The whole business of "seeing a pattern" needs to be looked at much more carefully. In our original report and subsequent reports on the adults, I suggested that between 90 and 95% of students saw a pattern in that they could produce the 5th dot rectangle. I now think that close to 100% saw a pattern. These students even saw a pattern of missing dot rectangles in the second student's pattern invention. Our job is to get them to see mathematically acceptable patterns and from those to select the mathematically useful. Even the student who could not correctly copy the dot rectangle problem on the test saw a pattern, and though "a bunch of dots in borders" may be good enough in some situations, it is not precise enough in mathematics. How many dots? Exactly where? ... are what we want to know. "Dots between some lines" might be precise enough for a fabric shop but not for an algebra course. A large part of the unacknowledged learning in mathematics has to do with this question of knowing the questions one is expected to answer to, the nature of the precision required (for instance, we rarely ask for precision of color as one would in a fabric shop).

The effect of concentrating on the generalizing tasks and treating the syntax as a secondary by-product in these sessions was interesting; it links to some recent experimental results involving the use of computers or calculators in algebra courses (e.g., Kieran, Boileau, & Garançon, 1989; Rojano & Sutherland, 1993) to focus attention on the semantics rather than the syntax of algebra. When our students did deal with syntax problems, it was on their own initiative in response to a particular need. At these times there was a great investment of energy in "understanding" the particular algorithm (if for no other reason than that one's credibility in the group depended on it). Like others who have tried alternate methods, I am left rethinking what we have come to accept as "cognitive obstacles" in the learning of algebra.

4.7. *Reflections on the Teaching Experiment as a Whole*

It is hard to describe the group dynamic that dominated these and the other sessions. If one listens to the tape the most remarkable feature is that it is the students who are doing all of the talking. There were times when I felt they had forgotten I was there. Almost from the beginning, the students behaved like a community of mathematicians. They were there to talk about algebra, and they talked about nothing else. They seemed reluctant to quit when our two-and-a-quarter-hour sessions were over, and when I reminded them that our sixth and last session would be on a certain date, there was an immediate and unanimous request to continue meeting. None of the students knew each other before, and I don't think any friendships were established through our meetings. They had neither age nor culture nor interests in common. As I mentioned earlier, we met in my crowded office Friday evenings at the coldest time of the year on a remote campus, which was almost empty and always dark. These were working adults spending evenings and weekends studying and trying to raise families as well. What we were doing was certainly of little or no immediate help in the prerequisite math courses they were struggling through. I think you could say that we had found the algebraic equivalent of American football: They were coming out in the snow, hail, and sleet to play algebra and nothing was going to stop them. There was no possible incentive except the joy of the game.

In an interview with one of the students, one gets an idea of what the experience meant for her:

I really like the whole idea of using math as not just a means of playing games with something but as a means of making generalizations. That's a neat way of thinking of math. I always thought of it as a teacher trying to get me somehow. And that means all the power is in the teacher and none of the power is in me. If I ever thought of it and this is my whole, ... if the power gets to me and I can start playing around with that and making statements that I can do things with, then math is ... much more empowering rather than disempowering, and to me that's going to be the difference in terms of ...?... and wanting to continue with it. That's a big, to me that's a big change.

She came back to this theme of empowerment again in the interview but from the angle of being in control:

And I liked the fact that you turned over the control. The teacher in me liked the fact that you gave us control of it sometimes. Alright you came up with the problems. Which made it seem like it was not only in your ability to do this. You gave us the full control over the problems by giving us a chance to come up with something. So we had multiple ways of approaching. First we were playing with it in our own rather limited way. You broadened our way of looking at it and then said, "Now you come up with a problem and try to stump each other." But we were now taking over control and to me that was one of the most powerful things. But then that's not enough alone until you've got the understanding. That once, you carry that understanding on and not just run off to another subject, another concept and go on to that. To be able to really fully master it and say you've got it and could come up with a problem as well as an answer. And that was also going backwards sometimes. So that was good too, just to see it from another perspective. Cause sometimes when we had to come up with it we couldn't come up with it immediately unless we figured out some other way. So that was seeing it from many, you know, from every angle. That was pretty useful.

Coming up with problems themselves seemed to be a new experience for these students and one that they appreciated. Seeing someone else's pattern was quite a different task from creating one's own pattern and presenting it to others. And solving pattern problems posed by one's colleagues has a whole different dynamic than solving ones coming out of a textbook or from a teacher. Students did not really try to "stump" each other with their problems. They did take great care to produce problems that were neither too easy nor too hard and yet interesting and do-able. An appreciation of what was a "good problem" seemed to develop in the group, and they delighted in being able to contribute one.

5. GENERALIZATION AND THE INTRODUCTION OF ALGEBRA

I wrote earlier in this chapter of my own introduction to algebra and tried to describe it as a process of acculturation. I think it is possible to make a case for introducing algebra through functions, and through modeling, and through problem solving, quite as honestly as it is to make the case that generalizing activities are the only way to initiate students into the algebraic culture. Nor is it much of a challenge to demonstrate that functions, modeling, and problem solving are all types of generalizing activities, that algebra and indeed all of mathematics is about generalizing patterns. I could quote Whitehead (1947):

> The history of the science of algebra is the story of the growth of a technique for representation of finite patterns.
>
> The notion of the importance of pattern is as old as civilization. Every art is founded on the study of pattern.
>
> Mathematics is the most powerful technique for the understanding of pattern, and for the analysis of the relationships of patterns.
>
> Now in algebra, the restriction of thought to particular numbers is avoided. We write $x + y = y + x$ where x and y are any two numbers. Thus the emphasis on pattern as distinct from the special entities involved in the pattern, is increased. Thus algebra in its initiation involved an immense advance in the study of pattern.

or, more recently, Sfard (1995):

> The majority of authors seem to be quite unanimous as to the early origins of algebra because they spot algebraic thinking wherever an attempt is made to treat computational processes in a somehow general way. Generality is one of these salient characteristics that make algebra different from arithmetic. (p. 18)

Sfard herself defines algebra in terms of generalization:

> I use the term algebra with respect to any kind of mathematical endeavor concerned with generalized computational processes, whatever the tools used to convey this generality. (p. 18)

Looking at algebra as a culture, I think we would have to at least agree that pattern generalization is a central activity and that the symbolic language of algebra certainly facilitates this task. Generalization is one of the important things we "do" in algebra and therefore something students should be initiated into fairly early on.

The teaching experiment suggests to me that introducing algebra through work with patterns is not only possible, it has some very exciting elements both for the teacher and the learners. The two sessions that we looked at in some detail were typical of the series of six. Student involvement was high, and a number of major breakthroughs in algebraic thinking took place. There was a lot of room for student creativity in spite of very weak technical skills. Students felt part of the mathematical community they created and felt that what they were doing was significant in that community. They developed communication and even technical skills with remarkable rapidity because the social need for them was there.

One objection to this kind of experiment is that it does not reflect what can be done in the classroom. I was curious about this and with another researcher took on a group of 50 Concordia students in an introductory algebra course the following year with the aim of testing out some of these activities. Although I did not have time to tape sessions, and students had to write a common final exam--which meant covering all the "stuff" in high school algebra in one fast term--I did manage to spend a number of classes repeating the kinds of pattern generalizing activities I had done with the experimental group. In fact, algebra was introduced gradually through various pattern activities inspired by the experimental course. Breaking students into small groups allowed us to reproduce some of the group dynamic, and soon a core of students began to meet regularly on their own to discuss their work. The results on students' thinking were harder to monitor; however, the low drop-out and absentee rate and their very impressive results on the common final exam written by about 200 students confirmed the hypothesis that this sort of pattern generalization work is possible in the "standard classroom." The question remains as to whether or not the results would be similar with 14-year-olds. My guess is that they would.

6. CONCLUSION

6.1. *Concluding Reflections on Teaching Algebra*

At the beginning of this chapter, I gave a few examples of algebra problems drawn from an old textbook, and I would like to return to one of those problems in order to reflect on the more general topic of concern in this book: How should we introduce high school algebra in order that students may, if they wish, continue to learn and grow mathematically in and beyond that culture?

The question I have chosen is the one that involved fractions: "Find a fraction equal to 3/4 such that 1/3 of the denominator exceeds 2/9 of the numerator by 8" (Crawford et al., 1954, p. 49). When we set a/b down as the unknown fraction, that fraction was definitely an algebraic one. In the former arithmetic world it might be reducible to 3/4 by dividing both numerator and denominator by the same number but in the algebraic world this was obviously not possible. Nor was there any point in using our knowledge of equivalent fractions to express the fraction as $3x/4x$ in order to bring out the reducibility because the next few words tell as that we must consider one third of the denominator, $4x$, which only complicates the situation. Arithmetic had to be left behind and these newer, richer algebraic objects had to be dealt with using their own procedures and operations. The whole exercise was pointless in the world of arithmetic, dramatically so in the world of the marketplace. Yet this was a meaningful, interesting, and demanding activity within the algebraic culture.

Building meaning for algebra has been the *raison d'être* for most teaching experiments of the last decade and certainly is the inspiration for our careful look at the introduction of algebra. I think we all recognize that the meanings students build will depend a great deal on the algebra environment we direct them towards, the aspects of the algebraic culture we draw their attention to, as well as their first experiences within it. We should therefore look closely at what the words *meaning in algebra* have come to mean today. I think that there are two aspects to meaning and that the first, meaning within the culture or meaning from within, has been replaced by meaning outside the culture. When we say today that students, who do not connect their algebraic learning to arithmetic or to solving problems, have been engaged in "meaningless" learning, we are, I think, expressing a modern bias toward the immediate, the pragmatic, the "real world" of the marketplace. We tend to see the old teaching methods of total algebraic immersion as instruction in meaningless techniques, syntax without semantics (as if they can be disentangled), and so on. What we are dealing with in this book is the issue of leading students into the algebraic culture while maintaining the meaning links with the outside culture. I predict that in a very few years we will be discussing the issue of building meaning within algebra (attention to form, developing a feel for the culture, skill in the selection and handling of tools, etc.). Both kinds of meanings (or understandings) are important--functioning comfortably within the new culture and making the links outside algebra--in order to be able to see arithmetic from this new vantage point and to move forward from arithmetic algebra into abstract algebra where the symbols may be matrices or functions or arbitrarily defined objects.

On the other hand, it is very demanding on students to ask them to participate in the algebraic culture, while maintaining all the old links with the arithmetic one, particularly in a branch of mathematics, symbolic algebra, that is founded on a temporary suspension of meaning and context. This requirement of keeping one foot solidly in the world of arithmetic is bound to lead to all the difficulties and obstacles enumerated in the literature. Students are bound to consider the equal sign as a herald of "the answer," a number. They will persist in using letters as a shorthand for words; they will try to continue with their informal, non explicit, and context bound "undoing methods" to solve problems. And they will have to find some sort of cover story for the totally new procedures of simplifying expressions, solving equations, graphing functions, and so on.

6.2. *Concluding Remarks on a Generalization Approach to Algebra*

But since we are committed to maintaining the "arithmetic connection," then let us look at the generalization approach briefly in the light of some of the questions raised for each of the approaches to algebra presented in this book: What characteristics of algebraic thinking does it promote? What are the major conceptual changes and adjustments it demands? What are the difficulties that might arise subsequently in the learning of algebra?

The generalization approach in our teaching experiment immediately threw students into using letters as variables, though the exercises focused on can be easily adapted to allow students to work with letters as hidden numbers. Think-of-a-number activities were used in the experiment to focus particularly on the latter aspect. The infinite series generated by many of the activities can lead students into reflections on infinite processes, limits, and the calculus. Generalizing and specializing, the yin and yang of algebra, are the daily fare of this kind of approach. A meaning that can be given to equivalent expressions is that of generating the same pattern. Solving equations and proofs might be a little more difficult if one insisted on maintaining a pattern generalization approach.

The approach is not without difficulties or adjustments, however. Although generalizing is a basic human activity that we demonstrate when we formulate our first words, the type of generalizing activities suggested here are not easy. As we saw, there were obstacles at the perceptual level (seeing the intended pattern), at the verbalizing level (expressing the pattern clearly), and at the symbolization level (using n to represent the nth array or number and then expressing the number of dots in terms of this). Flexibility, almost unnecessary in arithmetic, must be developed since, as was shown, not all pattern perceptions are mathematically useful. The conceptual obstacles involved in moving into algebra while maintaining the meanings of arithmetic, as mentioned above, cannot be avoided entirely; though focusing on different kinds of patterns beyond strict number patterns does diminish their impact.

Since a strict generalizing approach has never been sustained throughout high school algebra, it is hard to predict the difficulties that might surface later on in algebra. Would the move to abstract algebra prove difficult or would it just be another step in the generalizing process? a generalizing of arithmetic algebra? How would these students conceptualize functions when the input was other than a whole number? What meaning could be given to the number of dots in the 1.56th rectangle? No one, however, is proposing that generalizing activities be forced beyond the introductory stage. As an introduction to algebra, an entry into the culture, I think a generalizing approach is grounded historically, philosophically, and psychologically and has proven its merits pedagogically wherever it has been tried.

The focus in our research was on very particular generalizing activities that involved the use of algebraic symbolism. We tested and taught students essentially using four different generalizing activities, based mainly on generalizing number and geometric[11] patterns. These activities were largely inspired by the Shell Center and the Open University work on generalization. Our very limited experimental work indicates that the rewards of such an approach are many with perhaps the greatest being the opportunity for beginning students to function as creative members of the algebraic community from their arrival rather than standing back like tourists to watch others perform and create. The problems they were dealing with from day one were far from trivial and they sensed this. And although some people may think abstractions of dot patterns or reflections on the nature of consecutive number are irrelevant or insignificant, in the algebra culture we know they are at the very core of the discipline.

NOTES

1 Culture shock is defined in the Webster's Collegiate Dictionary as "a sense of confusion and uncertainty sometimes with feelings of anxiety that may affect people exposed to an alien culture without adequate preparation."

2 From the Webster's dictionary definition of culture.

3 That is, two groups fighting almost to the death over a single ball that doesn't even have the roundness to bounce and which none of them seems particularly to want to keep.

4 Could we, for instance, get teenage boys out in the snow, hail and sleet to do algebra? And how do the coaches get them to memorize schematic diagrams of dozens of plays to the point that they actually visualize them in their mind's eye in the heat of the game? Why, for us, can't they remember a single algebra algorithm? The quarterback signals "play number 3" and everyone knows his role.

5 *A New Algebra for High Schools* by Crawford, Dean, and Jackson (1954).

6 She was in fact so good at learning game rules that she subsequently managed to pass her first three prerequisite courses, including calculus--perhaps, by finding "answer patterns.."

7 Had we lost all credibility by insisting that • and • • be considered rectangles?

8 Outside of mathematics, are rows always straight? Rows of seats in a cinema, for example, usually tend to curve.

9 ... with the researchers having the most restricted, though mathematically useful, number of pattern views.

10 In fact my insistence on the role of "+ 1" in $2x + 1$ might even have reinforced their assumption.

11 Dot patterns were chiefly used, although some work was done with polygons in the experimental group.

CHAPTER 7

SOME REFLECTIONS ON TEACHING ALGEBRA THROUGH GENERALIZATION

LUIS RADFORD

This brief commentary chapter devoted to issues suggested by the Mason and Lee chapters raises a number of fundamental questions concerning generalization: the epistemological status of generalization and the nature and complexities of generalization as it is manifested in the didactic context of the algebra classroom.

1. THE EPISTEMOLOGICAL STATUS OF GENERALIZATION

Mrs. Smith, sitting in her living room, hears the doorbell ring. She gets up to see who is at the door. No one is there, and she returns to the room. Mr. Smith pursues their conversation when once again the doorbell rings. Mrs. Smith gets up once again to answer, but no one is there. This scene repeats itself a third time. The fourth time the doorbell rings, Mrs. Smith exclaims to her husband: "Do not send me to open the door! You have seen that it is useless! Experience has shown us that when we hear the doorbell, it implies that no one is there!"

You surely remember the above scene from *Cantatrice chauve* by Eugène Ionesco. What is so captivating about this scene is that it reflects in an impeccable way the dynamics of a procedure of generalization (in fact, Mrs. Smith remains faithful to the "observed facts") and surely the conclusion is absurd (for us).

The fragile status of knowledge obtained through a generalization process brings us to the question of what constitutes a "good" or "bad" generalization. The answer to this question, which has puzzled mathematicians and philosophers for the last 25 centuries, has been given to us in various forms: normative logic, inductive or probabilistic logic, statistics, and so on. Surely generalization is not specific to mathematics: From a certain point of view, it is perhaps one of the deepest characteristics of the whole of scientific knowledge and even, perhaps, of daily non-scientific knowledge, as shown by the little extract of Ionesco's theater piece quoted above.

From a mathematical teaching perspective that favors generalizing activities, it may be convenient to try to answer the question: Why, in the construction of his/her knowledge, does the cognizer make generalizations? The "why" should be understood, of course, in its deeper meaning, so that we may specify the epistemological role of the generalization as well as the nature of the relation between generalization and the resulting knowledge.

From the same perspective, other interesting questions are:

- What is the significance of generalization in mathematics? and more specifically:
- What are the kinds and the characteristics of generalizations involved in algebra?
- What are the algebraic concepts that we can reach through numeric generalizations?

N. Bednarz et al. (eds.), Approaches to Algebra, 107-111.

These questions are not addressed directly by Mason and by Lee (this volume). The reason is, it seems to me, that they give to generalization a particular epistemological status. Mason, for instance, says that "generalization is the life-blood, the heart of mathematics." Lee states that the most important activities related to algebra can be seen as generalizing activities; "nor is it much of a challenge," she says, "to demonstrate that functions, modeling and problem solving are all types of generalizing activities, that algebra and indeed all of mathematics is about generalizing patterns." The basic (more or less implicit) argument of both authors is that this "inner" developmental characteristic of mathematical activity, shown in particular by the history of mathematics (see references to history by both authors), can be translated in the field of education and used as a didactic device when mathematics is seen as a subject to be taught. I think that the hypothesis that generalization can be seen as an *epistemic norm* needs to be studied in greater detail and that the consequences that it has for the teaching of mathematics need to be specified. I believe that the answers to the above questions depend on the way in which we interpret the development of mathematics and the way in which we conceive the development of mathematical knowledge. A superficial look at the history of mathematics leads us to the impression that all mathematics is about generalizing. A closer look suggests that, if we accept generalization as an epistemic norm, it could not function alone but may be related to another probable epistemic norm, namely the problem-solving epistemic norm. Put roughly, I think that the latter functions as a primary need for knowledge, while the former functions as a driven-norm. Certainly, this is a point in mathematical cognition that requires deeper study.

2. WHAT ARE THE KINDS AND THE CHARACTERISTICS OF GENERALIZATIONS INVOLVED IN ALGEBRA?

In considering generalization from a didactic point of view, we should take into account that generalization depends on the mathematical objects we are generalizing. Generalization is not a context-free activity. There are many kinds of generalizations that can all be very different.

What are the characteristics of generalization based on geometric-numerical patterns? I would like to point out two specific elements in these kinds of generalizations. The first deals with a logical aspect and the second is related to the role played by external representations in generalizing geometric-numeric patterns.

2.1. *The Problem of Validity in Generalizing Results*

A goal in generalizing geometric-numeric patterns is to obtain a new result. Conceived in this form, generalization is not a concept. It is a procedure allowing for the generation, within a theory and beginning with certain results, of new results.

A generalization procedure g arrives at a conclusion α, starting from a sequence of "observed facts," $a_1, a_2, \ldots, a_n$. We can write this as:

$a_1, a_2, \ldots, a_n \dashrightarrow \alpha$ (α is derived from $a_1, a_2, \ldots, a_n$)

The facts $a_1, a_2, ..., a_n$ are interpreted according to a certain way of thinking (Fleck, 1981, refers to a style of thinking), depending on the knowledge and purposes of the observer.[1] This way of thinking results from the observer's conceptualization of the mathematical objects and relations involved in and between the facts a_i, and leads to a particular form of mathematical thinking.

What is important here is that one of the most significant characteristics of generalization is its logical nature, which makes possible the conclusion α. The underlying logic of generalization can be very different, depending on the student's mathematical thinking. For instance, many students think that some examples (even one or two) are sufficient to justify the conclusion α (Radford & Berges, 1988). Other students think that guessing the result from the first terms a_1, a_2, a_3 of a sequence is sufficient to justify the conclusion α. Other students think that the validity of a conclusion α is accomplished by testing it with a special term of the sequence, let's say the 100th term, or even the 1000th term. After all, is it necessary to prove α, when it appears as an obvious statement?[2] Who decides about validity?

Generalization as a didactic device cannot avoid the problem of validity, and validity is in itself a very complex idea. This does not mean that generalization cannot be a useful bridge to algebra. I want to point out that using generalization supposes that we should be prepared to work with this additional (logical) element in the classroom.

2.2. *External Representations as Symbols*

The use of external representations as symbols in generalization will be different from their use in elementary arithmetic. This is, in my opinion, another aspect to take into consideration. In order to explain this, let $e_1, e_2, ..., e_n$ be the symbolic expressions of the facts $a_1, a_2, ..., a_n$. The e_is are "sentences" in a symbolic system L_1 (take as an example a certain arithmetic symbolic system, e.g., the modern arithmetic symbolic system or the ancient Greek arithmetic symbolic system) and let ε be the symbolic expression of α in a certain symbolic system L_2 (eventually the same symbolic system L_1 or another one, e.g., our modern algebraic symbolic system), then a generalization procedure can be seen as illustrated in Figure 1.

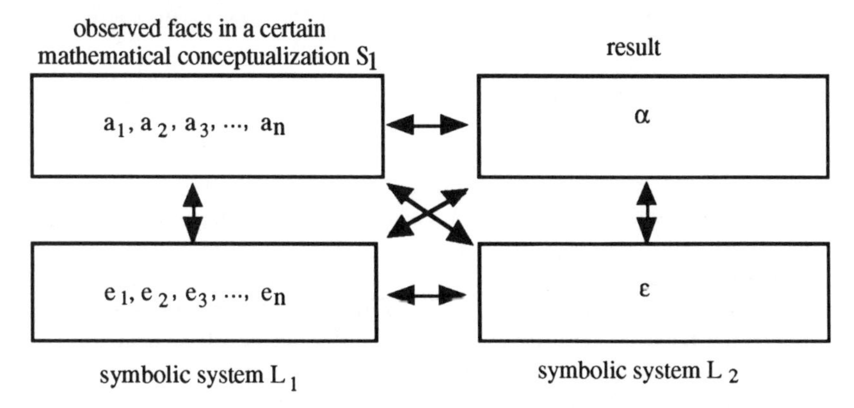

Figure 1.

The diagram shown in Figure 1 displays the many functioning modes of a generalization procedure. In the *How Many?* activity of Mason's appendix (this volume), he places emphasis on the double relation S_1 <--> L_1, in order to obtain the sequence of e_is: 1, 1 + 3 × 1, 1 + 3 × 2, 1 + 3 × 3. Of course, in doing so, one of the most important difficulties a student must face is to understand this very special way of counting, in which we do not write the actual number of rectangles (i.e., 1, 4, 7, 10, ...), but new expressions of these numbers. It requires "seeing" the "facts" a_i in a different way. Representations as mathematical symbols are not independent of the goal. They require a certain anticipation of the goal. The problem that now arises is that of knowing which facets of the object should be kept in its representation.[3]

We will now try to see, in "slow motion," the mental jump to larger terms in the sequence from this activity.

When the question is asked: how many squares will be used to make the 137th picture, the generalization is done from the invariance of the syntactic structure of the e_is (the student can then get the answer 1 + 3 × 136). But how can she be sure that this invariance is an acceptable argument?

We can take advantage of the above diagram to see an important difference between the questions asked by Lee and those found in Mason's appendix. When referring to the "dot rectangle problem," the first two questions are asked within arithmetic. The last question cannot be asked within arithmetic. In fact, in arithmetic we cannot make reference to the *n*th rectangle! This is a new expression related to a new concept. This last question is thus formulated within a domain where symbolic vehicles give precision to algebraic ideas. Even if both authors agree on the fact that algebraic symbolism is not the first goal of generalization, it is clear that this symbolism will be called upon to play a certain role.

For example, if in a certain pattern we see the sequence of terms in a certain way, we get a resulting expression $\varepsilon(n)$ of a formula α. But if we see this same sequence in another way, we get a different expression $\varepsilon'(n)$ for the same formula α. Take, for instance, the problem in Mason's appendix, where we get the formulas $\varepsilon(n) = 1 + 3(n - 1)$ and $\varepsilon'(n) = 3n - 2$. What arguments can we use to show the equivalence of these formulas? Will these be syntactic arguments?

3. WHAT ARE THE ALGEBRAIC CONCEPTS THAT WE CAN REACH THROUGH NUMERIC GENERALIZATIONS?

Let's now consider the kind of concepts that can be reached in generalizing geometrical-numeric patterns. The goal of these generalizations is to find an expression ε representing the conclusion α. The expression ε is in fact a formula and is constructed on the basis, not of the concrete numbers (like 1, 4, 7) involved in the first facts observed, but on the idea of a general number. General numbers appear as preconcepts to the concept of variable.

On the other hand, the goal in algebraic problem solving (where "problem" designates a word problem) is not to find a formula, but a number (i.e., unknown) through an equation. We therefore have a different situation than the case of generalizing a pattern. But this difference is not uniquely at the word level, which would lead us to believe that an unknown is but a variable, and an equation is only a type of formula. The difference is in fact a fundamental difference--a conceptual difference--that often goes unnoticed (school books and even school guidelines frequently do not recognize this difference).

In fact, the logical base underlying generalization is that of justifying the conclusion. It is a proof-process, which moves from empirical knowledge (related to the facts a_i) to abstract knowledge that is beyond the empirical scope. Yet, the logical base of algebraic resolution is found in its analytic nature. This signifies that when solving an equation, or a word problem, we are supposing that we know the number we are looking for, and we handle this number as if it were known, so that we can reveal its identity in the end. Therefore, we must place ourselves in a hypothetical situation. We therefore realize that the logical bases are, in both cases, very different. The generalization way of thinking and the analytic way of thinking that characterizes algebraic word problem solving are independent and essentially irreducible, structured forms of algebraic thinking.[4]

The above discussion suggests that the algebraic concepts of *unknowns* and *equations* appear to be intrinsically bound to the problem-solving approach, and that the concepts of *variable* and *formula* appear to be intrinsically bound to the pattern generalization approach.[5] Thus, generalization and problem-solving approaches appear to be mutual complementary fields in teaching algebra. How can we connect these approaches in the classroom? I think this is an open question.

NOTES

1 Lee's chapter shows, in the "dot rectangle problem," several kinds of perceptions or interpretations of the facts a_i (a_i being the sequence of dot rectangles).

2 Take, for instance, Lee's first problem, where students had to show, using algebra, that the sum of two consecutive numbers is always an odd number.

3 The results obtained by Lee in the consecutive numbers problem are quite eloquent in this respect: "+1"is difficult to perceive as an even number, since the symbol "+1" suggests an excess that is not compatible with the idea of even numbers.

4 It does not mean that a generalization task cannot lead to the solving of an equation or vice versa. For instance, in the formula ε of Mason's problem $N = 1+3\ (n - 1)$, we can ask the following question: What is the rank of the figure with 598 squares? What we claim here is that when the student engages herself in an algebraic procedure in trying to solve the equation $598 = 1 + 3\ (n - 1)$, the intellectual process will be supported by a different logical basis using concepts belonging to a different "form" of thinking than that used in the process of obtaining the formula ε.

5 There is another element that points to this same conclusion. If we look at the emergence of symbolism from a historical perspective, we notice that the use of the unknown in problem solving has often led to the development of different algebraic languages (Diophantus, Chuquet, Viète, etc.). However, the symbolic representations for the concept of variable came much later: Historically, the mathematical objects of variable and equation come from different conceptualizations (see *The Roles of Geometry and Arithmetic in the Development of Algebra: Historical Remarks from a Didactic Perspective* in this volume).

PART III

A PROBLEM-SOLVING PERSPECTIVE ON THE INTRODUCTION OF ALGEBRA

CHAPTER 8

EMERGENCE AND DEVELOPMENT OF ALGEBRA AS A PROBLEM-SOLVING TOOL: CONTINUITIES AND DISCONTINUITIES WITH ARITHMETIC

NADINE BEDNARZ, BERNADETTE JANVIER

Questions concerning the emergence and development of algebraic thinking in a problem-solving context oblige us to reflect, firstly, on the nature of the problems that are given to students and on their relative difficulty, and secondly, on the repertory of procedures available for handling them. More specifically, given the arithmetic experience students have already acquired in problem solving when algebra is introduced, an analysis of the problems presented in the two domains (arithmetic and algebra) is essential. Confronting their spontaneous arithmetic reasoning in these problems with that which is normally expected in algebra, we can get a better understanding of the fundamental changes required of pupils in the passage from one mode of treatment to the other (conflicts, necessary adjustments, new constructions). Any didactic setting (choice of appropriate situations and interventions) necessarily relies upon this knowledge.

1. INTRODUCTION

The work we refer to in this chapter is part of a larger research program[1] that aims at clarifying the conditions under which algebraic reasoning emerges and evolves in a problem-solving context.

For many reasons, we must consider problem solving as a significant perspective through which to introduce students to algebra:

- Historically, problem solving made a major contribution to the development of algebra (see Charbonneau & Lefebvre, Part III of this volume; Radford, Part I of this volume; Rojano, Part I of this volume). It was at the heart of the work of both Diophantus and the Arabs and was explicit in the case of Viète whose aim was to find a general method for solving problems.
- Problem solving also made a major contribution to the teaching of algebra: For several centuries, the teaching of algebra was based on a certain corpus of problems for which arithmetic and algebra proposed different solution methods (see Chevallard, 1989). Algebra was presented as a new and more efficient tool for solving problems that had previously been solved by arithmetic, in fact, as an indispensable tool that allowed one to tackle problems that arithmetic could only treat locally. Gradually algebra emerged as a privileged tool for expressing a general method for a whole class of problems (see, e.g., Clairaut, 1760, cited in Chevallard, 1989, or *L'arithmétique des écoles*, 1927).[2] This problem-solving approach to algebra was articulated so that the passage from arithmetic to algebra would reveal the relevance of this new tool, as well as a certain dialectic, questioning, and relativizing of the contribution of the arithmetic and algebraic approaches.

N. Bednarz et al. (eds.), Approaches to Algebra, 115-136.

Thus Clairaut, starting with a problem like those the first algebraists might have proposed, "Divide an amount, for example 890 L, among three people, so that the first has 180 L more than the second, and the second 115 L more than the third", presented a first solution to this problem without algebra. Then he used an algebraic method that expressed and solved the same problem (here different solutions were suggested with different choices of unknowns). Variations of this problem (modification of the relationships, making the existing relationships more complex, change of context) were then given to show the power of the algebraic solution.

While the evolution of teaching continued to privilege problem solving to some extent, the preoccupation with the transition between the two domains of arithmetic and algebra fell by the wayside. Teaching favored familiarization with an algebra that was stripped of all relevance, of all significance, by linking itself more to a study of the algebraic language and the symbolic manipulations required for an eventual use of the algebraic tool in a problem solving context (see Bednarz & Janvier, 1995).

To consider problem solving as a possible approach to algebra is, of course, to question the latter position and to question the very nature of the first problems offered to students and their subsequent gradation in order to better decide on an eventual progression. This knowledge can help in the choice of pertinent situations and interventions.

But, much more must also be revealed about the repertory of procedures available to students for handling these first problems given to them in algebra so that we might better understand what is required of them in this passage from arithmetic to algebra. The student has, in fact, a whole past experience on which this new learning will be constructed.

1.1. *The Introduction of Algebra in a Problem-Solving Context: Building on a Well Established Arithmetic*

Students in Quebec have behind them a whole arithmetic past (6 years of primary school), which continues during the first year of high school (ages 12 to 13), thus furnishing them with a long experience in this area when they start algebra in the second year of high school (Secondary II, 13-14 years old). Problem solving is not new to them.

During their arithmetic experience they were confronted with a certain body of problems. What was the nature of these problems? Are they different from the body of problems that are presented to them in algebra? In what way? Do these new problems, given to them in the introduction to algebra, provide the motivation to use a new and more powerful tool?

In arithmetic students also developed a set of strategies, implicit models, and convictions that they used in solving problems and which they will quite naturally bring with them when they are faced with an algebraic problem. What are the conflicts and the obstacles that will arise in coming to grips with a new way of solving these problems? Didactic analysis of the introduction of algebra in a problem-solving context cannot ignore the following:

> The acquisition of new knowledge involves both the extension of previous knowledge, which supplies a referential framework for questioning and decoding new information, as well as a breaking with this past, or at least a detour, such that the new knowledge is the

result of a different structuring of the various cognitive elements called into play. (Giordan, 1989, p. 250--our translation).

From this perspective, the procedures and the conceptions students develop in arithmetic are the anchorage point from which new solutions will emerge. These conceptions can act as obstacles or serve as bridges in the construction of new knowledge. How smooth a passage is possible from the arithmetic culture to the algebraic one? Are there enough similarities to warrant some hope of smoothness? Or are there too many contradictions and conflicts requiring profound reconceptualizations?

1.2. *How to Tackle these Questions?*

We must establish criteria that will allow us to analyze this transition in order to better understand the passage we require of students from the solving of "arithmetic" problems to the solving of "algebraic" ones. Our analysis here is based on the research team's earlier work in the area of arithmetic:

- We underscored the different factors that might influence the solution process: numeric domain; nature of data (known and unknown); context; structure of relationships involved, for example, multiplicative problems involving rate or additive problems involving transformations (Bednarz, Schmidt, & Janvier, 1989).
- We examined the relative complexity of problems with respect to the relationships involved. These studies revealed, in particular, the complexity of certain classes of problems in arithmetic that involve the reconstruction of a transformation (Bednarz & Janvier, 1991; Poirier & Bednarz, 1991).
- We highlighted the characteristic, spontaneous strategies that pupils developed to solve these problems involving specific combinations of factors of complexity.

This detailed knowledge of the problem-solving domain in arithmetic constitutes an essential basis for the analysis that concerns us here in the transition from arithmetic to algebra. On what thinking and on what experiences can the learning of algebra rely? What reasoning, what experience must it work against?

We will now look at the arithmetic-algebra transition by reflecting, first, on the problems that are generally proposed in the two domains, and second, on the reasoning used by students in these problems, before any introduction to algebra. We will then be able to show the fundamental changes that mark the passage from one mode of treatment to the other.

2. PROBLEM SOLVING: WHICH PROBLEMS?

What problems do we find in algebra courses and in arithmetic courses? What are their characteristics with respect to an eventual solution?

In algebra, the analysis of problems and their complexity has often been undertaken with respect to a symbolic treatment of the problem. It is, moreover, an "equation lens" that guides the distribution of problems from one school level to another. This classification is based on certain presuppositions which are mainly tied to a particular problem-solving tool, the equation. The latter guides the choice of problems to be presented at one school level or another and, in particular, the choice of the first problems that impose a single variable solution on the student. Our study of "algebraic" problems, as well as our first experimental work with different problems (Bednarz, Radford, Janvier, & Lepage, 1992; Janvier & Bednarz, 1993), questions this a priori ordering of problems using an "equation lens."

It was from quite a different perspective, and in continuity with our past work on problem solving in arithmetic, that our research team's analytic framework was developed. We wanted to underscore the "relational calculus" (Vergnaud, 1982) involved in the representation and solving of such problems: the nature of the relationships between the quantities of the problem, known and unknown, and the linking of these relationships,[3] in order to highlight in this a priori analysis the eventual difficulty of the problem for students and their possible involvement in solving it. We systematically examined the different types of problems found in the arithmetic and algebra sections of textbooks, both past and current, at various grade levels. Through an improved categorization of these problems, we constructed a framework that allows us to highlight those elements that characterize the passage from solving "arithmetic" problems to solving "algebraic" problems.

2.1. *What are the Characteristics of Problems that are Generally Presented in Algebra? Why are some of these Problems so Difficult for Students?*

The research team studied various types of problems. The analysis identified three major classes of problems based on the nature of the quantities involved and the relationships between them: problems of unequal sharing (see Figure 1 for examples of such problems), problems involving a magnitude transformation, and problems involving non-homogeneous magnitudes and a rate.[4] In what follows we will restrict our discussion to a well defined class of problems that is generally found in introductory algebra. We will show what distinguishes these problems from those generally given in arithmetic and what makes them difficult for students to tackle. These same problems will be used later in our analysis of students' reasoning processes.

The general structure of a problem brings out the quantities, knowns and unknowns, their relationship to one another, and the type of relation involved. These relationships are given more or less explicitly in the problem statement and must be reconstructed by the student (with the help of the known quantities or other mathematical or contextual knowledge, prior to solving). For example, let us look at the following problem: "*380 students are registered in three sports activities offered during the season. Basketball has 3 times as many students as skating and swimming has 114 more students than basketball. How many students are registered in each of the activities?*" An analysis of the relationships brought into play in this problem indicates two types: an additive comparison (114 more) between two unknown quantities and a multiplicative one (3 times as many). The two relations join together and continue in the same direction: The number of students playing basketball is obtained from the number of students skating, and the number of students swimming from the number of basketball players. Another type of relationship between the quantities is also explicit in this problem: the whole quantity in relation to its parts (which should be given equal consideration). In this problem, the whole quantity is known (number of participants in the combined activities).

The schema shown in Figure 1 illustrates this analysis of the general structure of the problem (the nature of the relationships between the quantities, the linking of these relationships, etc.).

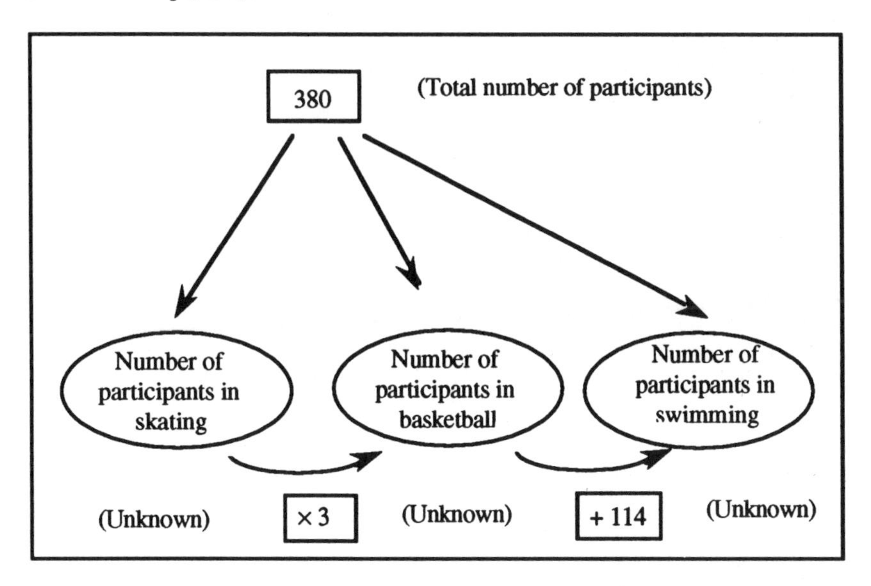

Figure 1.

By means of this schema, the criteria we use to characterize the problems, as well as their anticipated solving difficulties, are made more evident. For example, we can see that:

- No easy bridging can be made between two known data: Here the known is expressed in terms of relationships (3 times as many or 114 more) or as a state (number of participants in the combined activities), which is the only known state in the problem. For the students, this known state constitutes an entry point into the problem as we will see later. Arithmetic procedures are generally organized through the processing of known quantities, by attempts to create links between them in order to be able to operate on them. The unknown quantity then appears at the end of the process (see Bednarz, Radford, Janvier, & Lepage, 1992).
- This problem involves, as well, the simultaneous processing of two types of relationships (× and +) whose composition turns out to be complex (as we will see later) especially when it is approached arithmetically. The student must see that the same quantity (number of students skating) is repeated for basketball (repeated 3 times) and for swimming (once again repeated 3 times).

The structural diagrams of some of the problems used in our research with pre-algebra students (132 first year, high school students, aged 12 to 13) illustrate the difficulties the students have in building bridges between the data and thus engaging in the problems (see Table 1).

THE PROBLEMS	*PROBLEM STRUCTURE*	*SUCCESS RATE*
a) 380 students are registered in sports activities for the season. Basketball has 76 more students than skating and swimming has 114 more than basketball. How many students are there in each of the activities?	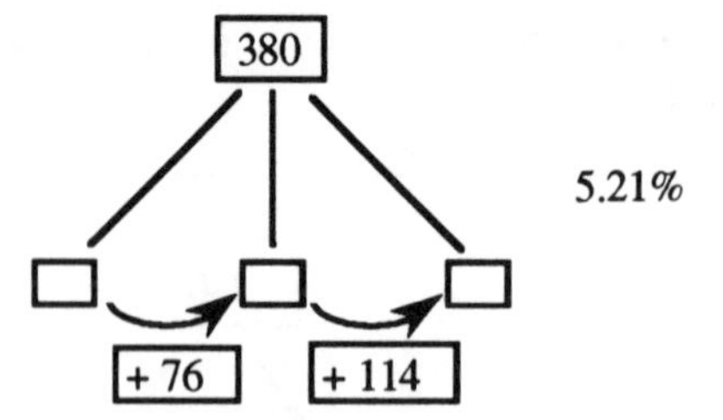	5.21%
b) 380 students are registered in sports activities for the season. Basketball has 3 times as many students as skating and swimming has twice as many as basketball. How many students are there in each of the activities?	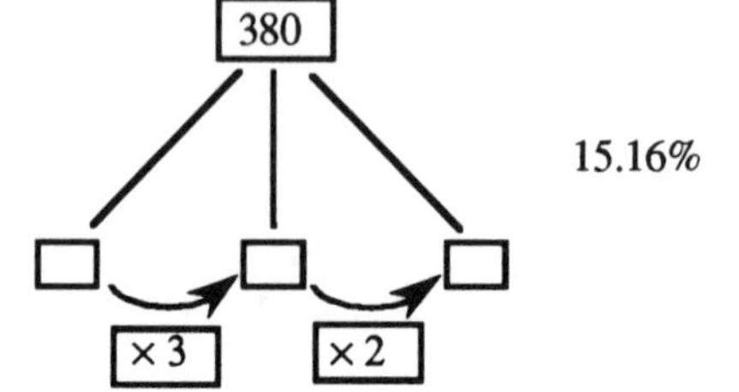	15.16%
c) 380 students are registered in sports activities for the season. Basketball has 3 times as many students as skating and swimming has 114 more students than basketball. How many students are there in each of the activities?	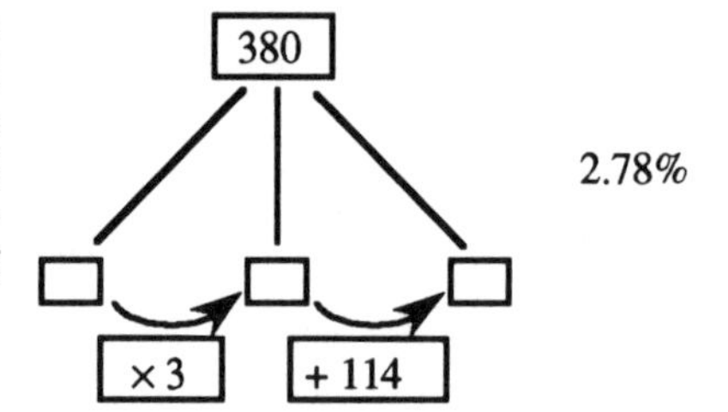	2.78%
d) 441 students are registered in sports activities for the season. Basketball has 27 more students than skating and swimming has 4 times as many students as basketball. How many students are there in each of the activities?	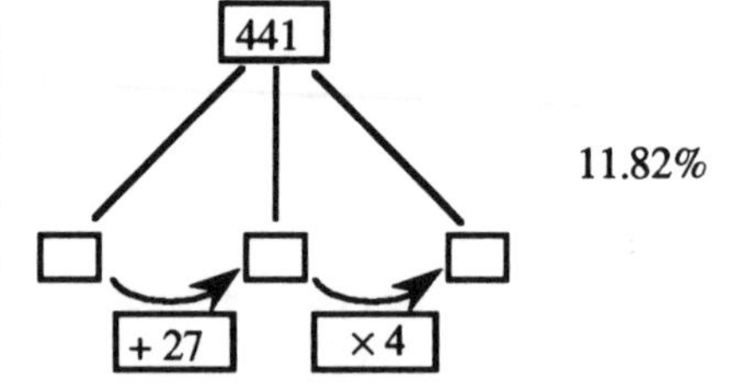	11.82%
e) Three children are playing marbles. Together they have 198 marbles. Pierre has 6 times as many marbles as Denis and 3 times as many as Georges. How many marbles does each child have?	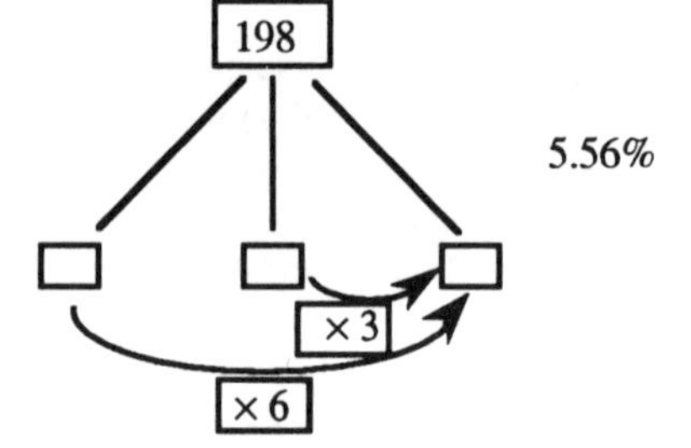	5.56%

f) Three children are playing marbles. Together they have 198 marbles. Pierre has 6 times as many marbles as Denis and Georges has 2 times as many as Denis. How many marbles does each child have?

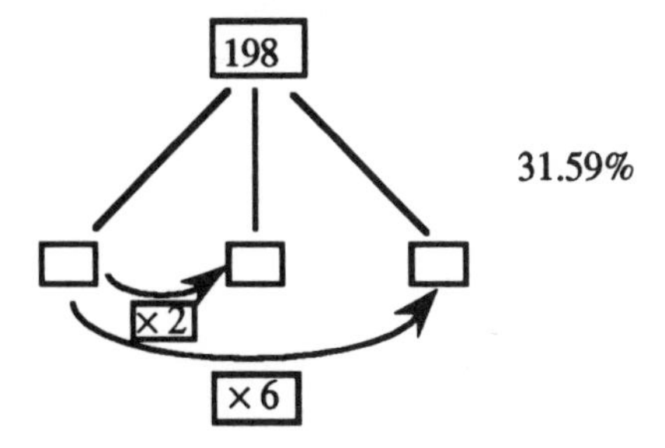

31.59%

g) Three children are playing marbles. Together they have 198 marbles. Georges has 2 times as many marbles as Denis and Pierre has 3 times as many as Georges. How many marbles does each child have?

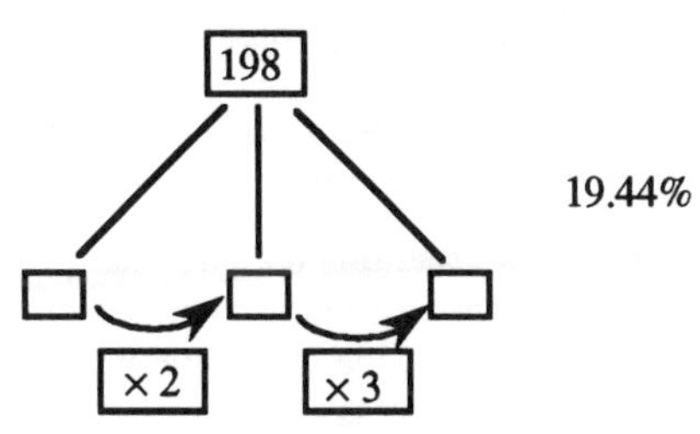

19.44%

Table 1.

Experimental results confirm the influence of those factors of complexity that were identified beforehand in the problem-analysis stage,[5] that is, the influence, for the same structure, of the nature of the relationships (homogeneous composition of two additive relationships vs. multiplicative, non-homogeneous composition of two relationships), of the number of relationships[6] (involving a more or less complicated management of the data), and of the linking of relationships involving a composition (direct or not) or no composition (Bednarz & Janvier, 1994).

Thus, as we had predicted, the problem discussed previously (problem c) was effectively more difficult for the students to solve (only 2.78% of students in their first year of high school, before any introduction to algebra, were successful).

2.2. *What are the Features of the Problems and What Makes them Complex? In What Ways are they Different from Problems Generally Given in Arithmetic?*

Our analytic framework allowed us to characterize the different types of problems generally encountered in algebra and helped us to shed some light on the differences between what we call an "arithmetic" problem as opposed to an "algebraic" one. For example, we can generate from the preceding problems (Table 1) some problems which could easily be presented in an arithmetic context.

PROBLEM # 1:

Steeve has 27 more hockey cards than Harold while Jacques has 3 times as many as Steeve. If Steeve has 138 hockey cards, how many cards do the three boys have altogether?

The semantic structure of this problem can be shown in this way:

?
KNOWN (state)
UNKNOWN (state)
+ 27
138
× 3

It is clear from this diagram that it is easy for the student to construct bridges between known data (expressed here in terms of states and relationships), thus allowing him to engage in a solution by working from the knowns to the unknown at the end of the process.

Starting with the known state (the number of cards Steeve has), it is possible to find the number Harold has ($138 - 27 = 111$) and the number Jacques has ($138 \times 3 = 414$). The total quantity can then be found: $111 + 138 + 414 = 663$ cards.

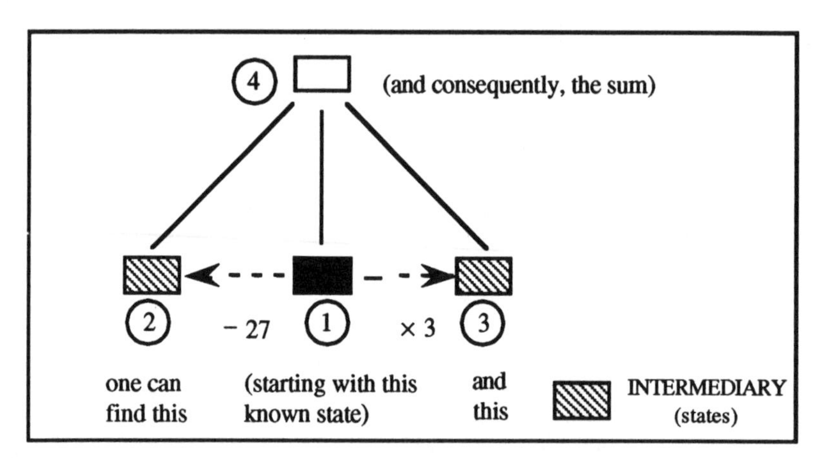

Problem # 2:

Albert has 4 times as many stamps as Judith and 7 times as many as Sophie. If Albert has 504 stamps, how many do the three children have altogether?

This problem can be illustrated by the following diagram:

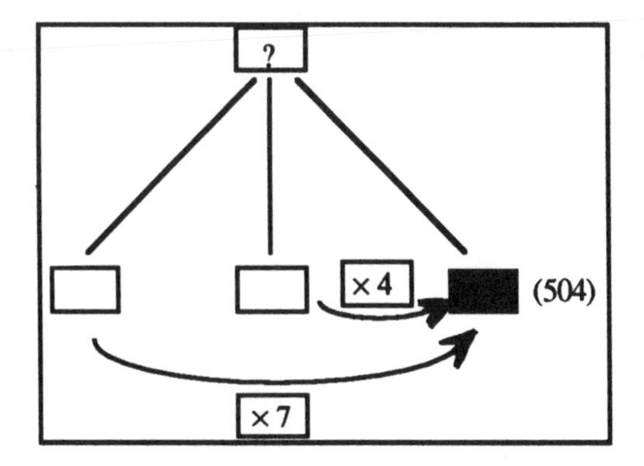

By using the inverse relationship, starting with the known quantity, the student can once again end up with the sought value:
$504 \div 4 = 126$ (Judith's number of stamps)
$504 \div 7 = 72$ (Sophie's number of stamps)
$126 + 72 + 504 = 702$ stamps (total quantity).

The preceding analysis of the nature of the relationships between quantities and their linking allows a better understanding of what distinguishes these two problems from the preceding ones (see Table 1) and sheds some light on a first facet of the passage from arithmetic to algebra:

- In arithmetic, the problems that are generally given to students are problems that we will label "connected": A relationship can easily be established between two known data (expressed here in terms of state and relationship), thus leading to the possibility of arithmetic reasoning (from the known to the unknown quantity at the end of the process) as in the preceding problems 1 and 2. Working in this way, there is no need for the student to deal with the three states at the same time. Starting with an initial known state, he can obtain a new state, and the third quantity can be found from this intermediary state. The composition of two types of relationships involved in some problems (see the a priori analysis of problem c), which implies that the student perceives that the same quantity (number of students skating) is repeated and included in the new quantity (the number of students skating is repeated 3 times for basketball and also for swimming), is not at all part of these two-step procedures.
- On the contrary, in algebra the problems that are generally given to students are ones we label "disconnected": no direct bridging can be established between the known data, as was the case in the problems in Table 1.

These results show the fundamental differences between the "relational calculus" underlying these two kinds of problems, problems generally found in arithmetic and in algebra (see Figure 2). The success rates on problems 1 and 2, 82% and 80% respectively, show that students have a greater facility with the "arithmetic" problems (connected) than with the preceding ones (see Table 1).

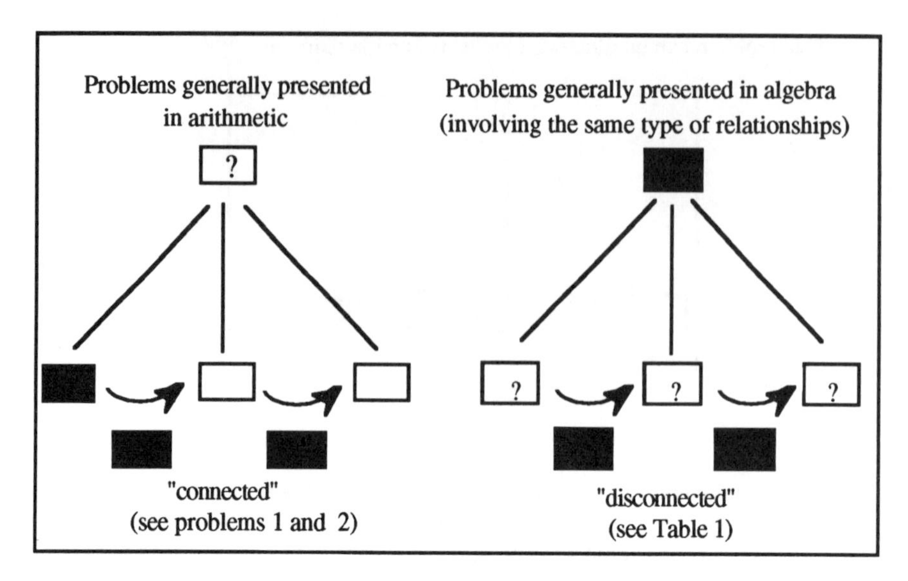

Figure 2

This does not mean that certain "disconnected" problems cannot be introduced in the arithmetic context.[7]

What can the student do in this latter case (disconnected problems) given his background of arithmetic experience? The results of the research undertaken with first year high school students (12-13 years old) throw an interesting light on the way students engage in these problems. We find various stable, arithmetic, reasoning patterns that are used by students, before any introduction to algebra, for this particular class of problems (reasoning that differs in some cases from the two-step sequential procedures presented previously, reasoning that starts from an initial state).

3. ARITHMETIC REASONING PROFILES USED BY STUDENTS FACED WITH PROBLEMS GENERALLY ENCOUNTERED IN ALGEBRA

An analysis of the procedures used by the students for this particular class of problems allows one to see some important differences in their ways of managing the quantities and relationships involved. We will rely on the preceding problems to illustrate these differences (see Table 1). The arithmetic procedures were generally organized, as we have already mentioned, according to the processing of the known data (states or relations) by attempting to create links between them in order to be able to operate on them. The student thereby tried to turn the problem into a "connected" one. Several approaches of this type can be identified. We have grouped these reasoning types into four major categories.

3.1. *Take the Known State in the Problem as the Starting Point*

In search of a state that will allow them to generate the successive relationships given in the problem and operate on them, some will use the only known state in the problem, which they will treat as an initial state, thereby enabling themselves to operate with and to generate the various unknown quantities they are looking for.

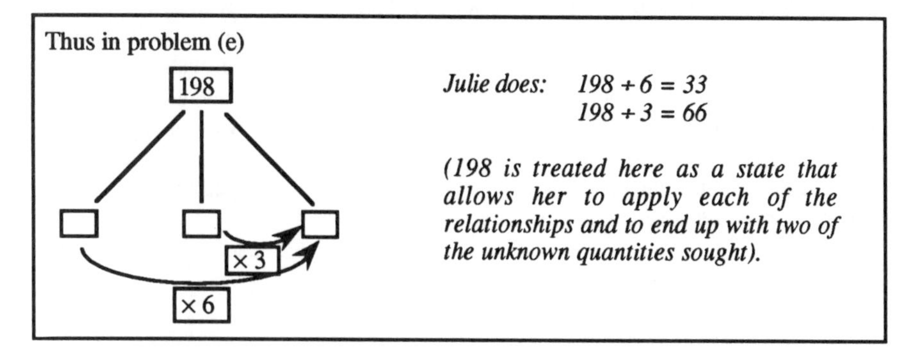

Looking for a known state, students who adopt this erroneous method connect the given problem by only considering one type of relationship present.

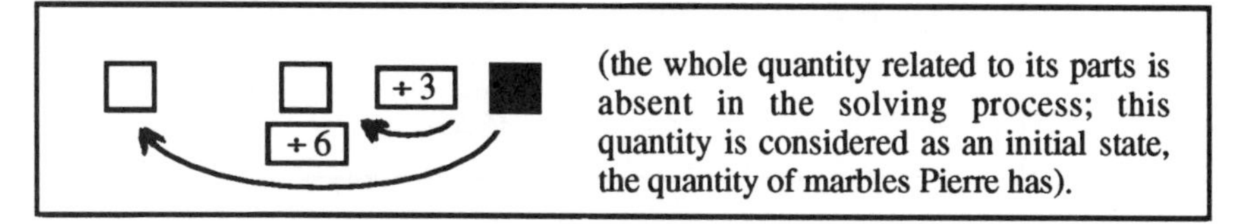

3.2. *Create a Starting State by Using a Fictional Number*

The search for a state that allows the student to carry out the given operations leads some to invent a fictitious number. Here we find procedures like "numeric trials," often mentioned in algebra research. Note, however, that this way of operating appears to be only one of the possible procedures (numeric trials are used relatively little by students).

Eric fits this reasoning profile. This is how he proceeds for the entire set of problems:

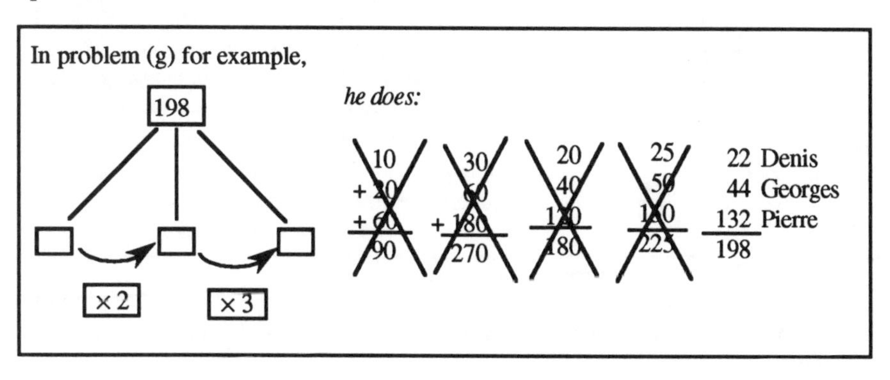

Here Eric fixes a starting value for one of the unknown quantities, a starting state from which everything else can be reconstructed. He then calculates the total number and compares it to the number given initially. He then begins a new trial.

These students who call upon procedures of the "numeric trial" type take into account, unlike the preceding students, the two types of relationships present (comparison and parts-whole relationship). However, like the students in the first category, they operate in a sequential manner starting from a known initial state and do this by passing through intermediary states and following the order of the relationships present. The problem's connections are thus constructed in a direct way from the bottom up (sticking very close to the situation as given).

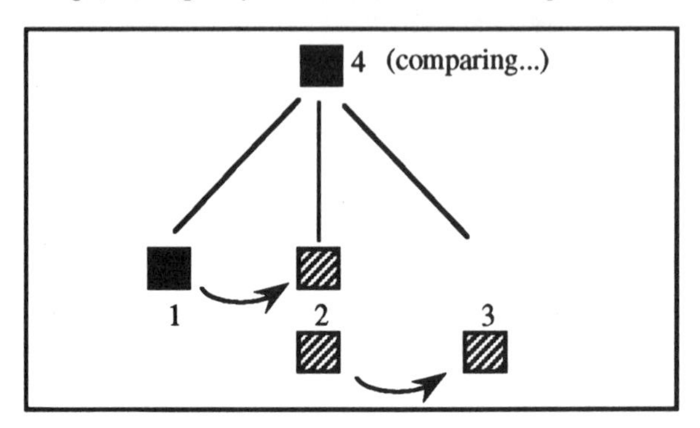

3.3. *Share and then Generate*

Other students create a known state, allowing them to generate the various relationships involved, by invoking a frequently used numeric strategy of equal sharing, which involves dividing the only known state of the problem (the whole) by the number of parts (composing this whole). The number thus obtained then serves as generator and allows for the reconstituting of the various unknown quantities.

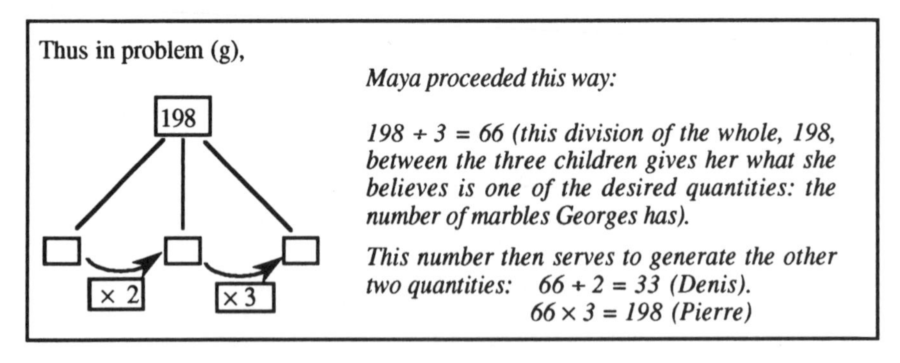

Thus in problem (g),

Maya proceeded this way:

198 ÷ 3 = 66 (this division of the whole, 198, between the three children gives her what she believes is one of the desired quantities: the number of marbles Georges has).

This number then serves to generate the other two quantities: 66 ÷ 2 = 33 (Denis).
66 × 3 = 198 (Pierre)

With this method students proceed in a sequential manner by a succession of steps starting from the known state, finding at the end of the process the quantities they were looking for.

3.4. Take the Structure into Account

Other solution methods demonstrate a superior level of mastery of the relationships involved and a more global representation of the problem's structure (linking of relationships involved).

Thus Éléonore, in problem (g)

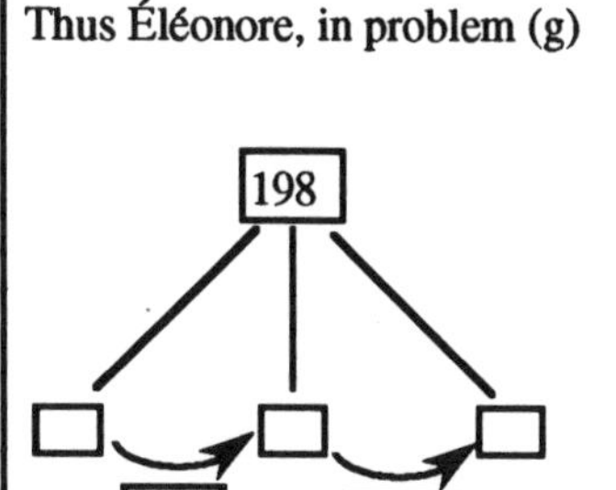

writes: $198 \div 9 = 22$
(she seems to see 1 share, 2 shares and 6 shares).

She thus obtains Denis' share and then figures out the two others:
$22 \times 2 = 44$ *(Georges)*
$44 \times 3 = 132$ *(Pierre)*

In problem (d)

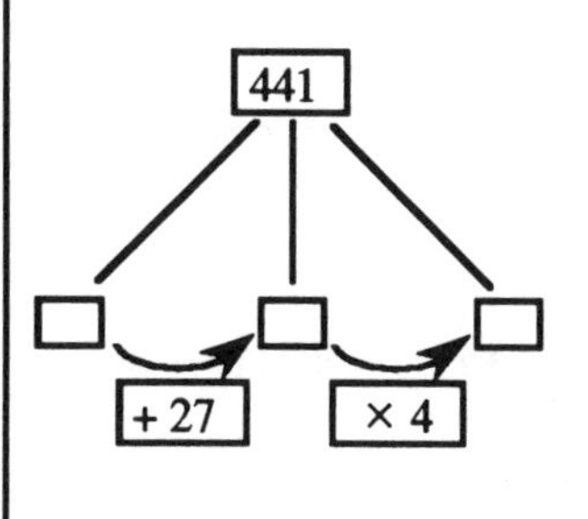

she writes: $27 \times 5 = 135$ *(here she seems to see the + 27 which is reproduced 4 times, thus perceiving the effect of the composition of the two relations).*

Then she calculates: $441 - 135 = 306$ *(taking away this surplus), and finds at last* $306 \div 6 = 51$ *(she divides the quantity obtained in 6, seeming to see 1 share, 1 share, 4 shares).*

Having got this amount, she uses it to calculate the other two: $51 + 27 = 78$ and $78 \times 4 = 312$

Answer:	*Basketball*	*78 students*
	Skating	*51 students*
	Swimming	*312 students*

The preceding strategy, representative of certain stable patterns used by students, illustrates the transformations that take place in the problem solving process for those who use this kind of reasoning (the student operates on relationships).

Thus Éléonore completely changes the initial problem (problem d) she was given, after removing the difference, into a problem of equal sharing (a problem bringing into play only one kind of multiplicative link).

In this example the student operates on the relations between the quantities and transforms the initial problem into a new global one that makes further calculations possible. This is in contrast to the preceding procedures presented, which endeavored to connect the problem state to the solution by operating sequentially, following the order of the relations and sticking closely to the original situation. Here, the calculations do not follow the problem as stated but appear more and more "distant" from it.

The preceding analysis shows different arithmetic reasoning profiles used by students, before any introduction of algebra, in dealing with problems traditionally presented in algebra. These reasoning profiles demonstrate, as we have seen, important distinctions in ways of representing and managing the quantities and relationships involved.

How will these students, depending on their reasoning profile, react to a new mode of treating these problems?

4. THE PASSAGE TO AN ALGEBRAIC MODE OF TREATING PROBLEMS: CONTINUITIES AND DISCONTINUITIES

4.1. *Hypotheses can be Formulated Concerning the (more or less easy) Passage to Algebraic Reasoning for these Children*

Before considering these hypotheses, we will try to describe the reasoning that is traditionally used in algebra when the work directly involves an unknown.

In operating on an unknown quantity, which we directly perceive as the one which will allow us to generate all the others, and doing it as if this quantity was known, the algebraic reasoning, just as with the arithmetic reasoning, "connects" the problem in some way.

The diagram for problem (c) will serve to illustrate the reasoning process in this example:

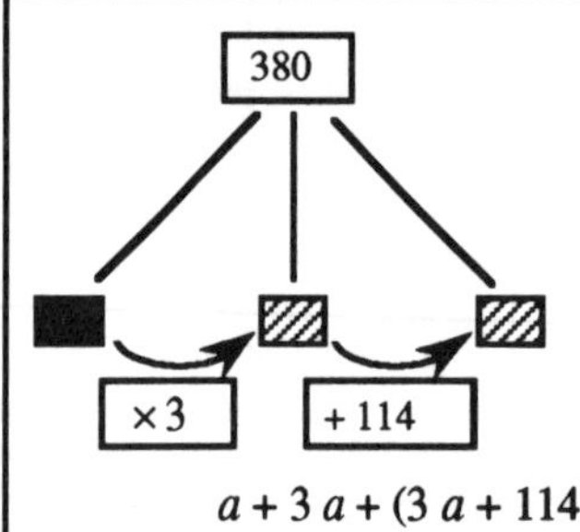

Let a be the number of students enrolled in skating (here we fix a state a).

$3\,a$ will be the number of students in basketball; and

$3\,a + 114$ the number of students in swimming.

$a + 3\,a + (3\,a + 114) = 380$ (the three states are tackled together here)

$7\,a + 114 = 380$

$7\,a = 380 - 114$

$7\,a = 266$

$a = \frac{266}{7} = 38$

The preceding analysis appears to link *a priori* this way of connecting the problem (algebraic solution) to the reasoning we called "numeric trials." In both cases, the initial states are fixed (by a fictitious number in numeric trial reasoning and by a symbol designating this state, whatever it may be, in algebra). We act "as if" in both cases, and we work "linearly" through the diagram (no modification is necessary). This seems to show that students adopting a "numeric trials" strategy are perhaps closer to algebraic reasoning, in their managing of the data, than students who adopt other types of arithmetic reasoning. We will see, however, in what follows that this is not the case and we will try to explain why. The preceding analysis already leads one to suspect certain differences: in one case (numeric trials), one operates in a sequential manner on the states, going through intermediate states, and there is no necessity to link the third state to the first one which generates it. In the other case (algebraic reasoning), the three states are tackled together in the equation.

Certain arithmetic reasoning (Type 4 reasoning: taking the structure into account) appears, on the contrary, to be *a priori* more removed from algebraic reasoning in the way the data are handled (students operate here on relations or transformations directly without state). How will students who have a global perception of the relationships involved react to an algebraic solution? Are they ready to model the problem using only one unknown?

In order to shed some light on these questions, we shall return to the principal results of the interviews with students having different arithmetic reasoning profiles; but first we take a global look at the aims of the interviews and the way in which they were constructed.

4.2. *The Nature and Aims of the Interview*

The subjects (12-13 years) represented different arithmetic solution patterns (they were selected using a written test of more or less complex "algebraic" word problems). Some of the specific questions addressed by the interview were the following:

- How does the student spontaneously envisage the solution of an "algebraic" problem: Does he see one unknown or several? (The student has a chance here to express his opinion as to whether or not the problem is solvable.)
- How does he interpret the algebraic reasoning he is presented with? How, in particular, are certain key aspects of this reasoning interpreted? Does he accept them? Can he use them? Is he convinced that the simulation of the reasoning (done in front of him) will allow him to actually solve the problem? Is he a long way from this approach? What are the difficulties that arise? Why?

4.3. *The Interview: Its Principal Stages*

During the interview three "algebraic" problems were given to the students. These problems were based on the same context: an inventory in a warehouse for sports equipment where three articles were considered (tennis rackets, balls, hockey sticks). The student had to find the number of articles in each category from the relationships between the respective quantities (comparative type relations) and the total quantity (whole related to its parts).

The three problems given were of different complexity:

- The first (the simplest but not classical) involved two multiplicative relationships, linking the quantities of balls, rackets, and hockey sticks. One of these quantities was not directly connected to the sum (here the quantity of balls is not necessary in order to solve the problem but can be found subsequently). The problem's structure can be illustrated by the following diagram:

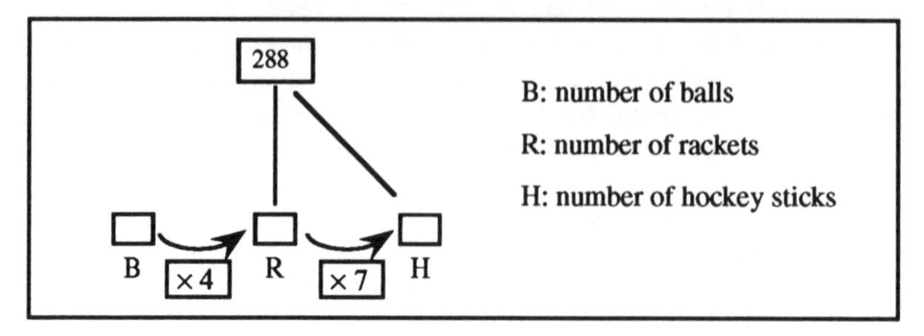

- The second problem, of the type given on the test (see Table 1) brought into play a composition of two different relationships (× and +) and appeared more complex (it involved a composition of two relations):

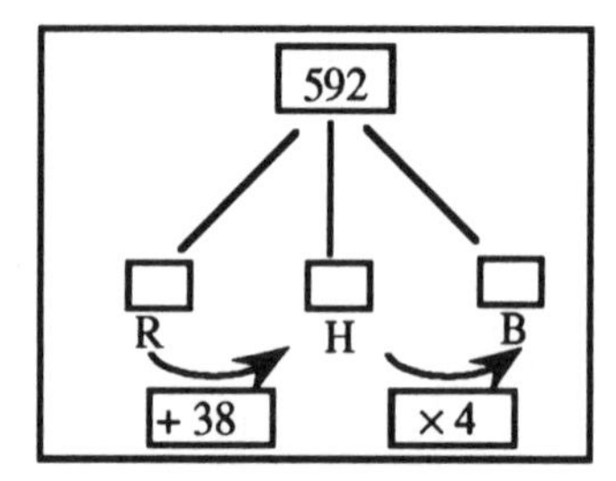

- The third problem presented also appeared more complex to the students, as we have seen (see Table 1), and as in the preceding case, could motivate, if they were to see the relevance, an eventual passage to algebra:

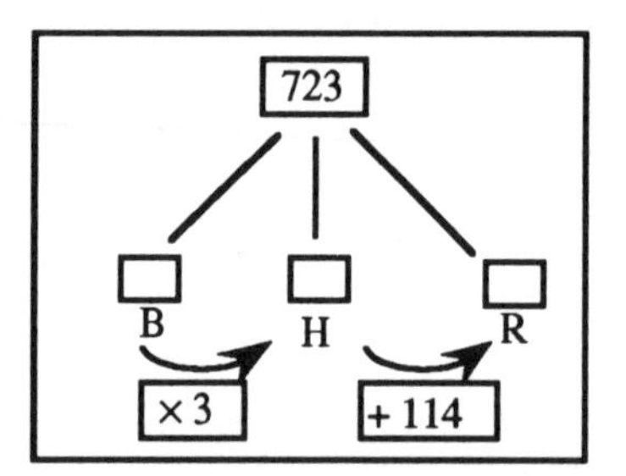

In what follows, we will focus primarily on the protocol followed for the first problem in order to better understand what is involved.

In presenting this problem, we hoped, firstly, to see how the student perceived it and to grasp how he spontaneously foresaw its solution.

The following situation was given with this in mind: "A son (hired by his father to do an inventory) left him the following message: 'Three types of articles were counted. The rackets and hockey sticks in all make 288. There are 4 times more rackets than balls and 7 times more hockey sticks than rackets'." The student was then asked the following question: "Do you think the father can get by with the message his son left him in order to find what is in the warehouse or, in your opinion, must he go to the warehouse to count anything?"

Using this question, we can, in fact, see if the student perceived the problem as involving unknowns: a single unknown (number of rackets, or balls, that must be recounted), two unknowns (number of balls and number of rackets that must be recounted), or if in fact he saw no unknown in this problem (he could adequately get by with the message that was left him).

Secondly, two models of the problem were suggested to the student: "The father thinks he can get by with the message without returning to the warehouse. So look, he writes this ... and this .... What do you think of that? If I told you to take one of the two, which, in your opinion, would be the most useful for finding what is in the warehouse?" Two equations were presented to the student at the same time: a "standard" equation[8] expressed in terms of the number of balls:

4 × + 7 × 4 × = 288

and another equation, also expressed in terms of balls but based on the student's reasoning on the written test (i.e., a model induced from his reasoning).

By this question we got a first indication of how the "standard" equation was perceived by the student: Was it close to his way of representing the problem or not?

Using a legend, we next observed whether students could directly give the number of hockey sticks from the number of balls. This stage of the interview, before getting into the algebraic solution as such, allowed us to clarify the meaning of the symbols used and the links between the quantities, and to gauge the ease or difficulty with which the students could consider the problem in terms of one of the quantities (the number of balls), and could use this to generate the other two quantities (resorting to a composition of the two relationships).

The algebraic solution was then simulated for the student, getting his reactions and interpretation at each step.

- "Look at what the father did" (we show the student the following equation):

"What do you think of this? Do you agree?" (this first equation that was presented to the student referred to two unknowns).

- "Now look at what he did" (the label representing the number of rackets is lifted to reveal 4 × the number of balls):

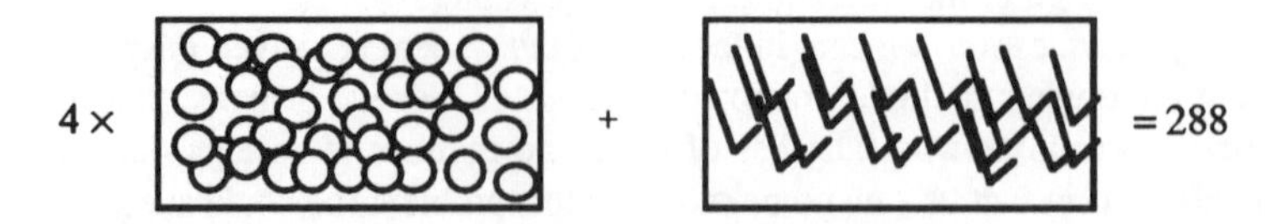

(How will the student react to this substitution of one of the unknowns?)

- Then we go on to the other quantity (we lift the other label to reveal the number of hockey sticks in terms of the number of balls: "Now look at what he did. Do you agree?"

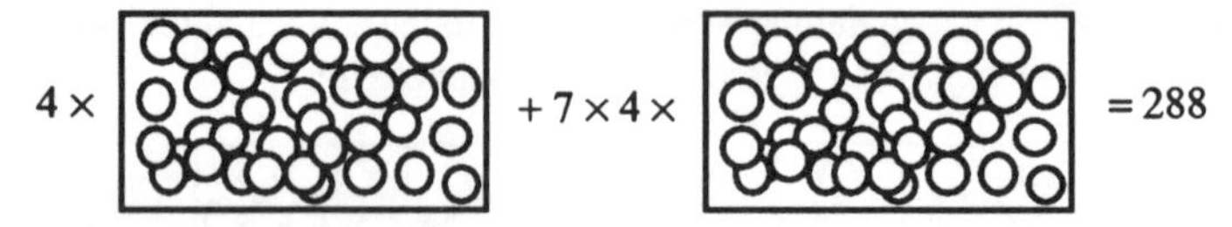

(What difficulties will the student experience at this stage in dealing with the composition of relationships?)

- We then ask the student to continue the solution: "Now it's as if I converted my entire warehouse into balls, do you agree? How many times more balls do I have in the warehouse now?" (Can the student at this stage operate on the unknown quantity? Will the student accept doing so?)

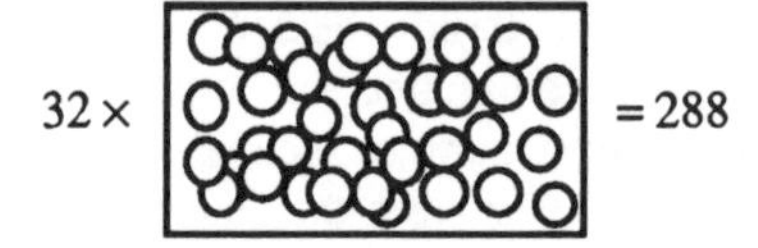

- "Now can you find how many balls are in the warehouse? And the number of rackets? hockey sticks?" (Is the student really convinced that the problem can be solved in this way?)

It is to be noted that the reasoning that is favored in this simulation attempts to maintain a certain meaning throughout the solving process.[9] The proposed solution method, with its constant reference to a certain meaning, is quite removed from those habitually found in algebra. Voluntarily suggested in this case, it allows one to better understand the distance that separates the student from the solution at each stage and to highlight the interpretation difficulties he encounters.

We will now return to some of the important results revealed by this study of the passage from arithmetic to algebra in problem solving.

4.4. *Fundamental Changes in the Passage from Arithmetic to Algebra*

4.4.1. *Conceptualization of the problem*

Three different attitudes are found in the ways students spontaneously envisage the problem's solution:

1) Some of them, as they did on the written test (reasoning Types 1 and 2, built on an initial state), look for a known state as a starting point for the solution

(taking the whole, the whole which they divide in 3, or a fictitious number) and then initiate a sequence of unrelated successive actions: The number of balls, for example, allows one to find the number of rackets and the number of hockey sticks then follows from this new number (a two-step procedure). This way they are far from the algebraic approach as we will subsequently see.

2) Others quickly perceive the structure of the problem (the balls constitute a false generator here). Thus, they solve the two-branch problem, as it is presented (by transforming it into a new, unified configuration): they see 8 parts, and divide into eighths. The quantity of hockey sticks is then generated from the number obtained, then the number of balls (here we find children having Type 4 reasoning: taking the structure into account). Thus, Éléonore rapidly replies: "Oh, he can manage with that ... because it gives an exact number for the rackets and then the hockey sticks. And then they say 7 times more hockey sticks than rackets so you can divide that number by 8, the number of times plus 7 times there because that makes nevertheless 1 times the rackets. Then that would give a number for the rackets, and then he could make 7 times ugh! the rackets, that would give a number for the hockey sticks .... Then they say it's 4 times as many rackets as balls, then the number of rackets you divide it by 4 to find the number of balls."
This way of functioning appears to be far removed from algebraic reasoning which operates in terms of a single generator: the number of balls. In the approach established by these students, there is no common generator.

3) A very few students say they must go and recount something in the warehouse (spontaneously perceiving an unknown in the problem). These *search for a generator* which, if they knew it, would allow them to generate the other quantities. In this way they may appear to be closer to the expected algebraic reasoning (these are the students with Type 3 reasoning: share and generate). Thus, Julie says: "I think he'll have to recount the balls." Then, when the Interviewer asks, "And with that will he be able to find everything in the warehouse?", Julie responds: "Yes, he'll count the number of balls times 4, then he'll count the number of rackets times 7 and he will have the hockey sticks" (we find a two-step procedure here). Elsa answers in a similar fashion: "Well I think he has to recount one of the articles, then ... when he knows what the number is, then he can get the others."

This first part of the interview, through the students' statements about the feasibility of the problem, shows the distance that separates them (for the majority of students) from the expected algebraic reasoning. Their choice, as to the most useful equation for solving the problem, confirms these first results. The majority reject the "classic" equation, choosing the one that is closest to their reasoning.

4.4.2. Obstacles in interpreting algebraic reasoning

The rest of the interview allows us to bring out the obstacles the students encounter in interpreting the algebraic reasoning, and to better understand the distance that separates them from this reasoning. We find that:

- To directly generate using a single unknown is a process that appears to be extremely complex for these students.

In their actions (when we ask them if, using the legend, they could find the number of hockey sticks knowing the number of balls, or later when we ask them to react to the solution), the students never take on the composition of the two relationships: they go through an intermediary state, the different states being generated by a sequence of two-step actions. "He will take the balls times 4 to find the rackets; then after that, he will take the number of rackets times 7 to find the hockey sticks."

- Substitution, which requires the passage to a single unknown, appears to be equally difficult for the student.
  The student's remark, "We can't do 4 times the balls plus 7 times 4 times the balls, that doesn't equal 288 which are the rackets and the hockey sticks, that would make that 288," clearly shows the difficulty encountered in accepting the substitution of one quantity by another (288 represents the total of hockey sticks and rackets, and not the total balls!). Thus "thinking with one unknown" requires that the students maintain a certain distance from the problem (one has to accept here imagining the entire warehouse full of balls).
- For certain students, these difficulties are accompanied by a refusal to operate on the unknown.
  Thus, Julie who, as we saw earlier, appeared to be nearer to algebraic reasoning in recognizing a generator from which the different quantities could be determined, is in fact far from it since the generator must *be known* to be able to operate: "We don't know how many balls there are, we can't take 4 times the balls plus 7 times 4 times the balls, that doesn't equal 288 which are the rackets and the hockey sticks."
- The symbolism used to present the relationships in the equation creates difficulty for some students.
  Thus, Valérie reacted to the symbolism:

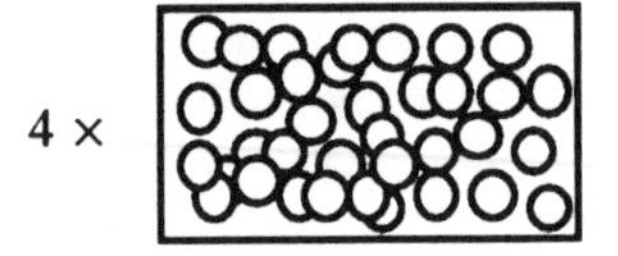

(to express the number of rackets), saying:

"No, it's the opposite." Why? "There are four rackets for one ball." One finds here a certain interpretation of the relationship 1 to 4 (see also Clement, 1982) that would lead to the symbolism:

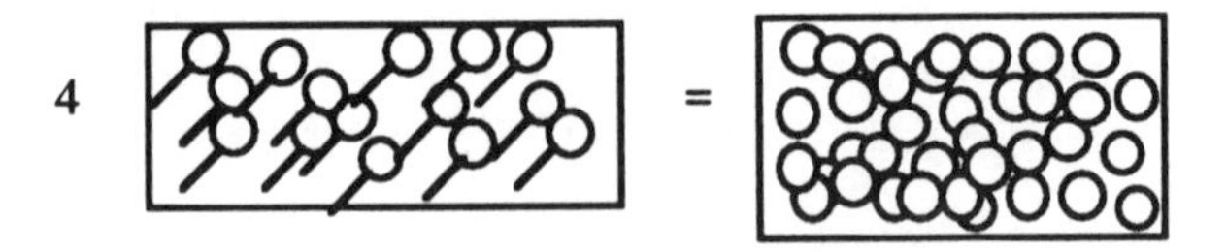

5. CONCLUSION

The preceding analysis helps us to better understand the passage required of students in going from solving "arithmetic" to "algebraic" problems. A first facet of this analysis leads us to reflect on the very nature of the problems we give to students in the two domains. Our analytic framework for these problems and our experimental results show the differences between the "relational calculus" on which an arithmetic mode of reasoning is founded and the "relational calculus" on which an algebraic mode of reasoning is based. We can distinguish here arithmetic profiles among student problem solvers on the basis of the consistent procedures they use in solving their first algebra problems and problems in the algebra corpus which "break" these regular procedures. These latter problems (e.g., problems involving composition of several relationships), which students find more difficult to solve arithmetically, may be interesting to use to motivate a transition to algebra. Moreover, the criteria that were developed by the team to analyze the relative complexity of problems in algebra enable us to anticipate a possible gradation in these problems for the purpose of teaching.

Finally, the contrast between the spontaneous arithmetic reasoning used by the interview students and the reasoning that is normally expected in algebra shows the fundamental changes that mark the passage from one mode of treatment to the other and the distance that separates certain previous arithmetic reasoning profiles (Types 1, 2, 3) from the expected algebraic reasoning. The analysis particularly shows the distance that separates numeric-trials reasoning from algebraic reasoning and calls into question those approaches that are generally promoted for encouraging the passage to algebra. Do these approaches not lead the student to think in terms of a sequence of successive actions, passing through intermediary states, hardly encouraging any consideration of the composition of relationships?

Structural-type students deserve particular attention. Although generally quite able to explain the key steps in algebraic reasoning when they are presented to them and to interpret the various equations that are proposed to them, their reasoning remains very removed from algebraic reasoning. They spontaneously revert to their own way of doing things, not seeing the pertinence of algebraic reasoning, which scarcely convinces them that one can solve in this way. Yet, their recognition of various equations and the ease of their interpretation of the different generators clearly show that they perceive the problem structure and master the play of relationships. Their interpretation is not limited, unlike that of the preceding arithmetic profiles, to a sequential treatment. Does the equation fulfill an entirely different function for these students, that of a model of a class of problems, one which could advantageously be exploited?

NOTES

1 This research was funded by FCAR-Québec and the Social Sciences and Humanities Research Council of Canada.

2 In the book, *L'arithmétique des écoles* (1927), we find the following definition of algebra which illustrates well the role it played in teaching: "Algebra is a science which simplifies problem solving and generalizes solutions by establishing formulas to solve problems of the same type." This role was put into practice in problems without numeric data and which were explicitly aimed at a general solution as, for example, the following problems:

"A sum of money is divided between two children such that one receives twice what the other gets. How much did each receive?"

"Two cyclists set off in the same direction with one traveling a certain number of kilometers more than the other. What is the distance between them after a certain number of hours?"

3 The set of criteria developed also involves other elements (wording, context,...).

4 For an analysis of problems involving different types of relations between the quantities (transformation problems, rate problems, comparison problems,...), see Bednarz et al., 1992; Bednarz & Janvier, 1994.
5 We worked at establishing criteria from which a grid emerged for the analysis of problems. This grid proved helpful in analyzing the relative complexity of problems. Experimental results in different grades (high school I, II, III, IV, V) showed that in spite of a higher success rate in the highest grades, some problems presented here appear to be difficult to solve. We will not get into this analysis here.
6 We varied other elements in this structure besides those presented in the table.
7 Such appears to be quite rare in the current textbooks we examined.
8 Obviously, the equation format (the symbolism employed) is not "standard."
9 The calculations, however, require the student to distance himself more and more from the problem situation.

CHAPTER 9

DEVELOPING ALGEBRAIC ASPECTS OF PROBLEM SOLVING WITHIN A SPREADSHEET ENVIRONMENT

TERESA ROJANO

Previous analyses of the historical development of algebra, which looked at problem solving as a force for change in the most significant processes of evolution, lead to thinking about the didactic role that this activity could play in the construction of algebraic knowledge on the part of the students. This idea, together with the growing body of evidence that spreadsheet environments may help pupils to develop fundamental aspects of algebraic thinking, are the bases upon which an empirical study with pre-algebraic children and with algebra resistant pupils is developed. In this chapter some relevant results of the study are discussed, in particular those emphasizing the shift that pupils make within this environment from proceeding from the known to the unknown (pre-algebraic process) to dealing with the unknown in order to determine its numerical value (algebraic process).

1. INTRODUCTION

There is frequently a tendency to think that the utilization of strategies, such as trial and error for solving algebraic word problems, can become obstacles to the progress of students who are learning algebraic methods. However, trial and error, together with other strategies considered informal and which are found in students beginning the study of algebra, are indeed a real foundation upon which the methods or strategies of algebraic thought are constructed. Thus, an important area of research is that of the role played by students' pre-algebraic antecedents with respect to problem solving as they learn new methods.

On the other hand, a common denominator in the results reported in studies on the approaches and strategies used by students when solving algebraic problems is the almost complete absence of school algebraic methods (see, e.g., Bednarz et al., 1992; Puig & Cerdán, 1990; Filloy & Rubio, 1993; Hall, Kibler, Wenger, & Truxaw, 1989). There are multiple explanations for this finding. As well as the intrinsic difficulties of problem solving, other possible causes of students' reticence to use algebra can be mentioned, such as the separation of the symbolic manipulation of algebra and its use in problem solving in introductory teaching programs. In addition to the latter, there is the inability shown by many students to achieve an integration of these two domains of knowledge.

The points mentioned above, together with the growing body of evidence that some of the difficulties encountered by students in the interpretation of algebraic symbols can be approached didactically by means of activities within certain computational environments (Sutherland, 1989), were the factors that led to the formulation of the Anglo/Mexican project, "Modeling algebra problems within a spreadsheet environment."

N. Bednarz et al. (eds.), Approaches to Algebra, 137-145.

The main objectives of the project can be summarized as follows:

- To investigate the way in which pupils use a spreadsheet environment to represent and solve algebra problems and to relate this to their previous arithmetical experiences and their evolving use of a symbolic language.
- To characterize pupils' problem-solving processes along the arithmetic/algebraic dimension as they evolve through working in a spreadsheet environment.
- To develop and evaluate a didactic sequence to help pupils make links between spreadsheet and traditional algebraic syntax.

The three stages of this project have been carried out over the last three years in parallel in Mexico and in Great Britain as a joint effort of the Centro de Investigación y Estudios Avanzados in Mexico and the Institute of Education of the University of London.[1] The first stage involved pre-algebraic children between the ages of 10 and 11 years; the second was carried out with students 14 to 15 years old who were reticent to use algebra; and the third stage dealt with average students of 10 to 11 years of age. In this chapter there is reference to some of the results obtained from the first two stages of the project, which are related to the role played by the spreadsheet in the development of "more algebraic" methods for solving word problems (proceeding from the unknown to the known), and which emphasize the influence of the students' "own" strategies and of their arithmetical antecedents in the development of these methods. An extensive and detailed description of the results and of the theoretical and methodological background to the studies is reported in recent publications by Rojano and Sutherland (1991b, 1992, 1993) and by Sutherland and Rojano (1993).

2. THE ANGLO/MEXICAN PROJECT: MODELING ALGEBRA PROBLEMS WITHIN A SPREADSHEET ENVIRONMENT

In addition to the studies carried out by Capponi and Balacheff (1989) and by Healy and Sutherland (1991), which show that spreadsheets can be a substantial support in the development of essential aspects of algebraic thinking, the Anglo/Mexican project has other antecedents which influence its central ideas. These are found in the history of the development of algebraic thought, and more specifically, in what we have called "lessons from history" in the chapter entitled, *The Role of Problems and Problem Solving in the Development of Algebra* (Rojano, this volume).

Three fundamental aspects of these "lessons" are considered: (a) the risk of preferring, as an object of knowledge, a merely syntactic and manipulative algebra, without a previous construction of meaning for the algebraic symbols (in this regard, didactic work from problem solving, aimed at providing meaning for symbols and general methods might be an answer); (b) the recognition of the significance of working with problems that reach the limits of the mathematical knowledge developed up to a certain moment in history and that lead us to the subsequent creation of new knowledge in the field of symbolic algebra (this idea is taken up in the didactic design of the spreadsheet activities, when non-routine problems are included as a challenge to the student); and (c) the importance of not denying the value of the knowledge, methods, and abilities that the students already possess, whether or not these are informal and primitive, with respect to the construction of new knowledge (as in the case of Viète who, in the 16th century, in order to achieve his aims in the formulation of "the new art of algebra," put to use knowledge and skills from classical Greek mathematics). This idea is utilized in the project when the students' own strategies are incorporated as the necessary starting point for the learning of systematic and general methods of problem solving.

An overview of the theoretical background to this study, which appeared in Rojano and Sutherland (1991b), can be summarized as including: (a) consideration of a didactic cut along the evolutionary line from the child's arithmetic to algebraic thought, a cut produced by the difficulty in operating on the unknown quantities (Filloy & Rojano, 1989), and (b) awareness that computer-based symbolic environments can support pupils in their learning to formulate mathematical generalizations (Sutherland, 1992a).

3. METHODOLOGY

The methodology used within the first stage of the project is essentially that of a longitudinal case study. Two groups of 10- to 11-year-old pupils were studied simultaneously, one in Mexico (7 pupils) and one in Britain (8 pupils). This age group was chosen because the pupils had not been involved in any formal algebra instruction. Both groups engaged in three blocks of spreadsheet activities over a period of five months with pupils working in pairs for approximately 12 hours of "hands on" computer time.[2]

- Block 1: Function and inverse function
  Within this sequence pupils were introduced to the following spreadsheet and mathematical ideas: entering a rule; replicating a rule; function and inverse function; symbolizing a general rule; decimal and negative numbers.
- Block 2: Equivalent algebraic expressions
  Within this sequence pupils were introduced to the ideas of equivalent algebraic expressions (e.g., $5n$ and $2n + 3n$).
- Block 3: Word problems (see, e.g., Figure 1)
  Within this sequence pupils were introduced to solving a word problem in a spreadsheet by representing the unknown with a spreadsheet cell, expressing the relationships within the problem in terms of this unknown, and varying the unknown to find a solution.

All pupils were given a pre- and post-test and interviewed individually on their responses to the questions in these tests. The interview items consisted of questions related to activities within the three blocks. The data collected included: tape or video recordings of individual interviews; printouts of computer work; paper and pencil work; detailed notes of teaching sessions; occasional video recordings of computer work.

4. RESULTS OF ANALYSIS FROM TWO WORD PROBLEMS

All the pupils became competent at entering and replicating a rule in the context of solving the spreadsheet problems in Blocks 1 and 2. In this section we discuss the results of analysis of Block 3 problems and the ways in which the Mexican and British pupils solved two word problems: the Measurement of a Field problem and the Chocolates problem (see Figure 1). The former problem was given to the pupils in the pre-interview before they carried out any spreadsheet activities. Both problems were given to the pupils in the post interview. Similar problems were presented to the pupils within Block 3.

MEASUREMENT OF A FIELD	CHOCOLATES
The perimeter of a field measures 102 meters. The length of the field is twice as much as the width of the field. How much does the length of the field measure? How much does the width of the field measure?	100 chocolates were distributed between three groups of children. The second group received 4 times the chocolates of the first group. The third group received 10 chocolates more than the second group. How many chocolates did the first, second, and third groups receive?

Figure 1. Word problems

When analyzing the pupil strategies to the word problems, we characterize as "algebraic" a solution in which the pupil works with the unknown within a problem, operating on it as if it were known. We characterize as "non-algebraic" a solution in which a pupil works from the known to the unknown.

The Measurement of a Field problem could be solved on paper, using the following algebraic approach, which involves working from the unknown to the known:

Let the width of the field = x *meters*
Let the length of the field = y *meters*
Then $y = 2x$ (1)
and $2y + 2x = 102$ (2)
So, by substituting (1) in (2),

$$4x + 2x = 102$$
$$6x = 102$$
$$x = 17 \text{ meters}$$

Many studies indicate that pupils are more likely to use a non-algebraic approach when solving this sort of problem (Bednarz et al, 1992; Lins, 1992).

One of the non-algebraic strategies used by some pupils of the study in the pre-interview is the *Whole/Parts* strategy, illustrated here by the way a pupil solved the Measurement of a Field problem: "I did 102 divided by 6 ... I just did two of the lengths to make it sensible ... I just thought there must be two of those in the length." This solution involves working with a known whole (the perimeter) and dividing this into parts to find the unknown lengths of the sides of the field. It can be considered that this method works in the opposite direction from that of the algebraic approach, which involves working with the unknown as starting point.

4.1. *Results from the Measurement of a Field Problem*

Table 1 presents an overview of the strategies used by pupils in the pre-interview when solving the Measurement of a Field problem (children worked with paper and pencil and without the computer). Most of the pupils could not give a correct solution. The correct answers involved non-algebraic whole/part approaches to solving the problem.

	MEXICAN PUPILS	BRITISH PUPILS
CORRECT SOLUTION		
102 ÷ 6; 17 × 2	0	1
102 ÷ 2; 51 ÷ 3	0	1
INCORRECT SOLUTION		
102 ÷ 2; 51 ÷ 2	1	2
102 × 2; 294 + 102	1	1
102; 102 × 2	3	0
Other	2	0
NO SOLUTION	0	3

Table 1. Overview of pupil strategies for the Measurement of a Field problem in the pre-interview (paper and pencil solutions)

During the Block 3 sessions, children worked on the following Rectangular Piece of Land problem: "In a rectangular piece of land the length is 4 times the width. The perimeter is 280 meters. What is the area of this piece of land?" This problem is similar to the Measurement of a Field problem. Throughout the spreadsheet word-problem sequence, pupils were taught a spreadsheet-algebraic approach (algebraic because it involves working from the unknown to the known; see Figure 2). The unknown is represented by a spreadsheet cell (this might be called x in an algebraic solution). Other mathematical relationships are then expressed in terms of this unknown. When the problem has been expressed in the spreadsheet symbolic language, pupils then vary the unknown either by copying down the rules or by changing the number in the cell representing the unknown (in a paper and pencil algebra approach, this would be the equation-solving part). In Figure 2 the spreadsheet formulas are shown in order to present the pupils' solution processes. However, once the formula is entered into the spreadsheet, a number is automatically computed and a table of numerical values is produced (although the user can ask to view the formulas).

	A	B	C	D
1		WIDTH (CM)	LENGTH (CM)	PERIMETER (CM)
2		25	= B2 * 4	= (B2*2) + (C2*2)
3		= B2 + 1	= B3 * 4	= (B3*2) + (C3*2)
4		= B3 + 1	= B4 * 4	= (B4*2) + (C4*2)
5				
6				
7		rule copied down	rule copied down	rule copied down

Figure 2. Carla and Laura's spreadsheet solution to the Rectangular Piece of Land problem.

There is evidence that both pointing with the mouse or with their finger and using spreadsheet language in their speech help pupils to express the problem constraints in the spreadsheet environment.

	MEXICAN PUPILS	BRITISH PUPILS
WITHOUT SPREADSHEET		
Non-algebraic Whole/Part solution.	1	2
Algebraic solution.	0	0
WITH SPREADSHEET		
Non-algebraic Whole/Part solution.	1	1
Algebraic solution.	5	3
NON COMPLETED	0	2

Table 2. Overview of strategies used by pupils to solve the Measurement of a Field problem in post-interview (spreadsheet available).

Both the Mexican and British pupils found this problem difficult in the post-interview, and for some pupils the interviewer provided support[3] with the solution process. The uncompleted solutions in the British sample were due to the time constraints of the interview session. However, comparing the results of the post-interview with the results of the pre-interview, we find an improvement in the pupils' ability to express symbolically the relationships of the problem. It is noted that, during the post-interview, some Mexican children still found it difficult to understand the relationship between the width and the length. So the teaching strategy used during the experimental work was reused at this time by the interviewer. This strategy consisted in asking the pupils (when necessary) to analyze particular cases (e.g., "If the width measures 3 cm, what does the length measure?"). In this way, these children could "jump" to the general case with no additional support on the part of the interviewer, that is, after analyzing a number of particular cases, they could produce a general spreadsheet formula to express the relationship. Most of the students in both groups needed a nudge to express the correct rule for the perimeter (many pupils spontaneously entered the rule for the area). Once the problem was expressed symbolically in the spreadsheet, all of the students were able to find the unknown values.

A number of methods were used by the Mexican and British pupils to express the formula for the perimeter of the rectangle (e.g., length + length + width + width; length*2 + width*2; length + width = s, length + width + s; 2* (length + width)). This variety of approaches suggests that the spreadsheet environment helps pupils to symbolize their informal ideas in a more formal way. The additive rule (length + length + width + width) relates more closely to the objects of the problem and seemed to be used by pupils who had difficulties with a multiplicative strategy. In this respect pupils were able to generalize and symbolize the arithmetic that was most familiar to them.

4.2. *Results from the Chocolates Problem*

Table 3 presents an overview of the strategies used by the pupils to solve the Chocolates problem in the post-interview (pupils were able to use a spreadsheet to solve the problem).

	MEXICAN PUPILS	BRITISH PUPILS
WITHOUT SPREADSHEET		
Non-algebraic Whole/Part solution.	0	0
Algebraic solution.	0	0
WITH SPREADSHEET		
Non-algebraic Whole/Part solution.	0	0
Algebraic solution.	7	7
NON COMPLETED	0	1

Table 3. Overview of strategies used by pupils to solve the Chocolates problem in the post-interview (spreadsheet available)

In this problem, some pupils started using a whole/parts, non-algebraic strategy as illustrated by the following solution on the part of Mike: He first performed the calculation 100/3 in a cell and then entered the result 33.33333333 in cell A5 (for the number of chocolates in the first group). He then entered the rule 33.333333*4 (for the number of chocolates in the second group) in cell A9 indicating that he was initially thinking with the specific value. He immediately changed this to the correct rule A5*4, indicating a shift from thinking with a specific object to thinking with a general object. Other pupils (in both Mexico and Britain) started with 100/3 in the cell that represented the unknown number of chocolates in the first group, indicating an integration of a non-algebraic and an algebraic approach. The majority of the pupils then varied this number 100/3 when the rule for the number of chocolates in the second group produced a number greater than 100.

5. SOME REMARKS ON THE ANGLO/MEXICAN SPREADSHEET STUDY

This study has shown that most of the pupils in this first stage of our project did not spontaneously think in terms of a general formula when first presented with the spreadsheet environment. The problems in Blocks 1 and 2 were designed to encourage pupils to move from thinking with a specific to a general object, and by the end of the study, the majority of the pupils in Mexico and Britain were able to do this. When solving the Block 3 word problems, pupils initially had difficulties in accepting the idea of working with a general unknown number and had to develop strategies for dealing with this. This could involve calculating a first approximation to the unknown and then placing this number in the cell that represented the unknown. The spreadsheet environment supported pupils in moving from thinking with the specific to the general, both in terms of the unknown and of the mathematical relationships expressed in the problem.

Informal non-algebraic approaches were found in Mexican and British pupils. With the exception of the British children, who had efficient informal strategies for solving word problems and who were the least motivated to learn the "spreadsheet method," the rest of the pupils seemed to be synthesizing their informal strategies with a more formal algebraic approach.

We have found that pupils can begin to use the spreadsheet to express their rather unformed ideas: "I think that it helps you because you put what you think in and then you can check to see if you are right, and it gives you more encouragement to continue because you know that even if you do it wrong you can use the spreadsheet to help you."

On the other hand, it can be said that the students who participated in this research progressed towards a "more algebraic" way of proceeding. In the majority of cases, both during the experimental work and in the post-interview, we found an awareness of the role played by the variation of the unknown in problem solving, once the rest of the unknowns and data have been expressed in terms of this unknown which varies. The awareness of the interdependent relationship between unknowns, the choice of an unknown for variation, and the work itself on an unknown quantity in order to find its numerical value, subject to the constraints of the problem, form the basis of the substantial differences between the variation of the unknown in the spreadsheet method and the informal strategy of trial and error based on numerical experience in the world of arithmetic.

In a second phase of the study, we worked with two groups of 14- and 15-year-old students (one in Mexico and the other in Britain) with a history of failure in mathematics (algebra-resistant pupils). The study involved the students in a sequence of activities with spreadsheets, almost identical to the sequence used with the 10- and 11-year-olds of the first phase. Some relevant results of the second stage reproduce those found in the first stage, and others reveal part of the nature of the reticence of these students to use algebraic methods and symbolism, as well as the high likelihood of making the connection between the representations of a spreadsheet and the algebraic representation of the relationships of a problem (Rojano & Sutherland, 1993).

The results of this research and of other studies carried out on the same themes (e.g., Filloy & Rubio, 1991; Garançon, Kieran, & Boileau, 1990) make the incorporation of problem solving into the teaching of symbolic algebra plausible, not as the concluding part but as the starting point and, suggest, as well, that we consider the processes of change that intervene in the students' passage from arithmetic to algebraic thought.

6. FINAL REMARKS

The lessons from the historical development of algebra regarding problem solving as a force for change in the most significant processes of evolution lead to thinking carefully about how algebra is taught, and in particular, about the role that this activity could play in the construction of algebraic knowledge on the part of the students. In this chapter, the empirical references are limited to reviewing a study that allows us to suggest the plausibility of introducing students to algebra through problem solving and, in this process, of recovering the students' own pre-algebraic strategies and informal methods (in some senses, the variation of the unknown within the spreadsheet evokes the trial-and-error informal strategy).

Aspects of fundamental importance to this issue have not been touched on, such as the question of maintaining algebraic syntactic manipulation as part of the curriculum, the teaching options that allow the introduction of the same--via "meaningful" approaches--and the communication to the students of the virtues of mastering the technique. Nor have we referred to the difficulties confronted by the students when, involved in learning approaches of this sort, they have to distill the syntactic rules using their own experience with "concrete" models or in "problem situations" where they have previously carried out actions analogous to algebraic ones (Filloy & Rojano, 1989).

The reason for making these minimal comments is that of indicating the presence of a dialectical relationship between these two components of algebraic thinking, that is, the development of operativity (purely syntactic component) and its use in modeling and solving problem situations (semantic component). One cannot speak of a significant evolution in one of these without a basis in the other. The details of this inextricable interdependence can be observed both at the level of individual thought (Filloy & Rojano, 1989; Rojano, 1985) and at the historical level (the case of the birth of Analytic Geometry is an example).

These considerations lead to the contemplation of teaching proposals that do not just include different ways of introducing the student to algebraic thought, but also maintain it in constant evolution through the above mentioned components, such that the student can become a competent user of algebraic language.

NOTES

1 Project funded by the National Program for Mathematics Teacher Training, the National Council for Science and Technology (Grant No: 1390–S9201) in Mexico, and the British Council. Part of this work derives from the ESRC funded Project, "The gap between arithmetic and algebraic thinking," (Grant No: R000232132) in England.

2 The 12 hours of computer activities were distributed in sessions of one hour a week. The five weeks of work included the period devoted to the final written test as well as the final individual interviews.

3 We categorized this support as: no support, a nudge, substantial support. The nature of this support was taken into account in the analysis.

CHAPTER 10

ROUGH OR SMOOTH? THE TRANSITION FROM ARITHMETIC TO ALGEBRA IN PROBLEM SOLVING

DAVID WHEELER

The two research studies just presented in this part of the book on the transition from arithmetic to algebra are situated within the historical "algebra as problem solving" tradition. The following comments mention the possibility of distinguishing algebra much more sharply from arithmetic, question the assumption that growth in learning should be gradual, and point out the influence of research designs on research "conclusions."

Reports built around empirical research, as the two chapters of Bednarz/Janvier and Rojano (this volume) are, should perhaps not be criticized too thoroughly or too ruthlessly. A presentation which tosses around theoretical ideas, and which may well have been written "off the top of the head," probably merits a response made in the same spirit. But reports of empirical research represent the fruit, very often, of months of hard and dedicated work; the particular research being reported is quite probably embedded in a sequence of work stretching over years.

Furthermore, any piece of empirical work in the educational field is inevitably confronted with more complexity than it can easily handle. The field lacks the clarity of problem definition and the consensus about appropriate problem-solving tools that have obtained in, for example, physics at particular periods. There are no "critical experiments" in education. Educational research is always very easy to attack. For sure, *all* researchers will have overlooked something, or extrapolated too extravagantly, or made some unwarranted assumptions.

I am not suggesting, of course, that empirical research should never be scrutinized and never be criticized, but I am saying that, particularly when we have in front of us only two examples out of the very many research studies that have been carried out in this area, I am reluctant to employ the full extent of possible critical fire-power. So I shall move immediately from the particularity of these studies to some general questions that they bring to my mind.

As several authors in both chapters claim, the evidence of history is definitely on the side of the problem-solving approach to algebra. If the mathematical developments that took place in history are trustworthy guides to the development of mathematical instruction, then it seems clear that the introduction of algebra should follow the problem-solving approach and focus on the solution of equations. My own view, which I think is shared by the authors of the chapters, is that the lessons of history are much less direct than this: They do not tell us to copy what happened, but to be informed by what happened. We must always remember, in any event, that our historical knowledge is inevitably partial and can rarely offer us a full and unbiased account of any part of the past, whether of mathematics or of anything else.

N. Bednarz et al. (eds.), Approaches to Algebra, 147-149.

Both chapters deliberately make much of the fact that algebra is introduced to students who already know some arithmetic, mainly the arithmetic of integers, and both are concerned with the parallels and the discrepancies between the arithmetical and algebraic strategies for handling similar problem types. An emphasis in research, or indeed in instruction, on the transition *from* arithmetic *to* algebra will always tend to stress the commonalities between arithmetic and algebra, the quasi-invariances of the transformation from arithmetic to algebra, and play down the very important differences. I am prompted to ask:

- Are we sufficiently clear about the ways in which ordinary symbolic algebra differs from the arithmetic of integers?
- Are there any reasons why we should not try to teach algebra to secondary school students as something totally fresh and different from anything they have learned before? (Biology, for example, is commonly taught as a "new" subject even though beginning students bring with them considerable amounts of relevant information and intuitions.)

Any pedagogical approach to the transition from arithmetic to algebra has to take account of the complementary requirements of (a) facilitating the transition while (b) not shirking the intrinsic obstacles. Getting the balance right is not easy. When one is attempting to make decisions about the structure of the curriculum, the production of teaching materials, and so on, the tendency is to work on all the components of the situation which might *ease* the way through the curriculum. But some psychological research, and certain philosophical analyses of mathematical and scientific development, suggest that in learning something new, the learner necessarily has to grapple with developmental "cuts," "ruptures," and intellectual "conflicts." If these are removed, learning doesn't take place or suffers impairment. The practical problem therefore, for the teacher or the curriculum developer, is to work out a compromise between the claims of smoothness and of necessary roughness.

The Bednarz/Janvier research starts out from a classification and articulation of "traditional" types of algebra "word" problems, whereas the Rojano/Sutherland research employs a recent technology to invent a new pedagogical approach to one type of traditional word problem. The first seems to begin "from below"--the style of the research is largely determined by the authors' extensive work on elementary arithmetical problems of a parallel type--whereas the second seems to start "from above," with the spreadsheet facility of the latest computer software. But from another perspective, the non-traditional representations, which the Bednarz/Janvier research introduces in order to clarify their classification categories, seem implicitly to suggest an alternative pedagogical possibility, whereas the innovative technological approach of the Rojano/Sutherland research only serves to disguise the familiar "test and adjust" strategy of solving numerical problems that dates from long before the development of algebra. I draw attention to these somewhat ambiguous features not to evaluate them, but only in order to make the point that, just as no pedagogical device can be mathematically neutral, neither can any research design. A research design, once employed, necessarily interacts with the mathematical content that is being examined. The achieved output is never quite the anticipated output. One begins a research intending to examine such-and-such a question, but the constraints of the procedure--whatever the procedure--ensure a degree of mismatch between the outcomes of the research and the assumptions embedded in the original question. The results from the research help to redefine the nature of the experiment itself *and* the terms in which it was first cast. In a world in flux, like the world in which educational experiments take place, one cannot rectify matters by repeating an experiment with the benefits of hindsight, but the educational researcher can utilize a sort of feedback-with-delay and improve the starting position for the next run.

My final remark draws attention to the way in which the study of pedagogical questions, and educational questions more generally, tends to embody unconscious assumptions of an evolutionary nature. Almost all educational research and development seems to be built on the assumption that learning is an evolutionary business, an evolution, moreover, which should be gentle, regular, and sustained. When I reflect on my own learning, I find few instances of such a gradual and continuous increase. On the contrary, learning appears to me to be very largely a discontinuous, non-linear, business. At almost the same time, I can be learning things that are totally unrelated to each other, and learning closely related things separated by vast tracts of time. The model of spiral learning fits experience no better. Learning seems to me certainly never additive and almost never convergent. It coalesces and disintegrates, it continually backtracks and recasts itself, as if it were some hugely-complex, on-going experiment that never concludes.

And if learning is not evolutionary, what becomes of the principles governing the drawing up of a curriculum?

The assumption of a steady evolutionary state may be derived from the estimable principle that what one invites students to learn should be connected to what they already know. But I think there may be a confusion here between two quite different and independent hypotheses. In the area of teaching and the curriculum, I suggest that "taking account of what the students know" is an imperative, whereas the use of an evolutionary model is merely one of several options, and not the most plausible one at that.

CHAPTER 11

ALGEBRAIC THOUGHT AND THE ROLE OF A MANIPULABLE SYMBOLIC LANGUAGE

ALAN BELL

This commentary chapter begins with a discussion of the question, "What is meant by algebraic thought?" and of its possible answers, and emphasizes the necessity of including a manipulable symbolic language. Then, the need for a fuller classification system for the analysis of word problems is argued. Such a system should take account not only of a larger variety of problems but also of the interactions and distinctions between formulating the problem equations and manipulating them to find the problem solutions.

Empirical studies, such as those of Bednarz/Janvier and Rojano (this volume), raise questions about the teaching of algebra, in particular about whether the teaching should follow the historical stages, or the psychological stages, or should intervene with tasks aimed at leapfrogging these stages and getting students to operate at a higher level, with the use of the intellectual and material technologies now available. These questions are introduced below. But the studies also raise the question that I should like to pose first; that is, *What is meant by algebraic thought?* This could mean any one or more of the following:

1) The resolution of complex arithmetic problems:
   (a) by step-by-step methods, working from given data to unknowns, or
   (b) by global perceptions and the use of multiple arithmetic relations;
2) The codification and use of systematic general methods for different types of problems;
3) The finding and proving of number (and geometric) generalizations;
4) The recognition and use of general properties of the number system and its operations;
5) The recognition, naming, and use of standard functions, for example, $y = kx^2$;
6) The use of a manipulable symbolic language to aid this work.

Of these, 1, 2, and 6 are the ones at issue in relation to the problem-solving theme that we are discussing here. All six are recognizable aspects of mathematical activity. To which of them should we ascribe the term *algebraic*? My own answer is, "to all of these," but only when 6 is also present, as this is the only one that is clearly distinct from other types of mathematical activity. However, other authors may disagree. Lins (1992) defines algebraic thinking essentially as 1b, and Bednarz and Janvier (this volume) describe four types of student approach to problems, of differing degrees of closeness to algebraic thinking. These are (using my terms):

- Arithmetic step-by-step problem solving (my 1a),
- Trial and adjustment,
- Intermediate (between the first and the fourth of these approaches),
- Global recognition of the problem structure.

N. Bednarz et al. (eds.), Approaches to Algebra, 151-154.

Previously, Bednarz, Radford, Janvier, and Lepage (1992) hypothesized that the second of these was close to an algebraic approach, into which it might be converted by starting with the assignment of a letter instead of a trial number; but they now suggest that the fourth is closer to algebra, since an algebraic approach presupposes this global recognition of structure prior to expressing the relations symbolically. I am not entirely convinced by this argument, since the classic (Cartesian) approach requires, at least at the formulation stage, only the ability to translate the verbal data into symbolic statements, and this can be done piecemeal. The coordination and combination of the resulting equations so as to reduce the number of variables to one, and to solve the final equation, need a global appreciation of the set of equations obtained, and of their meaning, that is, an ability to *work with the algebraic language*, rather than a global appreciation of the relations in themselves. In the Rackets, Sticks, and Balls example, the problem could be formulated by the three equations

$$r + h = 288, \quad r = 4b, \quad h = 7r,$$

which are then combined by substitution to give

$$4b + 28b = 288;$$

or this equation might alternatively be obtained directly by identifying the single variable b. A study of the properties of these different approaches would be important.

The Bednarz and Janvier diagrammatic classification of problems is of interest, and I have attempted to use it for some of the problems which I have used in recent work (see Figure 1). The results show, I think, the need for a fuller system!

The class of problems analyzed and compared in the chapter by Bednarz and Janvier is that in which there are three unknown quantities, and three relations connecting them, one of which is the total. This is a type of problem that I think began life as Piles of Stones and that has since been embodied as Towns and Populations, and Chocolates and Boxes. The facility values presented by Bednarz and Janvier show vividly the effect of different combinations of additive and multiplicative relations; some of these are new and unexpected (to me, at least). These results, like those of similar facility comparisons in other fields, do suggest hypotheses about students' views of the problems and methods of approach, and also demonstrate what variety of problem needs to be included in the curriculum.

A fuller classification, though, would include, in the first place, problems of the same type but in which the total was not given--thus giving rise to equations containing the unknown on both sides. But beyond this, we need a classification to include the other types shown in Figure 1. Some relevant criteria for such a classification include (a) problems that lend themselves to easy/hard formulation versus easy/hard solution of the equations, and (b) problems containing quantities that are intrinsically inter-related, additively or multiplicatively.

PAD & BIRO

A writing pad costs four times as much as a biro. The biro is 30 cents cheaper than the writing pad. What are the costs of a writing pad and of a biro?

RECORDS

Maggie and Sandra went to a record sale. Maggie took 67 dollars and Sandra took 85 dollars. Sandra spent four times as much as Maggie. On leaving the shop, they both had the same amount of money left. How much did they each spend?
(modified: 1.7 times as much / 0.4 times / 32 dollars less than / ...)

MONEY

A mother and her son and daughter spent a certain amount of money, with the mother and son together spending \$22, the son and daughter together \$15, and the mother and daughter together \$20. How many dollars did each of them spend separately?

RELAY

The track for a relay race is divided into four sections. We need to find the total length of the track, if the length of the first and second sections together is 175 m; of the second and third is 200 m; of the third and fourth is 175 m, and of the fourth together with twice the first is 200 m.

FISH

The tail of a fish weighs 4 kg. The head weighs as much as the tail and half the body. The body weighs as much as the head and the tail together. How much does the whole fish weigh?

LAKE

A boat travels from one end of a lake to the other at 3 km/hr, stops for an hour, and returns at 4 km/hr. The total time is 7 hrs. What is the length of the lake?

Figure 1.

The Money problem is easy to formulate using three variables:

$$m + s = 22, \quad s + d = 15, \quad m + d = 20.$$

Then the solution involves the more difficult process of successive substitution. Alternatively, it is possible, but more difficult, to formulate the problem with a single variable, and the solution is then more direct (as for the Rackets problem discussed above). The much harder, but similar, Relay problem is probably only feasible if, at first, formulated with several variables and then solved by matching and substituting in the algebraic language:

$f + s = 175$	(1)	from (1) & (2) $t - f = 25$
$s + t = 200$	(2)	from (3) & (4) $2f - t = 25$
$t + r = 175$	(3)	Hence, adding, $f = 50$ ...
$r + 2f = 200$	(4)	

The difficult step is seeing that the equations can be combined in pairs to give two equations in the same two variables, t and f. This is, I suggest, easier in the algebraic language than working from the verbal representation.

The Fish problem is a harder example also of a problem that is most easily formulated using several variables initially. It is clearly not true (cf. Rojano, this volume) that problems once formulated algebraically are solved *automatically* within the algebraic language. Rather it is the case that one has a *different* representation to work with, offering a different set of transformational possibilities (somewhat analogous to the difference between oral and written language).

The criterion (b) mentioned above is illustrated by the Records and Lake problems. In the former problem, the amount of money possessed at the start, the amount spent, and the amount left, for each girl, are related additively; in the latter, distance, time, and speed are related multiplicatively. One of the difficulties in solution is knowing which one of these quantities to use as the basis of the connection that will lead on to equations.

The sets of problems mentioned by Rojano, as distinguished in early algebraic work, were *distribution, work, cisterns, mixtures,* and *ages*. Consideration should be given to clarifying and extending this classification; some examples would be helpful. The question that arises at this point is to what extent are the problem classifications derived from historical study relevant pedagogically?

This is a specific version of the general question, "What can we learn from history?"; and one might also raise the more technical question, "How does the presentation of historical material in original or modern notation affect what we can learn from it?"

I have saved until last the most crucial question for algebra teaching. How should we combine purpose-giving problem solving--*Authentic Algebraic Activities*--and the learning of manipulative skills, and can/should the skill learning be meaningful? Enlightened pedagogy would give primacy to the authentic activities, and plan them so as to give rise to the need for specific transformations, which are then introduced, explained/discussed, and practiced. But in some earlier work, we made the observation that:

> Equations represented a quite different form of expression which was unlikely to be adopted by the pupils spontaneously unless they both recognised that this expression led to the possibility of algorithms for solution, and also that they had had some practice in the solution of such equations resulting in the solution of the corresponding problem, that is they needed to know that equations could be manipulated and solved in order to decide that it was worthwhile trying to formulate the problem in this way. (Galvin & Bell, 1977, p. II)

A way out of this dilemma may be to begin with situations in which the formulation stage is very easy, and the unknowns have a very concrete meaning--examples are Pyramids and Arithmagons (see Bell, this volume). Pyramids also illustrates the previous point: Expressions obtained can be made to require the distributive law and the subtracting bracket law simply by changing the defining rule from $C = A + B$ to $C = A + 2B$ or $C = A - B$. Other examples are provided in my main chapter found elsewhere in this volume.

CHAPTER 12

PLACEMENT AND FUNCTION OF PROBLEMS IN ALGEBRAIC TREATISES FROM DIOPHANTUS TO VIÈTE

LOUIS CHARBONNEAU, JACQUES LEFEBVRE

A natural question to ask historians is: "What were the problems that brought about the development of algebra?" A shift of attention leads to another question: "Where can we find the problems that brought about the development of algebra?" A first answer, not a bad one, may be: "In the beginning were the problems." A second answer may also be: "Problems are central in mathematics; they are at the core of the main texts of algebra." A third answer, another good one that will be presently justified, is: "The problems are at the end ... of the text."

1. INTRODUCTION

There is no need to emphasize the role of problems in mathematics teaching. Problem solving is now seen more and more frequently in curricula as the core of a healthy mathematical education. The success of the École Polytechnique of Paris, founded in 1794, may be seen as historical evidence for that viewpoint. It is the first educational institution to have systematically included in the curriculum practical work sessions in which students were asked to solve problems. The results were so impressive for the time that scientists from all over Europe came to see its students perform in mathematical fields considered elsewhere to be the domains of only the best mathematicians.

Taking into account the centrality of problems in mathematics teaching, it is natural for didacticians to examine the history of mathematics with the idea of looking for problems that brought about important developments in mathematics.

We, however, have tried to attack the question of the role of problems in the history of algebra from a somewhat different point of view. Considering that, before writing a book, a mathematician erases the traces of most of the difficulties he has encountered, we decided to focus our attention on the structure of some works of prime importance in the history of algebra. This is, of course, the historian's approach. We do not know the corpus of problems with which the algebraist dealt before writing a book, but problems that have been included in a book may be analyzed. This analysis should, however, take into account the role of these problems within the theory or theory-in-the-making, a role that is in part revealed by the placement of the problems within the structure of the books.

In so doing, we have given some importance to the wording used in relation to the concept of problem. A survey of some texts shows a wide variation in the words used to denote a problem. Robert of Chester uses *quæstio*, *interrogatio*, *exemplum*, *ænigma*, in his Latin translation (1145) of Al-Khwarizmi's *Compendium on Calculation by Completion and Reduction*. When he introduces an example, he writes *si dicas* or *ut dicas* which means "when you say." *Quæstio* and *exemplum* are also used by Chuquet (1484), while Cardano uses *quæstio* in his *Ars Magna* (1545).

N. Bednarz et al. (eds.), Approaches to Algebra, 155-165.

Bombelli (1572) uses *problema,* but only in section titles. Viète (1540–1603) uses two words, *zeteticum* and *problema*. Whatever the terminology used, more significant is the relative placement of "problems" within the structure of a book.

In the next section, we study four algebraists: Diophantus, Al-Khwarizmi, Cardano, and Viète. We see problems at the very heart of Diophantus' *Arithmetica,* while Al-Khwarizmi and Cardano relegate them mostly to the end of their treatises. In Viète, whose motto is "To solve every problem," problems are clearly divided into two categories, a consequence of a remarkable theorization of algebraic practices. In fact, they are concentrated in one of the three parts of his analysis, the one devoted to the actual resolution of problems, numerical or geometrical.

2. PROBLEMS AND THE STRUCTURE OF FOUR ALGEBRAIC TREATISES

2.1. *Diophantus and Al-Khwarizmi*

Diophantus' *Arithmetica* (circa 150 A.D.)[1] is composed essentially of different types of problems organized into 10 books.[2] Of these, only Books I and IV contain sections devoted to types of numbers and their rules of calculation. The problems are roughly organized not only by their nature, but also by their level of difficulty.[3] On the one hand, Books I to VII seem to form a whole.[4] The mode of resolution used in Book II is reused in Books IV to VII, although usually without any explicit reference. On the other hand, Books "IV" to "VI" contain more difficult problems which involve more sophisticated methods of resolution. Diophantus does not develop an explicit theoretical corpus in his *Arithmetica.*. The problems and their solutions, although aimed at developing the reader's know-how, are nonetheless never explicitly described.[5] We may say that the resolution of the first problems of a certain type, that is, the steps through which one must pass to solve the problems, present the effective core of this know-how. The subsequent problems of the same type, along with their solutions, not only allow the practitioner to better understand those steps, but also give him a way to improve his intuition on how to modify a problem so that it becomes solvable by a similar process. Diophantus states at the beginning of his Book I: "In order that you [referring to Dionysius to whom the book is dedicated] miss no thing, in treating which you would acquire ability in that science [of the resolution of problems], I consider it also appropriate to write, once again, for you, in what follows, many problems of this kind,"[6] and in the first sentence of Book VII: "Our intention is to expound in the present Book many arithmetical problems without departing from the type of problems seen previously in the fourth and fifth Books--even if they are different in species--in order that it be an opportunity for (acquiring) proficiency and an increase in experience and skill."[7]

The *Compendium on Calculation by Completion and Reduction* (circa 825 A.D.) of Al-Khwarizmi has a completely different pattern of organization. Its structure is as follows (the pages are those of Al-Khwarizmi, 1831):[8]

Part I:

1) Types of numbers (x, x^2, etc.) and rules of resolution of the six canonical forms of linear and quadratic equations. The rules are given by means of numerical

examples only. Geometrical demonstrations of the rules are given for the trinomials $x^2 + bx = c$, $x^2 + c = bx$, $bx + c = x^2$ (pp. 5-21).

2) "On multiplication": multiplication of first degree binomials (pp. 21-27).
3) "On addition and subtraction": operations (+, -, ×) on radicals with geometrical demonstrations (pp. 27-34).
4) "On the six problems": six problems of the six canonical forms, in the same order (pp. 35-41).
5) "Various questions": 34 miscellaneous problems (pp. 41-67).

Part II:

6) "On mercantile transactions": three types of rule of three, each illustrated by an example (pp. 68-70).
7) "Mensuration": areas and volumes (pp. 70-85); one instance of the use of algebra (p. 81).
8) "On legacies": 61 problems, all of the first degree (pp. 86-174).

Historians of algebra usually focus on the first five sections, those which differ the most from Diophantus' *Arithmetica* by their theoretical tone. Nevertheless, we deem it appropriate to point out that these sections add up to less than half of the book. On the other hand, the eighth section alone contains half of the book.

That last section is the only one without any theoretical treatment. In that regard it resembles Diophantus' *Arithmetica*. But the difficulties are of a different nature in Al-Khwarizmi. In Diophantus, indeed, the wording leads immediately to the writing of equations. In contrast, one of the difficulties of the problems of the master from Baghdad lies precisely in the making of an equation, a task that necessitates knowledge of the Koranic laws governing legacies and inheritances. The following two examples show that difference. The first one is taken from Diophantus, the second one from Al-Khwarizmi:

We wish to find a cubic number such that, when we diminish it by an arbitrary multiple of the square having the same side, the remainder is a square number. (Diophantus IV-11) [It leads to the equation $x^3 - kx^2$ = a square.]

If he (a father) leaves five sons, and bequeaths to some person a dirham, and as much as the share of one of the sons, and one-third of what remains from one-third, and again, one-fourth of what remains from the one-third after deduction of this, and one dirham more. (Al-Khwarizmi, 1831, pp. 118-120, 24th statement)

If the legacy is x, a son's share is v, C is the capital and ∂ represents dirham, the system of equations is:

$$C - x = 5v,$$

$$\frac{2}{3} C + \frac{1}{3} C - v - \frac{1}{3} [\frac{1}{3} C - v] - \partial - \frac{1}{4} [\frac{2}{3} [\frac{1}{3} C - v] - \partial] - \partial = 5v$$

One sees that the nature of the problems based upon inheritances is artificial, notwithstanding the concreteness of the meaning.

The first five sections of Al-Khwarizmi's work are more revealing. The placement of the problems is completely different from that of Diophantus. One can make out three types of statements. First, there are statements to clearly enunciate a rule. Those statements are numerical, with what we call the coefficient of x^2 being 1, for instance: "A square is equal to five roots of the same" (p. 6). With the exception of two instances, they are followed by similar statements where the coefficient of x^2 is not 1. For instance, "Five squares are equal to the roots" follows the previous example. That second type of statement is used to show the process needed to fall back exactly on 1 as the coefficient of x^2. The "six problems" (Section 4) and the "Various questions" (Section 5) are quite different. They constitute what we call the third type of statement. Al-Khwarizmi explains their relevance:

I now subjoin these problems, which will serve to bring the subject nearer to the understanding, to render its comprehension easier, and to make the arguments more perspicuous. (p. 35)

The problems are chosen according to that aim. They demand more complicated algebraic calculations: the use of the al-jabr (which means completion and serves to get rid of negative terms) and the al-muqabala (which means reduction and serves to get rid of similar terms appearing in both sides of an equation), or of the operations on the binomials, operations which have been dealt with in the second and the third sections. Here is an example of a problem of the third type:

I have divided ten into two parts; I have then multiplied each of them by itself, and when I had added the products together, the sum was fifty-eight dirhems. ("Fifth Problem" from Section 4, "On the six problems," p. 39)

To solve it one has to transform the equation $x^2 + (10 - x)^2 = 58$ into the fifth canonical form, $x^2 + 21 = 10x$.

The problems that we find in Sections 4 and 5 are much less varied than those of Diophantus. There may be many reasons for that. The intended reader must certainly be a major factor. The structural differences between the two works also plays a major part. Being organized upon a theoretical kernel, the first part of the work determines the scope of the diversity of problems. In contrast to what happens in a book like Diophantus' *Arithmetica*--where the domain of knowledge remains unclear, but accordingly, there is a vast and rich scope of matters to be touched upon--Al-Khwarizmi's domain of knowledge is narrow, but its limitations are clear. That which is really mastered gets clearly distinguished from that which is not yet mastered. Diophantus builds up "know-hows." In the second part of his treatise, Al-Khwarizmi also builds up know-hows, whereas he has exposed organized knowledge in his first part. Knowledge gives room to displays and discourses. Know-hows, like those developed in Diophantus' *Arithmetica* and in Al-Khwarizmi's eighth and last Section, are acquired through a practical mode of experiencing, a foundation which the discourse is partly unable to establish.

2.2. *Cardano (1501–1576)*

Chapters of *Ars Magna*[9] can be regrouped into three larger parts.[10] The first one consists of general considerations about algebraic transformations, changes of variables, and the number of roots or unknowns. It includes Chapters 1 to 4 and Chapters 6, 7, 9, and 10. The second part gives the algorithms for solving equations of the second and third degrees and some equations of higher degrees up to the ninth degree, and even one equation of the twenty-seventh degree. That second part includes Chapters 5, 8, and 11 to 24.[11] The third part is about various rules: the rule of the method, the golden rule, the rule of equal position, and so on. There is a total of more than twelve such rules aimed at making it easier for the practitioner to find the algebraic solutions of problems. For certain cases, an algorithmic solution is given.

In this book, which is remarkable in many regards, it is interesting to note where the problems are located.[12] Their distribution is far from being uniform, as is shown by the following table:

PART	PAGES	PROBLEMS	NUMERICAL*	MERCANTILE	GEOMETRICAL
I	71	4 **	4	0	0
II	82	14 ***	8	6	0
III	102	76	67 ****	5	4
Total	261	94	79	11	4

* By numerical problems we mean problems involving only pure numbers.
** All contained in Chapter X.
*** Chapter V alone contains 10 of these 14.
**** One of these problems is of a mythological nature.

How can such an asymmetrical concentration of the problems in the third part be explained?[13]

In order to answer such a question, one has to take into account the peculiar nature of the chapters in that part. These have a totally different structure from those of the previous part.

On the whole, the chapters of the second part are constructed according to one and the same framework. Each chapter's title formulates the type of equations to be solved. Immediately follows the geometrical proof[14] of the solving rule. Finally, the solving rule is then stated in an algorithmic way. The chapter ends with some numerical examples of solutions. The statements for those examples are simply equations corresponding to the type of equations discussed in the chapter.

In the third part, the chapters are organized differently. In some chapters, those which contain *quaestiones*,[15] a short introduction is followed by what constitutes the main feature of these chapters, the problems and their solutions. The introduction tries to give the scope of applicability of the rule described in the chapter. A statement of the rule is sometimes given, but it is not always easy to understand. For an uninformed reader, these introductions get their true meaning only after the various problems have been worked out. It seems to us that Cardano organizes these chapters around and upon their *problems*. In contrast to what he has been doing in the second part, he finds himself incapable of writing in general terms a method for translating these *problems* into equations whose form would be readily solvable. Perhaps he may have been unable to transfer ways of solving into algorithms for practitioners, or calculators. Cardano himself gives credence to such an interpretation in his introduction to Chapter XXXI, called "On the Great Rule." Regarding that rule, he indeed writes: "It requires an expert and is [best] taught by problems since it is many-faceted" (p. 186).

Structurally, Cardano's *Ars Magna* is similar to Al-Khwarizmi's *Compendium on Calculation by Completion and Reduction*. However, because its context is much richer, Cardano's book allows us to go a little further in perceiving the function of the problems in the book. Once again, most problems are found in the later parts of the book. Most of them remain numerical. Their placement shows how difficult it is to ensure that the search for an equation leads to a form whose solution is known. Which unknown is to be chosen? Which changes of variables have to be made, in what kinds of situations? Of course, in Diophantus, for instance, those questions have to be considered, but they do not constitute the basis of a discussion outside the solving of the problems. In contrast, Cardano identifies them clearly, although he cannot yet treat them in a systematic fashion.

2.3. *Viète (1540–1603)*

In 1591, in Tours, François Viète publishes his *In Artem Analyticem Isagoge*, a book establishing his program and containing the foundations for his *Algebrâ Novâ*. Even before the dedicatory letter, the author notes that his restored analysis, also called new algebra, is made of many treatises. The list he gives contains the *Isagoge* and the following[16] (we have added in parentheses the dates of first publication):

- Ad Logisticem speciosam Notæ priores (1631)
- Zeteticorum Libri quinque (1591 or 1593)
- De numerosâ potestatum ad Exegesim resolutione (1600)
- De recognitione Aequationum (1615)
- Ad Logisticem speciosam Notæ posteriores (unpublished)[17]
- Effectionum Geometricarum Canonica recensio (1593)
- Supplementum Geometriæ (1593)
- Analytica angularium sectionum in tres partes tributa (1615)
- Variorum de rebus Mathematicis responsorum, Libri septem (1593)

Through these various treatises, Viète describes the road to be covered in order to solve problems according to the analytic art. His method has three parts. In the *Introduction to the Analytic Art*, Viète compares his program to the analysis and the synthesis of the Ancients. The analysis of the Ancients is divided into two parts: the Zetetics and the Poristics. Viète adds a new analytic part which he calls Exegetics or Rhetics. Making use of almost exactly the same terms as did Pappus, Viète defines the analysis as "assuming that which is sought as if it were admitted [and working] through the consequences [of that assumption] to what is admittedly true."[18] His definitions of Zetetics, Poristics, and Exegetics suffer from imprecision. Taking into account the sum of information in the *Introduction to the Analytic Art* on the meaning to be given to the three parts of analysis, we propose the following definitions:[19]

Zetetics: To put the problem into an equation and to transform that equation into a form that can be related to a proportion. It could be a simple relation as with an equation of the form $P(x)\cdot Q(x) = CD$, where P and Q are polynomials and C and D are composed of known magnitudes. In that case, the related proportion is

$$\frac{P(x)}{C} = \frac{D}{Q(x)}$$

However, as Viète shows in *The Understanding of Equations*, the relation is usually a lot more complex. An equation that can be related to a proportion involving its coefficients is said to be ordered, and the proportion that corresponds to it is also called ordered.

Poristics: In view of the synthetic reverse reasoning to be done, to study in the analysis some delicate points whose reversibility either is not convincing or cannot immediately be taken for granted.

Exegetics: To find, in a form compatible with the numerical or geometrical nature of the problem, the root or roots of the ordered equation established through the Zetetics.

None of the treatises in the above list bears on the Poristics.[20] They can be associated with the other two parts of the analytic art. For instance, among the published books, only *Preliminary Notes of Symbolic Logistic*, *Five Books of Zetetica*, and *Two Treatises on the Understanding and Amendment of Equations* belong to the Zetetics.

The *Introduction to the Analytical Art* ends with seven sections on the Exegetics, the last sentence being "to solve every problem." In these sections, Viète distinguishes between two kinds of Exegetics, the numerical Exegetics and the geometrical Exegetics. The Zetetics has the symbolic logistics (or, in other words, the symbolic notations and transformations) as its main tool and, for that reason, remains outside the particular fields of arithmetic and geometry. In sharp opposition stands the Exegetics, which is rooted in those two mathematical fields. Indeed, when the statement of a problem is arithmetical, the solution, according to Viète, must be expressed in purely arithmetical terms, thus in numbers. On the other hand, if the statement is geometrical, the solution must correspond to a geometrical construction.

The treatise, *On the Numerical Resolution of Powers by Exegetics* (1600), deals with the numerical resolution of algebraic equations. It is divided into two parts, the first focusing on the resolution of pure powers ($x^n = k$, $n = 2, 3, 4, 5, 6$), the second on some affected powers, that is, equations with two or three nonconstant terms (more specifically, of these three forms: $x^n \pm ax^m = k$, with $n \leq 6$ and $m \leq 4$; $x^4 \pm bx^3 \pm cx = k$; $x^4 + bx^2 + cx = k$). As stated at the beginning of the treatise, an equation for which the roots are sought should be prepared first, that is, transformed into one of the canonical forms of the treatise. The way to do so is shown in the *Two Treatises on the Understanding and Amendment of Equations*.[21] This double treatise, especially its first part, also specifies how to interpret these equations in terms of proportions, and in so doing, how to prepare the way to translate them into geometrical constructions. They are then essential to the understanding of the two books on geometrical Exegetics: *A Canonical Survey of Geometric Constructions,* which gives geometrical solutions to the second degree equations, and *A Supplement to Geometry,* which does the same for third degree equations.

In that complex Vietan structure, where do we find the *problemæ* and *zetetica*? We have already mentioned that the word *problema* occurs mainly in the Exegetics. In fact, *problema* is rarely used by Viète in his algebraic works. It appears in the last chapter of the *Introduction to the Analytic Art* and mostly in *On the Numerical Resolution of Powers by Exegetics,* where examples are given of processes of numerical approximation for five *problemæ* of pure powers and twenty *problemæ* of affected powers. The style of presentation of these *problemæ* is reminiscent of the last part of Cardano's *Ars Magna*. The statement of a *problema* is always a general one. However, immediately after having stated it, the author falls back upon a numerical example and specifies one or more *paradigmæ*, that is models to follow in order to get to the numerical solution of a problem. These models are described using numerical examples, except at the very beginning and the very end of the book, where Viète gives a sketch of the method in very general terms. Also, it seems that Viète wants, as did Cardano before him, to develop know-hows. However, because Viète's treatise is a link in a chain of books, each of which has well-defined aims, the theoretical canvas is a lot more closely-woven than in Cardano's *Ars Magna*.

In treatises belonging to geometrical Exegetics, the word *problema* is seldom used. There are four *problemæ* in *Universal Theorems on the Analysis of Angular Sections*. As in *Two Treatises on the Understanding and Amendment of Equations,* the statements of those *problemæ* are general. For example, *Problema II*: "to construct one angle to another as one number is to another." A *problema* may have sub-problems called *problemations*. For example, *Problema II* has 3 such *problemations*. *Problemation I* is as follows: "To divide a given angle into three equal parts." In the other two books of the geometrical Exegetics, *A Canonical Survey of Geometric Constructions* and *A Supplement to Geometry*, there are only propositions, some of which, such as Proposition XI of *A Canonical Survey of Geometric Constructions*, are followed by a "Corollary on the Geometrical Solution," showing the real objective of the proposition. The fact that Viète does not use the word *problema* in these books should not be considered unusual because those two books describe tools allowing one to pass from symbolic language to geometry.

Considering our affirmation that the word *problema* is related to Exegetics, it may appear surprising to find some *problemæ* in *Two Treatises on the Understanding and Amendment of Equations,* the double treatise that globally belongs to Zetetics. Those *problemæ* are, as before, stated in very general terms.[22] But in *On the Numerical Resolution of Powers by Exegetics,* Viète explicitly refers to one of those *problemæ*.[23] The other occurrences of *problemæ* are also seen to be directly related to the preparation of equations for numerical or geometrical resolution, thus clearly linked to Exegetics.

The use of *zeteticum* is more circumscribed. It is used almost solely in *Five Books of Zetetics*. We have quoted Zetetics II-4 and III-2. In many ways, *zetetica* are more limited than *problemæ*. However, one must recall Viète's objectives. As already mentioned, to resolve a *zeteticum* is to translate it into symbolic terms and transform these symbolic expressions in a way in which they may be suitable to Exegetics. One has a confirmation of this interpretation in *Universal Theorems on the Analysis of Angular Sections.* After Theorem IIII, Viète states the *zeteticum*: "To construct in a circle two arcs in a given multiple ratio, which ratio also holds between the squares of the straight lines by which the arcs are subtended." This statement is similar to the *problem* in the book. But the *zeteticum* is dealt with symbolically without any geometrical manipulation; while the *problem* in that book are resolved geometrically.

3. CONCLUSION

We have seen four important algebraists each presenting his own algebra. We can roughly characterize them in regard to their attitudes towards problems as follows.

Diophantus appears as a brilliant artist. Here, artist has the meaning it had in the 16th century, that is, a person who has special know-hows, which were consciously developed and mastered, and which are used toward a specific aim. For Diophantus, each problem is a new challenge. In a way, he is the epitome of the problem solver. Problem solving is at the very heart of his algebra. But, nevertheless, almost every problem is independent of the others. There are patterns of resolution, but those patterns are not clearly described. Truly new tools, a non-operational symbolism for example, are used. But one cannot speak of a method to approach problems, at least not an explicit method.

Al-Khwarizmi is what we would nowadays call a "handbook writer," carefully distinguishing and explaining general cases and giving algorithmic procedures (justified by "proofs"[24]) to be applied in "exercises." The tools he uses are not actually different from those of Diophantus. He does not use any symbolism. But Al-Khwarizmi has a theory, a theory that is well circumscribed, with geometrical demonstrations of rules that can be applied blindly. He is also a technician, subtle enough to mathematize complicated legacy problems. His approach to problem solving is different from that of Diophantus. First, the problems he deals with are of a more limited scope. Second, he is more precise than Diophantus about the tools he may use. The problems on legacies are real problems for artists. But Al-Khwarizmi, who is also an artist, does not start from scratch, he has a theory. His artistic skills are used as long as the problem is not translated, mathematized, into the language of his theory. As soon as the problem can be related to a mathematized model, the technician comes in to take charge of the process.

What we have said of Al-Khwarizmi may also be said of Cardano. Cardano is, however, a subtle technician of a different type. In the third part of his treatise, not only does he want to produce new rules blindly applicable to certain problems, but he also wants to show how to produce those rules. That would transform the technician into an engineer. This implicit aim is clearly not present in Al-Khwarizmi.

Viète is a complete mathematician: artist and theoretician. As artist, he goes further than any artist before him. He solves problems not by using algorithms but by skillfully putting together general results. He is an ultimate artist because he is also a theoretician. Trying to look at problems with a new vision, he has a program of which the motto is *To leave no problem unsolved.* His analytical program is based on a completely[25] new vision of a mathematical theory: the *logistica speciosa*, the calculation with letters. Letters do not have a meaning by themselves; so, to calculate with them, rules of manipulation have to be given. This is almost an axiomatic approach. For Viète, problems are geometrical or arithmetical, but the models for solving them are neither geometrical nor arithmetical. Algebra now has a status by itself, one which is explicitly at a meta-level.

The present study is limited to algebraic treatises considered as separate works, at best compared with one another. Though far from giving a detailed story of the influence of problems in the evolution of algebra through the centuries, it does show that there was an evolution in the context in which problems were presented in these major works. More precisely, algebraists had changing general attitudes towards problems: problems for a technician, as examples of technical applications of rules (Babylonians and Al-Khwarizmi); problems for an artist, as a field in which to exhibit one's ability to solve them (Diophantus and Al-Khwarizmi); problems for an artist-engineer, as a way to expand the set of rules and, therefore, the set of solvable problems (Cardano); and problems for the ultimate artist and theoretician, such that problems as a whole are taken as a meta-problem to be solved according to an inventive method that makes use of a universal language (Viète).

NOTES

1 We have consulted Ver Eecke (1959) and Sesiano (1982).
2 In his Introduction, Diophantus tells us about 13 books in his *Arithmetica*. Nevertheless, the text of only 10 of those books is known today. A symbolic description of the problems of these books is given in Sesiano (1982, pp. 461-483). The numbering of the books is somewhat problematic. We will use the numbering of Sesiano, that is, the three first books of the Greek manuscripts are numbered I, II, III; the four books of the Arabic manuscripts published by Sesiano are numbered IV, V, VI, VII; and finally the last three books of the Greek version are numbered "IV," "V," "VI."
3 For example, Book III deals with systems of three equations in three unknowns. In his introduction, Diophantus explicitly says that he presents the problems "as it is normal to do, from the simplest to the most difficult" [our translation to English]. (See Ver Eecke, 1959, p. 9.)
4 See Sesiano, 1982, pp. 4-8.
5 The expression know-how is consciously chosen here to emphasize the fact that what is to be done and when it is to be done is not very explicitly taken care of. The practitioner has to develop his intuition to know when to apply a certain type of rule.
6 Sesiano, 1982, p. 86.
7 Sesiano, 1982, p. 156.
8 The division in Parts I and II and the numbering of the sections are ours.
9 The complete title of Cardano's treatise published in 1545 reads as: *Artis magnae, sive de regulis algebraicis, lib. unus. Qui & totius operis de Arithmetica, quod Opus Perfectum inscripsit, est in ordine Decimus*. We make use of Witmer's translation (Cardano, 1545/1968).
10 These parts are not explicitly identified by Cardano.
11 Chapter 8 could be placed either in the first or in the second part.
12 From now on, in this section, we will use the word "problem" only where Cardano's translator has used it. To be quite rigorous, one would have to examine the examples given everywhere throughout the book. Let us simply note here that the examples are linked very directly to the theoretical context immediately preceding them.
13 In the third part, the distribution of problems is almost uniform.
14 The proofs in the first chapters are truly geometrical. Afterwards, they refer to results given in previous chapters.
15 Ten of the sixteen chapters.
16 The titles given here are those appearing in the first edition of *In Artem Analyticem Isagoge* (*Introduction to the Analytic Art*). When published at the indicated dates, some of the books had titles different from the originally announced ones. In Viète (1983, note A), the titles as translated by Witmer are in the following order: *Preliminary Notes of Symbolic Logistic, Five Books of Zetetica* (regarding the uncertainty of the date, see Viète, 1983, p. 9, note 31), *On the Numerical Resolution of Powers by Exegetics, Two Treatises on the Understanding and Amendment of Equations, A Canonical Survey of Geometric Constructions, A Supplement to Geometry, Universal Theorems on the Analysis of Angular Sections*. In Viète (1646/1970), the last book has as title *Variorum de rebus Mathematicis responsorum, Libri VIII*. It has not been translated by Witmer. From now on, we will use Witmer's titles. Unless

otherwise stated, quotations will also be taken from Witmer's, with the sole difference that we have rewritten the equations in a language closer to Viète's way of writing, whereas Witmer describes them in a more modern language.

17 According to Ritter, this book is probably the *Amendment of Equations,* which was published with *Understanding of Equations* in 1615 (Ritter, 1895, p. 374).

18 Viète, 1983, p. 11.

19 For a more detailed discussion of these definitions, see Charbonneau and Lefebvre (1991, 1992).

20 On the difficult task of correctly defining the Poristics, see Viète (1983, pp. 12-13, note 6) and Ferrier (1980, pp. 134-158).

21 Viète (1983, p. 323, note 25). This remark is in the original edition of 1600, but not in Van Schooten's edition of Viète's *Opera Mathematica* (Viète, 1646/1970).

22 The placement of the word *problema* in the double treatise is as follows: 1) *Understanding of Equations*: once in the second paragraph of Chapter I in connection with geometrical problems, and three times in Chapter XVI in connection with the constitution of equations; 2) *Amendment of Equations*: four *problemæ* in Chapter III, three in Chapter VI, and two in Chapter VII (see Viète, 1646, p. 173; 1983, p. 322).

23 The reference is to the problems of Chapter VI of the second treatise of *Two Treatises on the Understanding and Amendment of Equations.*

24 On the origins of Al-Khwarizmi's geometrical demonstrations, see Hoyrup (1989, pp. 77-80).

25 The only other major a-semantic mathematical theory before Viète's seems to be the Eudoxian theory of proportions as seen in Euclid's *Elements*, Book V.

CHAPTER 13

PROBLEM-SOLVING APPROACHES TO ALGEBRA: TWO ASPECTS

ALAN BELL

Problem solving by forming and solving equations is a historic route into algebra. Now it is, along with generalizing and working with functions and formulas, one of the main modes of algebraic activity that need to be learned. But the exploration of generic problems and their extensions that provides a mathematically authentic and motivating mode of activity through which all of algebra can be learned. The chapter reviews the nature of algebra and outlines such an approach, giving examples of typical tasks and of students work.

1. INTRODUCTION

The two aspects of the title refer to broad and narrow senses of the phrase *problem solving*. As one of four options for introducing and developing algebra, I understand problem solving to refer to the solving of problems by the forming and solving of equations. This is the narrow sense of the term. But the essential mathematical activity is that of exploring problems in an open way, extending and developing them in the search for more results and more general ones. Hence the learning of *all* the options should be based on such problem explorations. This is the broad sense of the term.

I also need to remark that I do not see algebra as an identifiable course, separate from other branches of mathematics, but as appearing throughout the mathematics course, its symbolism, concepts, and methods being used wherever appropriate in the other fields--in expressing arithmetic generalizations, solving geometric problems, denoting unknown elements, solving equations and establishing relations in trigonometry, as formulas for use in statistics and mensuration, and so on. This is, I believe, correct from both epistemological and pedagogical standpoints. Thus, what I propose to discuss and exemplify in this chapter is material for an algebra course in which all four aspects are developed through generic problems, but I shall give particular attention to the strand of *problem solving by forming and solving equations.*

I shall begin with a few remarks about the nature of mathematics, and of algebra in particular, then address the immediate topic in more detail.

2. THE NATURE OF MATHEMATICS AND OF ALGEBRA

2.1. *Applied and Pure Mathematical Processes*

Mathematics has two aspects, roughly fitting the traditional labels applied and pure. First, it is a means of gaining insight into some aspect of the environment. For example, the exponential or compound growth function gives us insight into the way

N. Bednarz et al. (eds.), Approaches to Algebra, 167-185.

in which a population with a given growth rate expands over time--first slowly and then with an increasingly rapid rate of increase. Some of these properties are encapsulated in well known puzzles--such as that of the water lily, doubling its area each day; how large will it be the day before it covers the whole pond?--or in the frequent exhortations of our financial salesmen to consider how a modest investment might grow. Another related example is that of the decrease of the rate of inflation--which many people believe means that prices are coming down. In the home environment, a little knowledge of the symmetry group of the rectangular block will tell us when we have turned the mattress on the bed as many ways round as we can; and a modest knowledge of probability and statistics will help us to interpret advertising claims about what toothpaste seven out of ten movies stars use, and not to be excessively hopeful that our next child will be a boy if we have already produced three girls. These are all "useful" aspects of mathematics--and note, by the way, that they all depend on the application of conceptual awareness, not on any technical skill; they are useful in the same way as is the knowledge gained in most of the subjects of the curriculum--history, geography, literature, science--that is, deriving from knowledge of some key facts and explanatory concepts.

The second aspect of mathematics is somewhat less loudly commended in public nowadays. It is the pure mathematical aspect that it shares with art and music, the solution and construction of puzzles and problems, and the enjoyment of recognizing and making patterns. Mathematical problems in newspapers and magazines still attract a following, and we might speculate that the capacity to appreciate mathematics as an art to enjoy is initially present in most people, though it often gets suppressed by distasteful school experiences.

These two modes of interaction of people with mathematics, representing the applied and the pure mathematical approaches, have been identifiable throughout history as the mainsprings of mathematical activity.

Freudenthal (1968) distinguishes applied and pure mathematical processes:

> Arithmetic and geometry have sprung from mathematizing part of reality. But soon, at least from the Greek antiquity onwards mathematics itself has become the object of mathematizing.... What humans have to learn is not mathematics as a closed system, but rather as an activity, the process of mathematizing reality and, if possible, even that of mathematizing mathematics.

More briefly, Peel (1971) says:

> Mathematics concerns the properties of the operations by which the individual orders, organizes, and controls his environment.

2.2. *Algebra and Mathematics*

What is the place of algebra in mathematics? At least two viewpoints are possible. First, as Freudenthal indicates, the raw material of mathematics consists of number and space. Mathematical concepts arise from the recognition of sameness and difference, repetition, and other relational aspects of our experience. These we might call the primary conceptual systems. Algebra then consists of the conceptual tools with which we operate in these areas to produce further hierarchies of abstractions.

The tools are classifying, comparing, combining, transforming, reversing (the actions associated with the basic structures of set, relation, function, composition, and so on). From this point of view, there are algebras of numbers, in particular the common algebra of the rational number field, and algebras of geometry, in which symbols denote points, lines, transformations or other geometrical elements and their compositions. And although, historically, mathematics was rooted in number and space, other fields of primitive experience have been mathematized: for example, propositions, logic and sets, and probability. Some examples of these algebras are discussed below.

The second point of view is that algebra and geometry correspond to the two distinct modes of reasoning, corresponding to the left and right hemispheres of the brain: algebra representing the sequential, logical, verbal reasoning component; and geometry, the holistic, spatial component. The contrast between these two modes of processing may be seen when a geometrical problem, for example, determining a locus, is solved either by direct Euclidian reasoning or by coordinatization and recourse to algebra. The two viewpoints are compatible if we recognize that both number and space may be manipulated either sequentially or holistically.

Is algebra a language? Or, rather, in what ways are algebraic expressions like or unlike ordinary language? Processes of comprehension, of meaning making, are similar in the two fields, although mathematical expressions tend to be much more dense, much more lacking in redundancy than those of natural language. There are, however, other specialized ways of using language, for example, in legal material, political material, poetry, and in the expression of scientific hypotheses for testing. Some of these demand the same precision of reading as mathematics and have an almost similar density. The formulation of propositions for voting in meetings is another example.

There seems to be a real difference in the fact that mathematical language, and algebra in particular, are not only for expression but also for manipulation. Syntactic transformations of symbolic expressions can be made mechanically and used to expose genuine conceptual equivalences without the degree of mental effort required to establish those equivalences by working with the concepts themselves. Of course, the syntactic rules have to be derived from one's knowledge of the concepts denoted.

2.3. *The Four Approaches to Algebra*

The four possible approaches to introducing and developing school algebra proposed for this book are generalizing, problem solving, modeling, and functions.

2.3.1. *The algebra of numbers*

Generalizing as a pervasive mathematical activity is discussed by Mason in another chapter of this book. An example of a specific problem calling for generalization, in which the use of algebra is explicit, is the following:

Show that the sum of a number of four digits and the number formed by reversing the digits is always divisible by 11.

This requires the representation of the problem algebraically, using knowledge of place value, as

$$1000a + 100b + 10c + d + (1000d + 100c + 10b + a)$$

which leads, in two steps, to

$$1001\,(a + d) + 110\,(b + c)$$

which is easily verified to be a multiple of 11.

This shows the full algebraic process in a pure mathematics context, of representing the given number by an algebraic expression, and manipulating it into a form which displays the desired property.

Forming and Solving Equations is the present day form of the historic tradition in algebra, which began with the solution of relatively complex problems concerning numbers or quantities and the four operations. The mode of representation was initially verbal, but developed through primitive semi-verbal symbolisms to the present algebraic language. There is no sharp line of distinction between problems soluble by arithmetic reasoning alone and those which require algebraic solution, and many of the easier problems offered for solution by algebra are also soluble by arithmetic reasoning. Consider the following Records problem from Lins (1992):

Maggie and Sandra went to a record sale. Maggie took $67 and Sandra took $85. Sandra spent four times as much as Maggie. On leaving the shop, they both had the same amount of money left. How much did they each spend?

Arithmetically, one might reason that Maggie's amount spent represents the difference between $67 and the amount left; and Sandra's amount spent is the difference between $85 and the amount left. Since this is four times as much as Maggie's amount, the difference $85 – $67, or $18, must be three times Maggie's amount, which is therefore $6, while Sandra's is $24. Here we have implicitly worked with two unknowns, the common amount left and Maggie's spending. The corresponding symbolic solution might be written as follows:

$67 - L = M$ $\quad$ $85 - L = S = 4M$

Comparing these, $\quad$ $85 - 67 = 3M$, $\quad$ whence $M = 6$

Lins (1992) defines algebraic thinking independently of the use of algebraic symbolism, as follows: To think algebraically is (a) to think arithmetically, and (b) to think internally, and (c) to think analytically.

The first two criteria imply working with the semantic field of numbers and arithmetic operations, which would agree with the characterization discussed above. The third, *analysis*, is defined following Pappus:

Now, *analysis* is a method of taking that which is sought as though it were admitted and passing from it through its consequences in order to something which is admitted as a result of synthesis; for in analysis we suppose that which is sought to be already done, and we inquire what it is from which this comes about, and again what is the antecedent cause of the latter, and so on until, by retracing our steps, we light upon something already known or ranking as a first principle; and such a method we call analysis, as being a reverse solution. But in *synthesis*, proceeding in the opposite way, we suppose to be already done that which was last reached in the analysis, and arranging in their natural order as consequents what were formerly antecedents and linking them one with another, we finally arrive at the construction of what was sought; and this we call synthesis. (Fauvel & Gray, 1987, p. 209).

Lins adds:

> In *synthesis*, one deals only with "what is known and true," and through a chain of logical deductions, other true statements are obtained; it is the method exclusively used in the whole of Euclid's *Elements*. In analysis, on the other hand, what is "unknown" has to be taken as "known," with the "unknown" elements being used "as if they were known," as part of the relationships which are to be manipulated until one arrives at "something already known."

The verbal solution of the Records problem would appear not to qualify as arithmetic because it does not use the full semantic field of numbers and operations but only part-whole reasoning; nor is it internal because its unknowns are not pure numbers, but quantities with concrete meanings. It is, however, analytic, in that it operates with the unknowns.

A problem that is less easily solved without algebraic representation is the following Rowing problem:

> A boater rows a certain distance upstream at 2 mph, stops for an hour, and returns at 4 mph. The total time, including the stop, is 3.5 hours. What is the distance each way?

Arithmetic approaches to this can hardly avoid the error of taking the average speed to be 3 mph. A solution by trial and adjustment might assume the distance to be 10 miles, giving journey times of 5 hrs and 2.5 hrs and hence a total of 7.5 hrs, just 3 times what is required (2.5 hours). If this factor is applied to the distance, giving 3 1/3 miles, we can check that this is the correct solution. This method resembles an early Egyptian method; it depends on the correctness of the assumption of proportionality. If this were false, it would be shown to be so by the check, and different adjustments could be tried, but success could not be guaranteed. The symbolic method, using the equation

$$d/2 + d/4 = 2.5$$
$$d(3/4) = 2.5$$
$$d = 10/3 = 3\ 1/3$$

is clearly more effective, and also more general.

A problem involving both *functions* and *modeling* is the following:

Pizza prices

Size	*Diameter*	*Cost*
Mini	20 cm	$ 4.00
Small	25 cm	$ 5.00
Medium	27.5 cm	$ 7.50
Large	30.5 cm	$ 8.60
Family	38 cm	$10.50

Explore the relations between diameter and cost, and discuss what is good value.

Here we need to start by considering what functions of the diameter might be reasonably expected to be proportional to the cost. If the thickness of the pizzas is constant, the amount of pizza dough is proportional to d^2; if the thickness varies as the diameter, the amount is proportional to d^3. The amount of topping, in either case, is proportional to d^2. Some sample calculations (approximate) are illustrated in Table 1. Graphs of d^2 and d^3 against cost are shown in Figure 1.

d cm	20	25	27.5	30.5	38
d^2 cm^2	400	625	756	930	1444
d^3 cm^3	8000	15600	20800	28400	54900
Cost $	$4	$5	$7.50	$8.60	$10.50

Table 1.

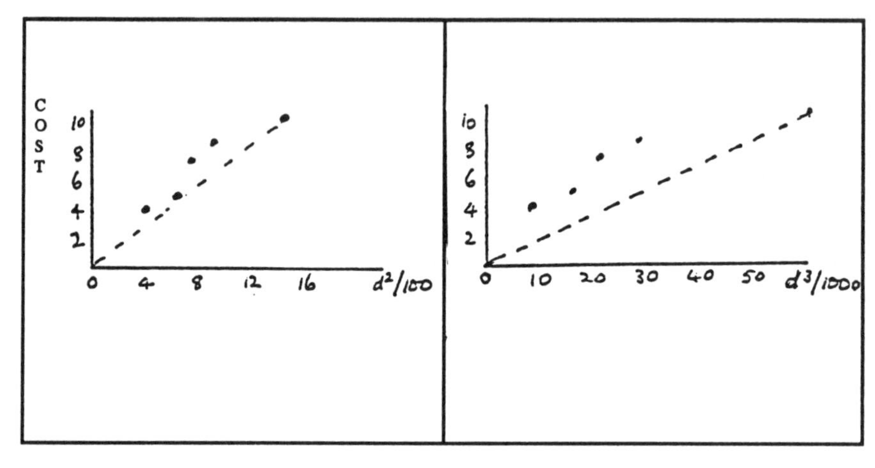

Figure 1.

The largest (Family) pizza is the best value by either criterion. The Small size is almost equally good value if d^2 is taken as the measure of amount of pizza. This problem requires an awareness of the square and cube functions and their relevance to surface and volume and also some competence in the modeling process.

These problems each tap somewhat different motivations. The number generalization probably arouses some interest in itself; in the Rowing problem, the attractiveness arises more from the power of the algebraic method in resolving a complex and otherwise difficult problem; while, in the Pizza situation, the satisfaction derives from the success of the mathematical method in giving some insight into a practical situation.

The classes of problems represented by these examples all merit a place in the algebra curriculum, both on the grounds that they provoke authentic mathematical activities and that they embody important and different algebraic strategies that are valid over a wide field. Moreover, they all concern the algebra of numbers.

There are, however, other algebras and other symbolisms in mathematics which need to be considered. They can illuminate number algebra through their similarities to it and their differences from it.

2.3.2. *Other algebras*

First, there are geometric algebras in which the elements are geometric transformations and the (single) operation is function composition. An example is the *Triangle Algebra* task (see Figure 2). The elements here are half-turns about specified points.

You need some triangular spotty paper and some tracing paper.
Copy the triangle, I, below onto the middle of your spotty paper.
Label the mid-points of the sides as shown:

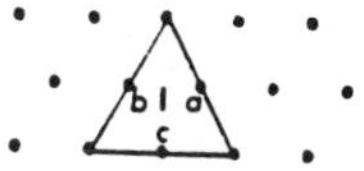

Trace the triangle and its labels on your tracing paper.
Now use your tracing paper to give I a half-turn about *a*.
Prick through the tracing paper to show the new position of the triangle.
Draw it and label it A.
Do the same for *b* and *c*.

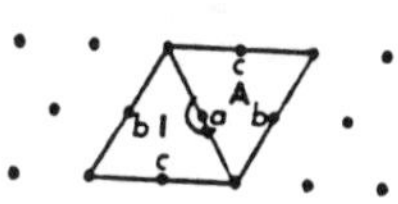

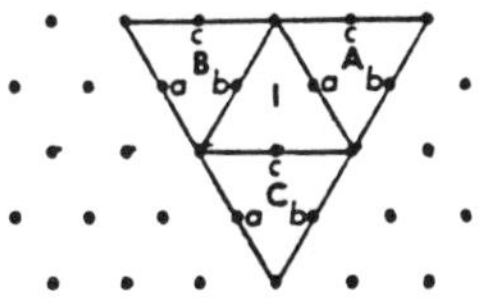

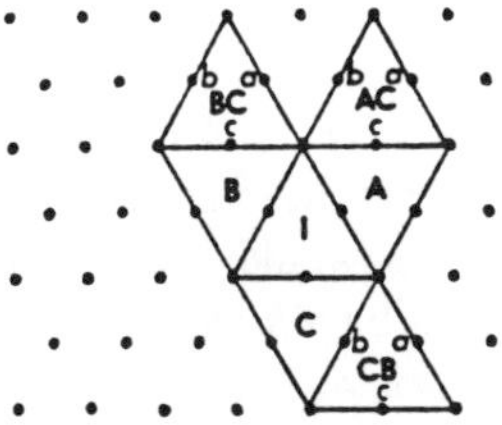

Now give A a half-turn about *c*, B a half-turn about its point *c* and C a half-turn about its point *b*.
Label them as shown.
Continue choosing triangles and giving them half-turns about one of their points *a*, *b*, or *c*.
Try to fill in all the gaps. Label all the triangles.
Remember that ABAC means a half-turn about *a*, then *b*, then *a*, then *c*.

Figure 2. Triangle Algebra

You need your tesselation of triangles. Find the triangles labelled AB, ABAB, ABABAB... What do you notice? If you add AB to the label of a triangle, what does it do to the triangle? Does this work with *all* triangles? Does the same thing happen with AC? What about BC? or CA...?

Study of this pattern leads to the observation that AB, ABAB, ABABAB, ... denote successive translations in a certain direction; and BA, ... in the opposite direction; and BC, ... in a different direction. Also, some positions can have multiple labels; we may write BCA = ACB; and ABCABC = I. In fact, XYZXYZ = I, where

X, Y, Z are any of A, B, C (generalization). Thus we may reduce a long "word," such as ABACACBAC, and show what position it represents, as follows:

$$\begin{aligned} \mathrm{ABACAC(BAC)} &= \mathrm{ABACA\underline{CC}AB} &&= \mathrm{ABAC\underline{AA}B} \\ &= \mathrm{AB\underline{ACB}} \\ &= \mathrm{A\underline{BB}CA} \\ &= \mathrm{ACA} \end{aligned}$$

Other geometric algebras arise from other sets of transformations. For example, the set of symmetries of an equilateral triangle--reflections in the three altitudes (p, q, r), rotations of 1/3 and 2/3 turn (w, w^2)--gives an algebra which contains relations such as $p^2 = q^2 = r^2 = I$, $w^3 = I$, $pq = w$, $qp = w^2$--these being relations among specifics, analogous to $5 \times 3 = 15$; and $x^6 = I$ as a generalization, true when x is any of the six symmetry elements.

Boolean Algebra, where the letters may denote propositions, and the operations are *and* and *or,* has just two "numbers," 0 and 1, and a set of laws somewhat similar to those of a field, but with, in particular, two distributive laws.

The set of residue classes modulo p (p prime) is a (non-ordered) field; and, even if p is not prime, is a ring. These are small number systems in which some or all of the normal "laws of algebra" hold.

Boolean algebra has serious uses in some areas of circuit design, and the finite fields of residues have applications in the design of randomized plots for agricultural experiments. Although these two topics could probably not command a place in the mainstream of school algebra, they provide valuable enrichment material and help to show something of the diversity of mathematics (Bell, 1964a, 1964b).

If these examples are all to be included--and there seems no good reason why they should not be--one would need to adopt a definition of an algebra as *any manipulable language by which relations or compositions are handled in the conceptual fields of space, number, or elsewhere in mathematics.*

3. WHAT NEEDS TO BE LEARNED IN ALGEBRA?

The algebraic processes needing development are:

1) using the algebraic language to express relationships and to work with the representation,
2) by manipulating the symbolic expressions into a different form to expose fresh aspects of these relationships, and
3) doing this in some characteristic ways, of which the most important are *forming and solving equations, generalizing,* and *working with functions and formulas.*

We can identify four aspects here. The first consists of being *able and willing to operate with symbolic (algebraic) expressions.* Pre-algebraic students may solve a complex arithmetic problem successfully by working directly with the number relationships in it, finding quantities that lead eventually to the desired solution, but are unable or reluctant to adopt the algebraic route of first representing the situation symbolically, then making syntactic transformations, then reinterpreting in the original situation. Note that this could be a description of what they do with numerical symbols and algorithms; the algebraic step is to do this with quantities which are either not specified numerically (generalized numbers) or are as-yet unspecified (unknowns).

The second aspect is that of learning the linguistic aspects of algebra--learning to *write* and *read* the notation correctly and meaningfully.

The third is to learn to *manipulate* it correctly and fluently.

The fourth aspect consists of acquiring the *strategic know-how* needed actually to deploy this language in activities such as those of generalizing, forming and solving equations, and working with functions and formulas.

Some experimental work on this fourth aspect is reported in Bell, Malone, and Taylor (1988) and Bell (1988). Here I shall now illustrate how problem solving in the broad sense can form the basis of an algebra course, with a particular focus on the forming and solving of equations, but also referring inescapably to generalizing.

3.1. *Problem Solving, in the Broad Sense*

The type of problem solving that has the greatest potential as a learning experience is that in which a generic, developable problem is given. In this the students, by changing the elements and the structure, can generate a set of problems that together give good insight into the topic and into relevant methods. I give a few examples of this, involving *Generalizing* and *Functions.*

In *Line Patterns* (see Figure 3), the letters T, M, B stand for the top, middle, and bottom numbers in the chosen boxes. They are thus *generalized numbers*, but have concrete support. The cognitive demand is therefore low. But these letters are being used in symbolic algebraic statements of generalization. The students are using algebraic language in a situation where it forms a natural means of communication.

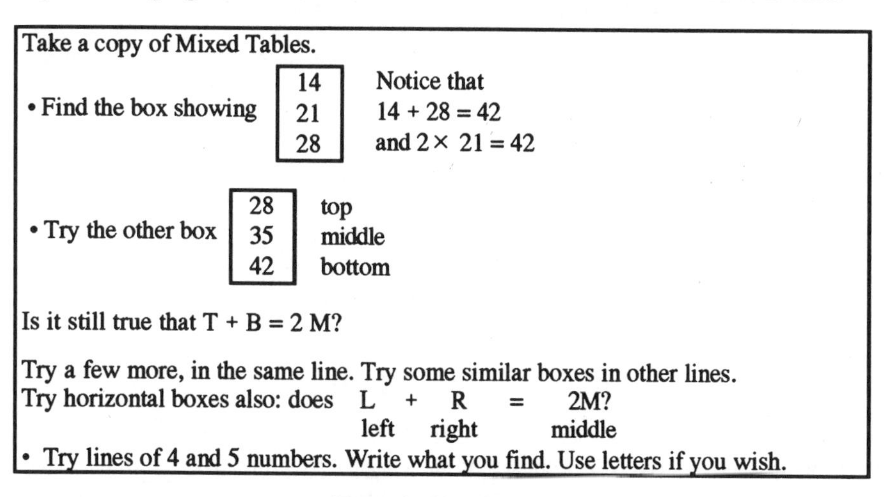

Take a copy of Mixed Tables.

- Find the box showing [14 / 21 / 28]   Notice that $14 + 28 = 42$ and $2 \times 21 = 42$

- Try the other box [28 / 35 / 42]   top / middle / bottom

Is it still true that $T + B = 2M$?

Try a few more, in the same line. Try some similar boxes in other lines.
Try horizontal boxes also: does L + R = 2M?
(L = left, R = right, M = middle)

- Try lines of 4 and 5 numbers. Write what you find. Use letters if you wish.

Figure 3. Line Patterns

Corners and Middles (see Figure 4) extends this use to a greater variety of expressions. Note that the approach is exploratory--after trying the initial suggestions, the students are asked to find, express, and test their own generalizations, which can become more complex as the students seek to challenge their colleagues. Thus, through these tasks, the students are gaining experience both in generalizing and in the use of algebraic symbolism as a language for expressing and working with their generalizations. They are also becoming familiar with the possibility of a conjecture being true always, sometimes, or never. It is an authentic algebraic experience, reflecting in this relatively simple situation the generalizing and symbolizing processes that continue to be important in mathematics at higher levels.

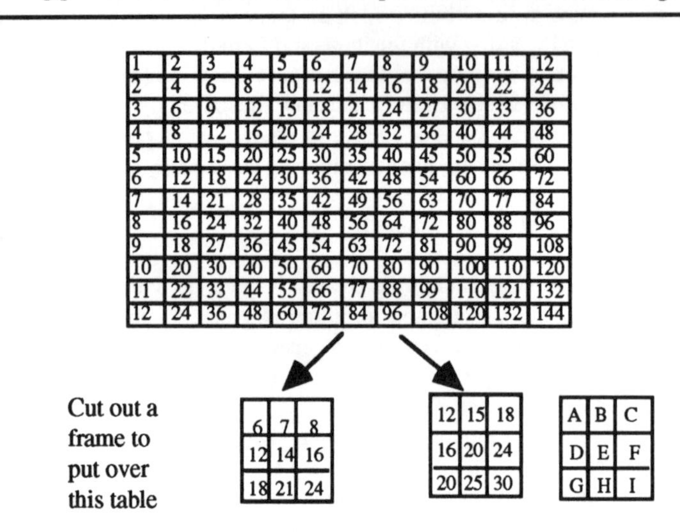

1	2	3	4	5	6	7	8	9	10	11	12
2	4	6	8	10	12	14	16	18	20	22	24
3	6	9	12	15	18	21	24	27	30	33	36
4	8	12	16	20	24	28	32	36	40	44	48
5	10	15	20	25	30	35	40	45	50	55	60
6	12	18	24	30	36	42	48	54	60	66	72
7	14	21	28	35	42	49	56	63	70	77	84
8	16	24	32	40	48	56	64	72	80	88	96
9	18	27	36	45	54	63	72	81	90	99	108
10	20	30	40	50	60	70	80	90	100	110	120
11	22	33	44	55	66	77	88	99	110	121	132
12	24	36	48	60	72	84	96	108	120	132	144

B - A = C - B
(the B number - the A number) = (the C number - the B number)

Check that this is true in both the boxes shown. Check other places on the table.
So B - A = C - B is always true.

- Consider A + D = F + I
  This is never true. Check this.
- Consider E = 2A
  This is sometimes true. Check this.
- With your partner, make a list of 6 relations and challenge another pair to decide whether they are always, never or sometimes true. Be sure of your answers first. Try to get some of each type.

Figure 4. Corners and Middles

These situations have been developed with 11- to 12-year-olds. A possible follow-up activity here would consist of the teacher collecting and recording on the blackboard a number of the relations discovered by students. With skillful choice from those offered, one could hope to have examples of the same relation expressed in different ways. Discussion of this could provide the beginnings of the awareness of the equivalences that form the basis of manipulations. In the task *Giving Clues,* discussed below, we show how this awareness was actually developed, in a different situation.

More extended individual explorations of additive and multiplicative squares, and of calendars, by 13- to 14-year-olds, have included the work illustrated in Figure 5. The students were encouraged to experiment and find their own such patterns for different shapes on the calendar and to include some that did not give a regular pattern. The example shown in Figure 5 is one in which the non-generality of the pattern is shown by the algebra; the difference $x - 3$ means, as the student says: "The answer isn't always 13, but, in the case I took, $x - 3 = 13$. So the difference will vary, depending on where the shape is."

(Heather)

I want to find the difference between the marked areas in ~~any~~ this shape anywhere in a calendar.

16	23	30
17	24	31
	25	

I will start with a simple case, by adding up each group. These came to 53 and 66 with a difference of 13. I will now put it into algebra to see if the difference will always be 13.

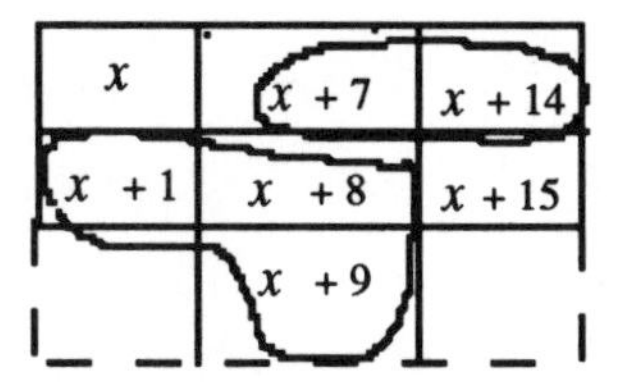

Again I added the two groups up and took the smallest away from the largest.

$x + 7 + x + 14 = 2x + 21$

$x + 1 + x + 8 + x + 9 = 3x + 18$

$3x + 18 - 2x + 21 = x - 3.$ (should be - 21)

This means that the answer isn't always 13, but will vary according to where the shape is.

Figure 5.

Here the student is explaining in her own words the process of testing a generalization using algebra. She assigns a single letter x to the number in the top left corner of her shape, and expresses the others in terms of it. She uses the formation and manipulation of her expressions not only to *express* her conjecture but also to prove or, in this case, to disprove its generality.

A multiplicative generalization is being tested by another student in the script shown in Figure 6. Here a higher level of manipulative competence is demanded. (It is not without error!) The need to learn a new manipulative procedure, in this case the distributive law, can be provoked by suggesting (if necessary) that students consider such multiplicative relations. An appropriate amount of lesson time could then be taken to discuss, establish, and practice the new skill, before returning to the exploratory work.

Proving with Algebra (James)

The problem

Taking a 5 by 2 rectangle from the number square: I ring the number in each corner, multiply the numbers in the two top corners together, and then the numbers in the bottom two corners together. I then find the difference between the two answers.

(76)	77	78	79	(80)
(86)	87	88	89	(90)

$= 76 \times 80 = 6080$

$= 86 \times 90 = 7740$

$7740 - 6080 = 1660$ or $6080 - 7740 = -1660$

I can prove this using algebra, for the example. I will use the same rectangle as above.

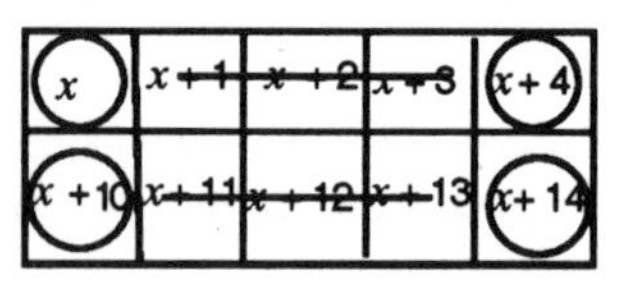

This is the same rectangle, converted into algebraic terms.

$(x + 10)(x + 14) - x(x + 4)$
$= x(x + 14) + 10(x + 14) - x^2 + 4x$
$= x^2 + 14x + 10x + 140 - x^2 + 4x$
$= (x^2 + 24x + 140) - (x^2 + 4x)$
$= 20x + 140$

In my rectangle, $x = 76$, so:
$(20 \times 76) + 140$
$= 1520 + 140$
$= 1660$

Conclusion: As the value of x is changed, the difference between the numbers will change. The answers will never be the same in 2 different positions on the 1-100 number square.

Figure 6.

These non-general relations show more vividly than the general ones how the use of a letter as a generalized number demonstrates the generality or not of the proposed pattern.

This is a point at which the broad pattern-finding activity was alternated with the more focused activity, in this case discussing how to multiply brackets. Throughout, periods of exploratory work alternated with lessons on specific manipulations. Subtracting brackets was discussed at an earlier point. These laws were approached first in an experimental mode, using cognitive conflict--expressions such as 13 – (6 – 2) being compared with 13 – 6 – 2 and the discrepancy discussed--before giving a lot of exercises for practicing the correct translation. Similarly, products such as 17 × 13 were discussed before $(a + b)(c + d)$.

The pedagogical principles here are that the x is introduced in such a way that it is clear that it stands for the number in a certain cell of the table and the purpose of the manipulation is to establish the generality of the pattern. Thus, we avoid the incomprehension that often arises when manipulative exercises are given in the absence of the meaning-giving formulation and interpretation stages. A similar rationale exists in the spreadsheet situation (Sutherland, 1992b).

3.2. *Problem Solving as Forming and Solving Equations*

In the early years of schooling, children solve missing number problems such as 8 + ? = 11 or ? – 7 = 5. Initially, these are solved by trial, but when larger numbers become accessible, a problem such as ? – 527 = 332 can provoke the realization that the needed operation is 527 + 332, and probably the awareness of the generality of this transformation is close. The question is whether the writing of $a - b = c \Rightarrow a = b + c$ is a suitable activity at this stage. Children could be taught to write this, but clearly it is not appropriate, unless this symbolic writing contains, or can lead to, some insight that is not already present in the mental perceptions. (Davydov, 1990, has experimented in this way with children aged, I believe, about 6 to 7 years).

Simple think-of a-number problems could be given to somewhat older children, who would be able, with practice, to reverse mentally such chains as add 5, multiply by 2, take 10, the answer is 4, to find the starting number. Some way of writing down the process would become helpful. This might be in the form

$$n \ +5 \ \times 2 \ -10 \ = 4,$$

and there might be some discussion about ambiguity and the need for brackets (e.g., "Is $n + 5 \times 2$ the same as $n + (5 \times 2)$?"). Noticing that the undoing sequence is "+10, ÷ 2, –5" might lead to the recognition that the rule is opposite signs, opposite order, and that this may be applied syntactically.

Some situations that lead directly to the forming and solving of equations (as distinct from the above operational sequences) are the following. These have been used with 11– to 12-year-old pupils with the aim of providing a gentle way of overcoming the first conceptual obstacle listed above, that of accepting the possibility and value of representing a situation by an algebraic expression and working with it. These diagrammatic situations have the advantage that (a) they are naturally self-checking, and (b) the unknowns have an obvious concrete existence, thus reducing difficulties associated with having to decide what quantity to denote by x and how to translate from verbal information into symbolic statements. The following are some examples.

In *Pyramids* (see Figure 7), the construction rule is that a lower number is the sum of the two adjacent ones above it.

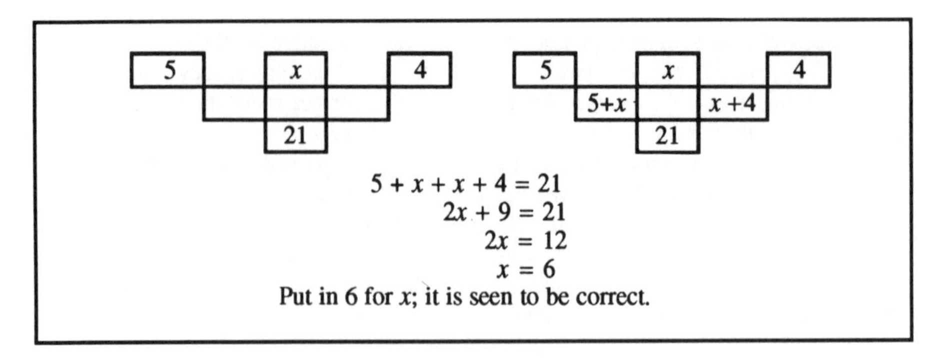

Figure 7.

The mode of writing the equation may be suggested by the teacher, but is normally readily adopted. The collection of terms and solution of the equation may similarly be the subject of discussion, but is usually accepted as self-evidently good by the pupils. By changing the construction rule from A + B to A + 2B, and then to A – B, more difficult manipulations can be made to arise. These would need dealing with by focused discussions, with experiment to show the validity of laws 2(A + B) = 2A + 2B and A – (B – C) = A – B + C. For other types of diagram that can act as a concrete setting for equations, see Bell (1994).

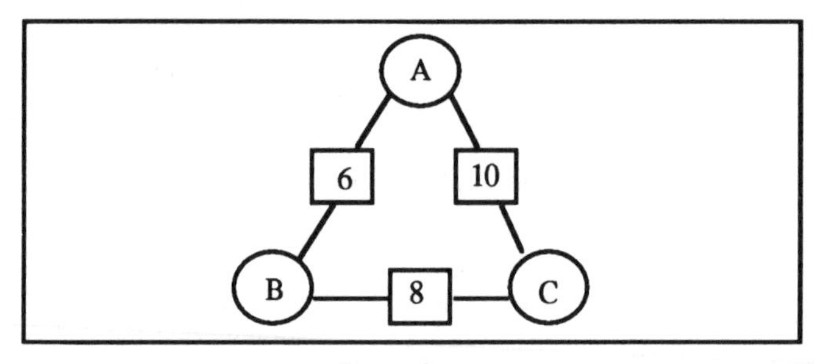

Figure 8.

In *Arithmagons* (see Figure 8), we have to put numbers in the circles that add to give the numbers in the adjacent square; for example, A + B must be 6. This example can be solved easily by trial, but when the numbers get big, say 65, 71, 80, it becomes better to denote one, say A, by x, get B = $65 - x$ and C = $71 - x$ and the equation

$$(65 - x) + (71 - x) = 80.$$

This is solved by thinking about its meaning, not by recourse to imposed rules.

Questions leading to generalizations can be posed, for example, is there always a solution--a positive solution? a whole number solution? The situation itself can be varied. This is a triangular arithmagon. Square, pentagonal, and larger arithmagons can be investigated; there are interesting properties to be found (and more generalizations) (Bell, 1995).

The *Piles of Stones* situation takes the step into forming equations from verbally described situations, but still in a restricted framework, to limit the difficulty of assigning the letter and constructing the equations.

Research has shown that many students regard the purpose of all questions in algebra as being the performance of some piece of manipulation. To avoid this danger in work on equations, it is important that students experience the full activity of beginning with a problem, forming the equation, then solving it, and interpreting the result. The activities discussed above have had letters embedded in them from the outset. We now need to consider cases where equations have to be formed from verbal situations. Students (aged about 13) were given two generic examples, which were discussed; they were then asked to make up (in groups) similar problems to be solved by another group. The first two problems given were the following.

1) There are two piles of stones. The second has 19 more stones than the first. There are 133 stones altogether. Find the number in each pile.
2) There are three piles of stones. The first has 5 less than the third, and the second has 15 more than the third. There are 31 altogether.

Students were asked to solve the first two problems, no method being specified. Most solved the first numerically: setting, correctly, 57 and 76 or, wrongly 47 1/2 and 66 1/2. I then showed them the algebraic method, taking x for the first pile, $x + 19$ for the second, obtaining $2x + 19 = 133$ and solving to give 57 and 76 for the two piles. They were then asked to take x for the second pile.

This gave $x - 19 + x = 133$, $x = 76$, and the same numbers 57 and 76 for the two piles. The aim here was to display the possibility of different x assignments, to observe the different expressions and equations that resulted, to note the appearance of the same solution for the size of the two piles, and thus, to get some insight into the relation between the algebra and the problem.

Following this they were asked to work, in groups of three, at solving Problem 2 above, taking in turn each of the three piles as x, and to compare their results. On the following day, each group was asked to make up and solve 3 similar problems, two easy and one hard, to be attempted by another group. This led to a lot of insight into the way different x-assignments affected the expressions, turning + into – and multiples into fractions. It also led to an unexpected degree of richness in the problem statements generated by the students. As well as four bean bags and the number of pupils in three rival schools, we had: "A nuclear scientist must complete 4 experiments to save the world, and has 23 days to do them in. The first will take twice as long as the second ..."

To conclude, we may say that the main initial difficulties lay in expressing relations such as "pile 3 has 15 more stones than pile 2," when pile 2 was x, making pile 3 "$x + 15$"; and more so when pile 3 was x, needing a reversal to make pile 2 "$x - 15$." "10 less than $x + 15$" was another step up in difficulty. However, although this was observed as a serious obstacle for some students in the early lessons, on being offered the answer, they soon picked up the idea; in the school examination question on this work, no student failed to formulate an equation, though there were some "reversal" mistakes.

3.3. *Recognizing Equivalent Equations*

Our past research shows a strong need for activities in which algebraic expressions and equations are treated by common sense and normal understanding rather than by a number of special rules, liable to misinterpretation, and not subject to normal critical appraisal. We sketch briefly one such activity; *Arithmagons* (above) provides another.

Giving Clues is another generic situation that leads naturally to the intuitive recognition of equivalent equations. It also provides a body of experience on which the study of sets of linear equations can be built.

$$(L - 3)/M = N \quad (1)$$
$$3M - L = 1 \quad (2)$$
$$4N - M = 2 \quad (3)$$

These equations were put on the board, with the statement that the teacher had chosen three numbers, denoted them by *L, M,* and *N,* and written down these three clues by which they might be found. The game was for each person to try to find the numbers (by trial and error or any other method). When a person had found them and had checked that they satisfied all three equations, she was not to say what they were, but was to make up a new clue and offer it to be added to the list.

It took some 5 minutes for the first person to find the numbers, and after that clues came along quite quickly. As more clues were added to the list, these provided additional help to those who had not yet found the numbers.

$L = 2M + N$	(4)	$L = M^2 + N$	(10)
$L - N = 2M$	(5)	$L - 4N = 1$	(11)
$L - 2M = 1$	(6)	$L + M + N = 8$	(12)
$2M = L - N$	(7)	$L = 5N$	(13)
$N + L = 3M$	(8)	$L - 4N = N$	(14)
$2L = 4M + 2N$	(9)	$2N = M$	(15)
		$L - N = 4N$	(16)

The discussion of this list included, as well as identifying duplicates, considering which were easy or useful clues--that is, those containing just two of the letters. One letter clues were rejected as "give-aways."

Subsequent work led to considering ways of starting with such a set and having a general routine for finding the hidden numbers. This consisted of combining equations so as to get a pair both of which contain the same letters (such as (11) and (13), then combining or comparing them. The method of "matching" consisted of working towards a pair such as

$$x + 4y = 9$$
$$x + 6y = 15$$

from which $2y = 6$ can be seen intuitively.

By this stage we have a general method for dealing with a number of equations in several variables: Combine equations in pairs so as to reduce the number of variables to one.

We close this section by giving some annotated examples of problems made up by pupils; they indicate the experimental approach that the pupils had and their sense of control over the situation (see Figure 9). Making up problems gives the pupils a strong sense of ownership of the mathematics and an outlet for creativity and humor. It also has a diagnostic value, in showing the teacher what aspects of the situation have been recognized by the pupils.

There are some fruit + veg. in a bowl. The 3 apples and 2 bananas have 39 maggots, 2 bananas and 1 carrot have 47 maggots, 1 carrot and 2 apples have 33 maggots. How many maggots does each fruit have.

$3a + 2b = 39$
* $2b + c = 47$
$c + 2a = 33$

Add together $5a + 4b + 2c = 119$
* Half them $2\,1/2\,a + 2b + 1c = 59\,1/2$

* and *

$2\,1/2\,a = 12\,1/2$ $a = 5$
$a = \frac{12.5}{2.5}$ $b = 12$
$= \frac{25}{5}$ $c = 23$
$= 5$

This pupil seems to be enjoying himself, but has not noticed, so far, that he has too few equations. There is an opportunity here for the teacher to discuss this point.

This is an interesting example, and successfully solved. But note the attraction for adding all the equations, which might not have been helpful had the second equation been different.

There once (was) a poet called Ben who owned one solid gold pen. He had 3 silver pencils, and 2 coloured stencils, and it only cost him £600.

He knew a man who lived down the road, who had 2 pens made of gold. He had one stencil, and only one pencil and it only cost him £300. Here's a question for you to do, it comes from us to you, how much does each item cost?

1. $p + 3s + 2c = 600$ 2. $2p + s + c = 300$
$p + 3s + 2c = 600$ - $2p + s + c = 300$
3. $-p + 2s + c = 300$

None of them seem to match with each other at the moment, so now I will try multiplying.

Figure 9.

3.4. *Progression*

The problem situations already discussed contain within themselves the possibilities of varying degrees of complexity. For example, the *Piles of Stones* situation--if the total number is given--leads to linear equations with the unknown on one side only (and so collectable and soluble without formal equation concepts); without a given total, the equation will contain x on both sides. That is, if in Problem 2 of *Piles of Stones*, the phrase "there are 31 altogether" is replaced by "the first and second together have three times as many as the third," the equation becomes $(x - 5) + (x + 15) = 3x$. The degree of abstraction of the given numbers (large numbers, decimals) affects the difficulty of recognizing and performing the correct operations and transformations.

Further problem types have been identified in recent studies aimed at examining the differences between algebraic and arithmetic approaches to problems. These studies may indicate how to choose a suitably varied and graded set of problems for a course. Examples are the following:

588 passengers must travel on 2 trains. One train has only 16-seat cars, the other has only 12-seat cars. The train with 16-seat cars will have 8 more cars than the other. How many cars will be in each train? (Bednarz, Radford, Janvier, & Lepage, 1992).

The tail of a fish weighs 4 kilograms; the head weighs as much as the tail and half of the body; the body weighs as much as the head and the tail together; how much does the fish weigh?

Because of its overlapping relations, the Fish problem is somewhat hard, whether attempted algebraically or arithmetically. There is an uncertainty about how many variables to employ. (It is good for students to see that this is not of major importance--if once the equations are formed they realize that it is easy to combine them to reduce the number of variables.)

Research aimed at categorizing problem types so as to ensure a good spread and to help the teacher manipulate the level of difficulty is of value here. Further examples of teaching material are contained in Bell, Hart, Love, and Swan (1980).

4. CONCLUSION

Through these problems, I have tried to show how generic problems can provide authentic algebraic experiences that not only cover the strategies for problem solving by forming and solving equations, but also develop the key algebraic abilities of writing, reading, and manipulating symbolic expressions.

Within such a problem-based course focused on forming and solving equations, it is clear that opportunities for generalization occur easily, provided the mode of working is exploratory. Functions can also arise when problems with insufficiently many constraints are offered. For example, in the *Pyramids* situation, if the contents of *two* cells are left unspecified, the functional relationship between the possible pairs of numbers in these two cells can be investigated, both as a table and as a formula. Similar extensions of other problems are possible. But the movement can also be the other way. A problem that begins as the study of a functional relationship given by a formula for y in terms of x gives rise not only to forward questions about what value of y corresponds to a given value of x, but also to the reverse question, what x corresponds to a given y, and this involves the solution of an equation; and the wish for a *general* expression for x in terms of y requires the transformation of the given formula.

I would not advocate any one approach as dominant or precedent. Regarding what aspects of algebraic awareness are fostered by these different approaches, I would say that experience of all the approaches--generalizing, functions, and equations--is essential. I would also assert the value of drawing students' attention *explicitly* to the nature of these three modes of algebraic activity and to the essential algebraic cycle itself--of representing, manipulating, and interpreting.

CHAPTER 14

"WHEN IS A PROBLEM?": QUESTIONS FROM HISTORY AND CLASSROOM PRACTICE IN ALGEBRA

JOHN MASON

Evolution in the role, classification, and treatment of problems in algebraic treatises from Diophante to Viète raises many questions concerning the didactic practice and intentions at the time as well as questions that arise with the attempt to extract some "lessons from history" for today's classroom. The problem solving approaches to algebra, situated in today's classroom, offer activities and obstacles that suggest different questions--some of which point back to the historical issues raised. The dichotomy of the general versus the particular and the question "What is a problem?" thread through the reflections in this commentary chapter.

Charbonneau and Lefebvre's chapter (this volume) focuses on the role of complex problems in the history of algebra. It begins by tracing the evolution of the term *problema*, and its relation to the term *quaestio* in early algebra texts by typical and well known authors such as Robert of Chester (1145), Nicolas Chuquet (1484), Gerolomo Cardano (1545), Rafael Bombelli (1572), and François Viète (1540-1603). Seeking specific meaning and uses of particular words is hampered by translators using different words according to their usage at the time of the translation; so the modern use of the word *problem* is extremely problematic, to put it mildly, with striking differences between use in Europe and North America. To appreciate the use of a word, original texts have to be consulted, and these are not always available. I have a suspicion also that medieval Latin was not always noted for its stability in word use, and in some cases authors naturally changed their ideas over the long period of manuscript preparation, which makes it hard to appreciate subtleties in old texts, and even leads sometimes to our introducing subtlety where none was intended. The Oxford English Dictionary gives *problema* as a thing thrown or put forward, hence a question propounded for solution. There seems to be a whiff of "throwing forward an educational or entertainment purpose." Wycliffe and Chaucer used it in terms of a riddle, but by the 16th century, it had become a question for academic discussion or dispute.

I was struck particularly by being reminded of Viète's classification into

zetetics: search (Polya--problems to find)
poristics: deduction (Polya--problems to prove)
exegetics: exposition, interpretation

and I hoped for further insight into how these distinctions were used to structure classification of mathematical questions and how these were then presented to students. There is much to be learned about the effects of didactic transposition (Chevallard, 1985), when expert awareness is turned into instruction, and historical study could provide insight into how people in the past have faced these challenges.

N. Bednarz et al. (eds.), Approaches to Algebra, 187-193.

For example, Diophantus of Alexandria is famous for his collection of "problems," but he did not develop an explicit body of theory for resolving them. However, the demonstration of each in turn may develop in the reader a growing sense of approach or method, and it would be fascinating to know how Diophantus and followers conceived the structure of their exposition. Diophantus is properly located in a long chain of authors, with earliest known representatives those of Babylonian tablets, Egyptian papyrus scrolls, Indian vedas, and Chinese manuscripts, in which questions are posed and solved in the particular, but in a form that is hard not to see as intended to be generic. There is still considerable debate as to the extent of generality perceived by those ancient scribes. My conjecture is that Diophantus and other authors were dwelling in the general and manifesting the particular, and either were not aware of the difficulties their audience had in generalizing (just as now in classrooms), or had a highly developed awareness of particular and general with cultural expectations that their readership would be seeing generality through the particular, or belonged to a tradition in which the general was part of the esoteric face-to-face teaching. There are certainly strong indications of awareness of generality, as evidenced by the end-of-solution comments distilled by Gillings (1982) from Egyptian manuscripts:

The producing of the same.
The manner of reckoning it.
Behold! Does one according to the like for every uneven fraction which may occur.
The doing as it occurs (or, that is how you do it).
Shalt do thou according to the like in relation to what is said to thee, all like example this.
The correct procedure for this [type of] problem.
These are the correct and proper proceedings. (p. 233)

Charbonneau and Lefebvre's analysis from history charts a development in the role played by word-problems in the early texts that they examined. With Diophantus, the "tasks" carry the exposition because the methods have to be reconstructed from the particular by the reader. In Al-Khwarizmi (translated by Robert of Chester), there is some explicit theoretical exposition of generality. Diophantus' use of complex problems is relegated to those sections where explanation of the general is not sufficient. This practice apparently reappears in Cardano, though his books have a richer classification of question-technique structure. Viète penetrated this gray zone, categorizing according to whether a solution is seen as belonging to a symbolic domain or to an arithmetic/geometric domain of construction.

A distinction attempted by Viète, between algebra and geometry, and more particularly, between the zetetics of algebra and of geometry, later merges when a classical geometrical construction involving geometrical means is used to resolve a quadratic equation. There are interesting parallels with the current developments of computational expressive environments providing symbolic manipulation on the one hand, and employing graphic screens and "mouse-mathematics," on the other. The overt supremacy of algebra is under attack, often being more useful when hidden under geometrical displays.

Christiansen and Walther (1986) point to a long philosophical tradition exploring what the concept of *problem* really means, and that

> whereas Leibniz did not consider the *aspect of difficulty* (Greek Aporie: doubt, indecision, indecidedness) as one of the characteristic features of a problem, exactly this (aporietic) aspect is emphasized in the line of development based on the tradition of Kant and N. Hartmann. (p. 271)

Brookes (1976), following the hermeneutic tradition, proposed the question "When is a problem?" to emphasize that problems are most appropriately seen as states experienced by people, when an action has resulted in a disturbance, so that a situation or question becomes problematic for them. Parthey (cited in Christiansen & Walther, 1986) makes similar observations. The situation may become problematic, but is not the problem as such. Christiansen and Walther go on to relate these considerations to Polya's (1966b, pp. 123-129) distinction between routine and non-routine tasks, which seems to be a modern version of the medieval use of *quaestio* and *problema,* and then to develop a sophisticated analysis of the relation between task and activity in classrooms.

Charbonneau and Lefebvre suggest that the origins of the algebra problem lie in the unresolved, the unsystematized, but that once a question belongs to a type and has a method of solution, classical authors no longer consider it a problem. Nevertheless, it may still become a problem for others. We have for too long treated mathematical tasks (problems, questions, exercises) as things having innate properties, such as interest, ease or difficulty, and relevance, when these adjectives are more properly assigned to the state of a person in the presence of and construing a text. In other words, a particular collection of words is not in itself a problem, merely a collection of words. If those words engender a state of *problématique*, then a problem has emerged, for the person. But that is the result of an action arising from an initial disturbance of the status quo and involving will/attention/orientation, past experience, and the present situation. Traditional textbook word problems rarely produce a disturbance that becomes a problem for the pupils. Even so, it is possible to re-experience problématique, and that depends on the teacher as much as the author (Movshovits-Hadar, 1988). It could be useful to consider the relation between problem and rule and the role of practice in the evolution of algebra (particularly in theoretical mathematics).

The importance of historical studies lies partly in the opportunities they afford to re-enter struggles to encompass the shifts of attention needed in order to make sense of, and to systematize, solutions to classes of problems. It can also provide access to root problems that drove mathematicians to develop what we now know as a topic to be taught in schools, however transformed through the ravages of the didactic transposition.

The role and influence of educational concerns in the development of a subject like algebra seem rather different now from the Middle Ages. Whereas the books being studied were the recent compilations of new mathematical ideas at a period, the influence now of educational concerns, particularly on algebra, is at best retrospective, perhaps raising a few side-issue mathematical questions, but the principal problématique is confined to re-thinking the nature of algebra in a pedagogic context that now includes symbol manipulators and spreadsheets. One current issue of concern is what a symbolic manipulator could be expected to do when solving a general problem that, in special cases, has special answers because of coincidences and alignments of particular values (Monagan, 1992, p. 31).

1. QUESTIONS FROM HISTORY

When did explicit exposition of generality first enter mathematics textbooks, and what did those authors say about this transition from the particular to the general? Are there two distinct lines of development to recent times, the particularity of arithmetic, and the generality of algebra, or is it an intertwined tale? If intertwined, how do authors choose between example and theory first, and with what justification? Has that justification altered over generations, or is our pedagogy basically medieval? What do authors think students are doing with examples, theory, and worked examples?

What is involved in the generation of a state of problemhood in an individual or group? How can a teacher act to bring about the requisite disturbance, not too little, not too strong, that generates this state?

What is involved in seeing the general through the particular when following a particular solution to a particular question?

What is the Chinese and Japanese experience of particular and general, of question and problem?

What connections are there between the use (and sense behind that use) of problems and what we now know as puzzles--and perhaps a shift in their purpose? Medieval puzzles became pretty arcane and ridiculous, and yet remained popular for many centuries. (I have in mind the Egyptian "St. Ives problem" and the many variations of resistance problems that employ harmonic means for their resolution, involving fountains spurting, animals eating, ships sailing, popes praying, ...)

2. CLASSROOM PRACTICE

I find myself, naturally enough given the similarity of our interests and context, in full harmony with Bell's observations and suggestions (this volume). I think he shares with me a difficulty in finding a way to express in words what is more to do with an ethos for the classroom, a way of thinking that could imbue and inform all aspects of a teacher's mathematical encounters with pupils, rather than a topic studied for a short while at some particular time. Bell has great breadth and depth of experience of constructing tasks for pupils, particularly connected with the emergence of algebra, and a few of these are offered in his chapter.

There is a direct link with Charbonneau and Lefebvre in that problem solving has different meanings in Europe and in North America, with Europe still tending to use the older meaning that Charbonneau and Lefebvre detect in the early algebra texts that they studied.

Bell's chapter begins with a succinct, but comprehensive, summary of the nature and place of algebra in mathematics: as a set of conceptual tools (classifying, comparing, combining, transforming, reversing) and as a way of thinking (language of expression and manipulation), complementary to geometry. Points are illustrated by tasks for pupils.

In the *Triangle Algebra* task, Bell has found something (a description of the effect of AB, using right-operators or Reverse Polish notation). What is the student to make of the directiveness of the task specification (i.e., is there an invitation to creativity and exploration?) How might a student discover that XYZXYZ = I for any substitution of X, Y, Z by members of {A, B, C}?

This example offers food for discussion of distinctions between abstraction and generalization: The use of *A*, *B*, *C* for rotations of the triangle is notational, rather than general, abstracting the essential aspect of the rotation as an operation related to a distinguished vertex, but ignoring details of its particular effect on a particular triangle; the word equation *ABCABC* = I expresses a generality because the effect is independent of the triangle to which it is applied; and the use of X, Y, Z involves yet further expression of generality.

Further examples of algebra are offered, including one arising from symbolizing geometric transformations. Tahta (1980) suggested that "the geometry that can be spoken is not geometry, but algebra," in trying to locate a boundary between algebra and geometry, and there are resonances with Viète's attempted distinction. The geometric context tasks can be seen as expressions of geometric generality, or as abstraction, as pulling back from the objects being operated on and focusing on transformations.

3. ALGEBRA IN THE CURRICULUM

Bell then turns to the algebraic content of the curriculum and suggests attention to: expressing relationships; manipulating to expose fresh relationships through forming and solving equations, generalizing, and working with functions and formulas.

The use of "working with ..." seems a bit weak, especially in view of the sophistication of the tasks he then offers to illustrate what he means. I would offer a stronger version along the lines of "formulating, testing, and employing functions and formulas in a variety of situations and contexts, with reflection on the role played by symbolic expression" in relation to four aspects identified.

- Aspect 1: Being able and willing to work with symbolic (algebraic) expressions.
  Can I find an example in my own experience of an unwillingness to work with symbolic expressions? Certainly I can, and it suggests a lack of manipulable confidence in the objects being symbolized. Hence the phases in a developing spiral, in which *manipulating* confidence-inspiring entities develops a *sense-of,* which may become *articulated* in increasingly succinct form, and thence an entity subject to further manipulation (Mason, this volume).
- Aspect 2: Reading and writing in the notation of algebra.
  I see competence and confidence in reading and writing with symbols as achieved through a natural enculturation process in which the teacher explicitly uses symbols to express generality and increasingly calls upon students to do the same (via diagrams and mixed word-symbol sequences). The difficulty experienced by pupils is the sudden introduction of a formal abstract language to no apparent purpose other than the doing of strange word questions.

- Aspect 3: Learning to manipulate symbols correctly and fluently.
  I see the rules of manipulation emerging naturally from the arising of multiple expressions for the same entity, coupled with a basic human desire to "avoid the middle stage" and be able to go directly from one expression to another. Fluency arrives from frequent experience of expressing one's own generality so that fluency tasks are seen as merely examples from a whole range of possible generalities. The Gattegno-Hewitt observation (Gattegno, 1978; Hewitt, thesis in preparation) is critical here: If you want to gain fluency and facility, you must be able to do something without fully attending to the details; so you have to engage in tasks that attract much of your attention *away from* the specific skill.
- Aspect 4: Developing strategic know-how.
  Surely this arises from attending to and reflecting on more significant mathematical tasks that require and exploit algebraic manipulation.

4. PROBLEM SOLVING AS A NORMAL MODE

Line Patterns and *Corners and Middles* offer an enculturation into the use of symbols to express relationships and the use of specializing for checking conjectures (though this language is not used explicitly). Bell provides an informative example of a student using symbols to demonstrate that a conjectured generality is false, and specific to the particular case tried.

Similar experiences using spreadsheets have been developed by Sutherland (1992b), suggesting that context is not terribly important, but rather that the process of expressing generality and then manipulating those generalities is endemic to mathematics.

Pyramids and *Arithmagons* offer further examples of tasks that can stimulate expression of generality and provide further enculturation into the uses of acknowledging one's ignorance (Mary Boole in Tahta, 1972).

In a tiny, throw-away paragraph, Bell remarks on the richness that arose from asking students to pose their own examples like the ones done the previous day. This is for me a particular case of an important general principle. If you want to support students in generalizing from their experience as part of their construal process, then engage them in creative as well as routine activities. In particular, when pupils can describe a class of problems, and even write down a very general question of that type, then they are in a position of power with respect to an examiner, for they know the source of tasks, and they can recognize and locate any particular task within that class. In the attempt to express what they mean by "this type of problem," and to pose what they consider to be an easy, a hard, and a general question of this type, students offer teachers insight into what features they are stressing and ignoring, which is the essence of generalization (Gattegno, 1990).

Pupils who have a sense of "types of questions" contrast markedly with pupils who are at the mercy of fresh tasks mysteriously coming as if from nowhere and each requiring a "new" technique. By explicitly invoking student taxonomizing and expression of generality, teachers can help students achieve a state that may approach the state of medieval authors who "have a method." They can be supported in becoming aware not just of having a method, but of the class of typical problems to which it applies. I am reminded of the observation of Charbonneau and Lefebvre, that the word *problema* was used by early writers for those questions that did not seem to fall into any particular class and required special methods. The history of the development of mathematics can be read as a search for techniques to resolve classes of problems, reducing them to mere questions. The didactic transposition then converts the experience of problematicity into instruction in techniques, without, often, offering any experience of the original problematicity.

5. COMMENTS ON CLASSROOM PRACTICE

As teachers, we want our pupils to think creatively, to master techniques that will enable them to tackle more challenging "problems." But we fall victim to the didactic transposition (Chevallard, 1985). It is more convenient to devise instruction, to the extent of treating pupils as ill-functioning computers to be programmed, than to invoke creative thinking. Any system for teaching algebra is almost bound to omit the very essence of algebra, for in trying to provide teacher-proof materials, or even teacher-supportive materials, the didactic tension (Mason, 1986) is breached at two levels:

> The more explicitly the teacher indicates the behavior sought, the easier it is for students to display that behavior without recourse to generating it from understanding and comprehension; the more explicitly educators indicate the behavior desired in the classroom, the easier it is for teachers to display that behavior without recourse to generating it through their own mathematical being, and so failing to be mathematical with and in front of their pupils. (p. 30)

If you think that latter conjecture is too strong, then observe that a growing number of teachers, bombarded by governmental statements of what they are supposed to do with their pupils, are putting their trust in the textbook: "My job is to make the doing of the text-tasks as pleasant as possible; they will learn as long as they do those tasks." Unfortunately, all the evidence over thousands of years runs contrary to that conjecture.

PART IV

A MODELING PERSPECTIVE
ON THE
INTRODUCTION OF ALGEBRA

CHAPTER 15

MATHEMATICAL NARRATIVES, MODELING, AND ALGEBRA [1]

RICARDO NEMIROVSKY

The construction of mathematical narratives is at the heart of modeling. This chapter explores the connections between algebra learning and the process of narrative construction. The first major section introduces the notion of mathematical narratives in the context of narratives in general. The next section describes and analyzes the work of high school students in episodes that exemplify three aspects of narrative construction: repair, playing out, and idealization. The last section focuses on the relationship between mathematical narratives and the early introduction of algebra.

1. INTRODUCTION

A shared goal among mathematics educators is having students come to be able to fluently use graphs and equations in the description and interpretation of events in the world. However, as currently taught, often these representations arise out of nothing--and so have to be imposed on students as notations devoid of personal meaning. To change this situation it is essential to identify and nurture the students' domains of everyday experience that may offer a fertile background for the growth of mathematical ideas. The domain of experience that constitutes the focus of this paper is story-making. This chapter is an argument for the productive and generative nature of narrative construction. Rather than being a marginal aspect of the learning of mathematics, it is suggested that the development and interpretation of narratives are a rich resource. The aim is not just to point out the importance of narratives, but also to elaborate on the dynamic processes involved in the production of narratives as a meaning-making activity.

To highlight the kind of story-construction that is specific and fundamental to mathematical modeling, the chapter deepens the notion of "mathematical narratives," that is, narratives that fuse aspects of events and situations with properties of symbols and notations. The same mathematical symbols--the same graph, equation, or number series--can always be interpreted within many different narratives. Often the term "mathematical model" is used to refer to an equation or a graph that is on paper or on the computer screen, collapsing all the possible ways in which it can be embedded in narratives, which reflect different modes of understanding and of recognizing connections between symbols and phenomena. Modeling involves at its core the construction of mathematical narratives.

Activities in which students develop and talk about mathematical narratives configure a productive territory for the early introduction of algebra. Young students can engage in rich conversations around the construction and sharing of mathematical narratives expressed with graphs or tables of numbers (diSessa, Hammer, Sherin, & Kolpakowski, 1991; Tierney, Weinberg, & Nemirovsky, 1992). Among the different avenues for the emergence of algebra--such as the generalization of

N. Bednarz et al. (eds.), Approaches to Algebra, 197-220.

arithmetical relations or of geometrical patterns--the development of mathematical narratives is a fruitful context for the exploration of algebra to describe the different ways in which things change and for the conceptualization of variables as expressions of continuous variation. The use of mathematical narratives in the introduction of algebra has been suggested or implemented by several educators, curriculum developers, and software designers (Barnes, 1991; Chazan & Bethell, 1994; Confrey, 1992a; Dugdale, 1993; Kaput, in press; Krabbendam, 1982; Lobato, Gamoran, & Magidson, 1993; Swan, 1985; Tierney, Weinberg, & Nemirovsky, 1994; Yerushalmy & Schternberg, 1992).

2. MATHEMATICAL NARRATIVES

The literature on narratives is, of course, huge. The first part of this section is not a literature review, but an articulation of a few central aspects that are especially relevant for the subsequent analysis of mathematical narratives, such as how narratives combine different temporalities, and the narrative use of episodic (or discontinuous) descriptions for events that change continuously.

2.1. *On Narratives in General*

The distinctive feature of a narrative, the one that distinguishes narratives from other forms of discourse, is that it is embedded in a sequence that is meant to reflect temporal order. Labov (1972), in a widely cited definition, says:

> We define narrative as one method of recapitulating past experiences by matching a verbal sequence of clauses to the sequence of events which (it is inferred) actually occurred. ... With this conception of narrative, we can define a minimal narrative as a sequence of two clauses which are temporally ordered. (p. 359)

Labov's example of a minimal narrative is the following:

(a) I know a boy named Harry.
(b) Another boy threw a bottle at him right in the head,
(c) and he had to get seven stitches.

Clause (a) is not part of the body of the narrative. It is a free clause that could be uttered at another point in the sequence. However, the fact that (c) comes after (b), using the temporal juncture "and," reflects that Harry had to get stitches *after* someone else hit his head.

Labov's analysis shows that, even though a temporal sequence is the skeleton, so to speak, of a narrative, the narrative cannot be reduced to a chronology. Ricoeur (1981) explained this crucial point:

> But the activity of narrating does not consist simply in adding episodes to one another; it also constructs meaningful totalities out of scattered events. The art of narrating, as well as the corresponding art of following a story, therefore require that we are able to extract a configuration from a succession. (p. 278)

Ricoeur distinguished between the episodic and configurational dimensions of a narrative. The episodic dimension refers to the sequential structure, whereas the configurational dimension alludes to the point of the narrative as a whole. Ricoeur stressed an issue that was analyzed by Aristotle: a narrative does not distinguish between fact and fiction. A narrative has the power to confer reality to a fiction, that is, to induce in the hearer the feelings of actually experiencing imaginary events. Ochs, Taylor, Rudolph, and Smith (1992) emphasized how this integration of fact and fiction takes place in the co-construction of narratives through conversation and social interaction.

Often a narrative incorporates parallel courses of time, embedded in characters and circumstances that relate to each other but keep their own passing duration; they converge and diverge at certain points. A narrative involves two temporalities, the temporality of the narration and the temporality of the events that are narrated. The relationship between the two can be quite complex. On this relationship, I highlight two aspects that are central for the analysis of mathematical narratives:

- The episodic structure of a narrative imposes some discontinuity in the narrated events even if they are continuous. Language offers many resources to neutralize the discreteness of the episodic structure. For example, suppose two contiguous episodes, A and B, concern a person called Mary, and the narrator says: "(A), and then, as Mary got increasingly upset, (B)." The expression "as Mary got increasingly upset" tells that episodes A and B are part of a continuum that is marked by the gradual change in Mary's feelings. The use of this type of resource mitigates, but does not eliminate, the elements of disjointedness resulting from the episodic structure of the narrative.
- The temporality of the narrative does not necessarily correspond to a sequence of events succeeding in time. A description of a landscape or a painting may take the form of a narrative even though neither the landscape nor the painting change over time. In this case the episodic structure of the narrative tells how the narrator is observing, and invites the listener to imagine a static object (e.g. " ... and more to the left ..."). Space can therefore be temporized by a narrative that incorporates the changing perspective of the narrator.

2.2. *On Mathematical Narratives in Particular*

I define a mathematical narrative as a narrative articulated with mathematical symbols. An example of a mathematical narrative is the following:

(a) First it rained more and more and it started to become steady (pointing to piece a of Graph 1)

(b) then it rained steadily (marking piece b of Graph 1)

(c) then it rained more and more (pointing to piece c of Graph 1).

Even though Graph 1 is continuous, the interpretation of this graph in a narrative is accomplished by splitting the event into phases or episodes. The narrative integrates events (e.g., raining more or less) with characteristics of the symbolic forms (e.g., the line on the graph going up or down, being curved or straight, etc.). There are many possible graphs to describe the same narrative (e.g., a graph in which section (c) is a straight line is also consistent with the narrative), and yet, at the same time, the narrative incorporates specific properties of Graph 1.

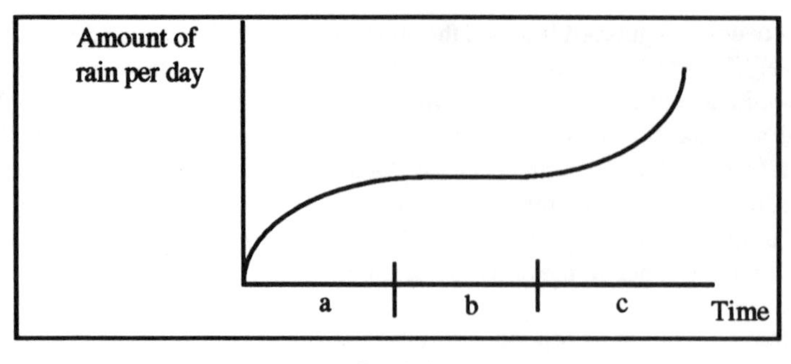

Graph 1.

For instance, "First it rained more and more and it started to become steady" describes the singular shape of the piece (a) of Graph 1, by combining an increase ("First it rained more and more") with a decreasing trend ("it started to become steady"). There are also many narratives to account for the same Graph 1 (e.g., another narrative might describe changes on the accumulated level of rain or on the average amount of rain as time passes). Some narratives are more complete or compelling than others depending on the overall purpose (e.g., to design a rain collector, to identify when it rains the most, to find patterns in a list of numbers showing amount of rain per day, etc.), but each narrative expresses a specific way of making sense of Graph 1.[2]

A mathematical narrative could also be an interpretation of a graph that is not over time, such as a graph showing the contour of an object over distance. Often a narrative is loaded with tensions because narrative construction is a confluence of different stories that struggle for dominance. When this struggle cannot settle down, the narrator acknowledges a feeling of not-making-sense. In other cases, the tension responds to the lack of connectedness and completeness. These are perceived as good, holistic qualities of a mathematical narrative. Students have expectations about completeness and connectedness. For instance, often they expect that a graph, in order to be useful, has to show what they conceive as the whole story; in a fourth grade class, after a number of activities based on the study of plant growth (Tierney et al., 1992), we asked students which of the two plants represented in this graph grew more:

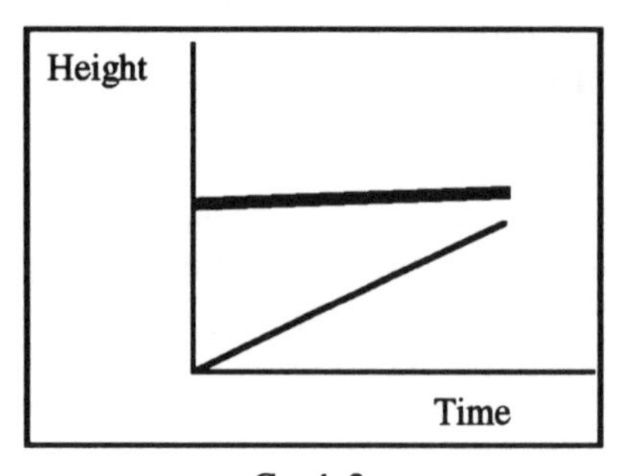

Graph 2.

One child said that the plant represented with a thicker line grew more. One might conclude that he interpreted "being above," instead of the difference between heights, as "growing more," but he went on showing that what really happened was like this:

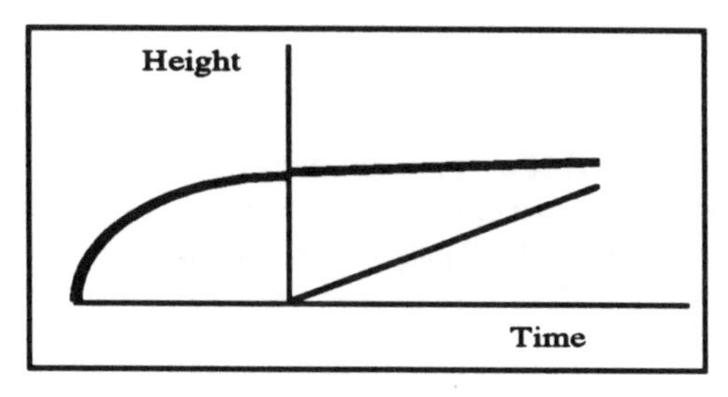

Graph 3.

Since the sharing and discussion of mathematical narratives is one of the ways in which mathematics is learned, a potential contribution is to clarify how mathematical narratives create tacit grounds of communication, language-use, and experimentation. Labov (1972) showed how the mastery of the English language is expressed in the richness of speakers' narratives. A parallel argument suggests that fluency with mathematical ideas should be reflected in the construction of mathematical narratives.

3. THREE EPISODES

This section includes three examples of students' construction of mathematical narratives. These episodes are part of a series of individual teaching experiments conducted at TERC. The main focus of the project was to study how high school students, who have not taken calculus courses, learn about qualitative differentiation and integration when the problems are posed graphically in the context of physical changes that students can control, predict, and measure.

The three students, Ken, Laura, and Steve, all used a computer-based motion detector that enabled them to produce graphs of a moving object in real time on the computer screen. The students moved a toy car along a straight path in order to generate graphs. By using tools to draw or to record movements, they expressed a narrative not only with linguistic means; drawings, gestures, and kinesthetic actions also became expressions of a narrative. Although the student-interviewer talk has a prominent role in our analysis, extra-linguistic elements, such as body motion and drawings, are also critical to elucidate narrative construction.

I include these three episodes not to show students' mistakes or obstacles but to explore some of the characteristics and functions of narrative construction. Mathematical narratives are not the exclusive domain of "right" or "wrong" ideas. They express the many ways in which people deal with mathematical problems and, for research purposes, they can be more or less illuminating regardless of their "correctness." Actually the three episodes are part of longer teaching experiments through which Ken, Laura, and Steve were able to overcome the difficulties shown in the next section. Rather than a description of stable narratives as-they-are, our

At the time of each teaching experiment, Ken was in 10th grade, Laura in 11th grade, and Steve in 11th grade.[3] They volunteered to participate and were paid a student fee. They studied at different high schools in the Boston area. All of them had taken Algebra I and Geometry.

3.1. *Ken*

This episode introduces a first example of mathematical narrative. It presents the notion of "repair" as a key aspect of narrative construction. Ken felt the need to repair his narrative to account for the continuity of the velocity versus time graph.

3.1.1. *Description*

I moved the toy car toward the motion detector and then away from the motion detector. A graph of distance versus time appeared on the computer screen.

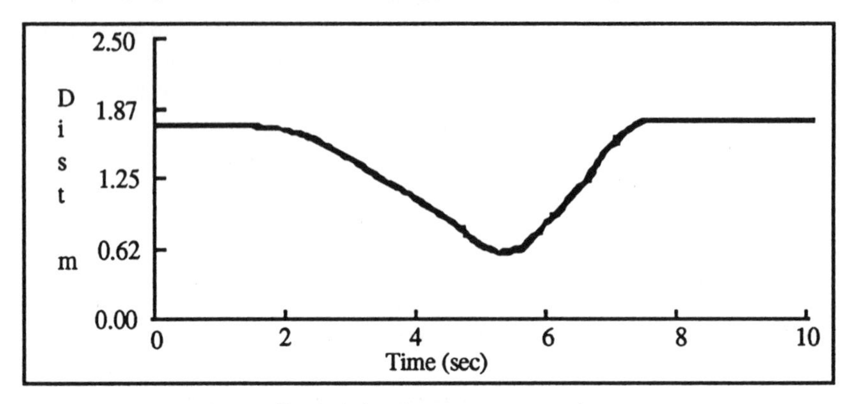

Graph 4. (on the computer screen)

Then I asked Ken to draw on paper a copy of the curve displayed on the computer screen. Ken drew Graph 5.

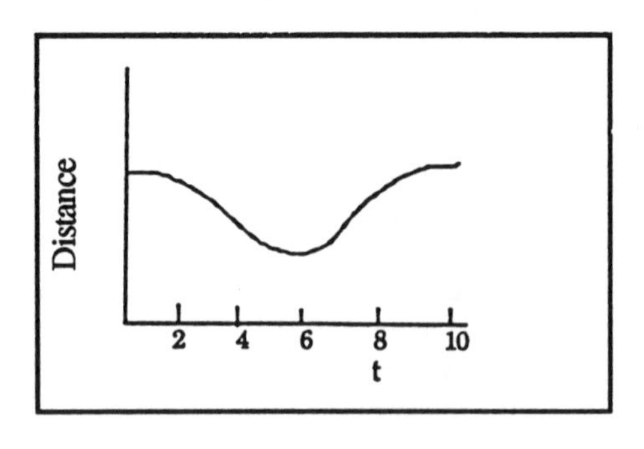

Graph 5.

1 Ken: *OK. OK. On this graph [Ken draws the axis, Graph 6a] ... Alright. OK.*
2 *Um, alright, um, I started off, um, with no, no speed for about two*
3 *seconds [adds a horizontal segment, Graph 6a]. And then, um, kind of, uh,*
4 *OK, yeah, this is going backwards, um, so, I'll do this (Graph 6b).*
5 *Alright, um, for two seconds there is no velocity until, um, it went*
6 *backwards for about, uh, until six, kind of a slow acceleration [Graph 6c].*
7 *Um, and then, OK, uh, actually, that's six. [Graph 6d] Then it slowed*
8 *down, slowed down about six, [Graph 6e] um, and started going forward*
9 *again until [Graph 6f] about, actually, until about eight [Graph 6g]. And*
10 *finished there. Kind of a, kind of a messy graph.*

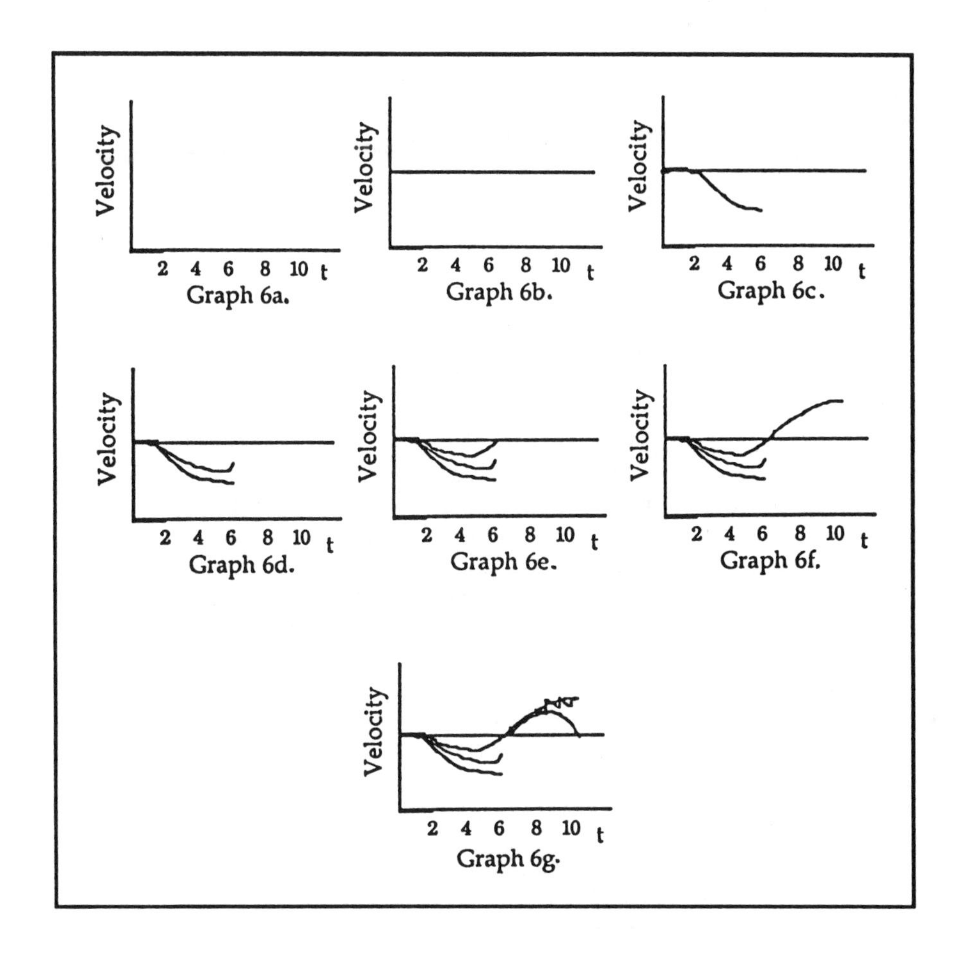

After Ken drew the prediction (Graph 6g), he asked for the computer graph of velocity versus time:

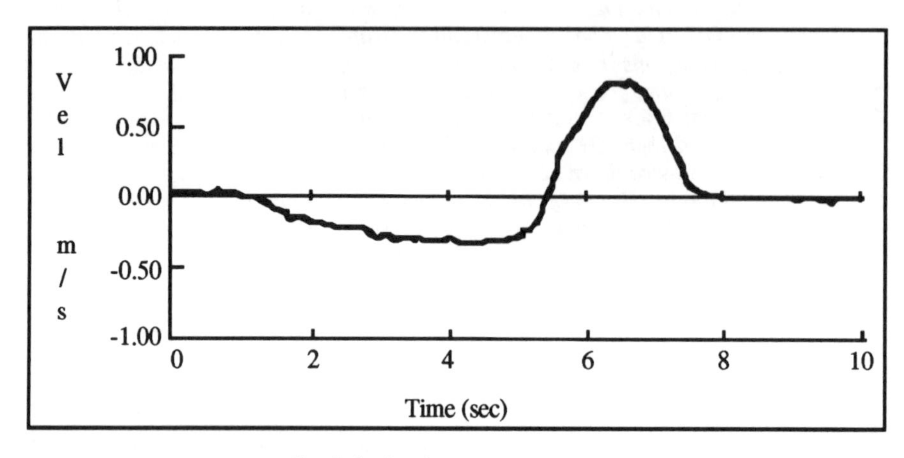

Graph 7. (on the computer screen)

Ken pondered the computer Graph 7:

11	*Ricardo:*	*What do you think?*
12	*Ken: ...*	*Um, it's kind of the same, same general direction, I guess, as*
13		*mine [Graph 6g].*

3.1.2. *Analysis*

In this episode, Ken constructed a narrative describing the motion that I had produced with the toy car. His narrative mediated both his interpretation of Graph 5 and his production of Graph 6. Lines 1-10 reflect Ken's constructive process. He structured the narrative in several phases; let us follow them.

- *Phase 1.* "I started off, um, with no, no speed for about two seconds" (lines 2-3). This is the first phase of the narrative, that is, the car is still. It corresponds to a horizontal piece on Graph 5.
- *Phase 2.* "And then, um, kind of, uh, OK, yeah, this is going backwards" (lines 3-4). This phase is about the car going backwards, towards the motion detector. Ken faced the problem that his Graph 6a did not include a space for negative velocity. He proceeded to redo the graph. I call this type of process of reorganization, *repair*. Repairs are different from adjustments because they respond to a felt need to reorganize the narrative as a whole. Whereas an adjustment might involve a local modification of the graph, such as to make it somewhat more steep or curved, a repair entails the revision of the graph/narrative as a totality.

  Ken repaired the lack of space for a negative velocity by drawing Graph 6b. Then he continued the narrative on Graph 6c: " ... it went backwards for about, uh, until six, kind of a slow acceleration" (lines 5-6). In Graph 6c, Ken expressed that the car continues to go backwards until 6 seconds. At this point, he found on Graph 5 that, right after 6 seconds, the car started to move forward (away from the motion detector), which meant positive velocity, but then he realized that he was far below the positive area. At 6 seconds, Ken needed to be closer to the positive side. He made two repairs in order to reach the positive side, which he expressed as, "it slowed down, slowed down about six" (lines 7-8), while drawing Graph 6e.

- *Phase 3.* "And started going forward again" (line 8). In Graph 6f, Ken showed a going forward until 8 seconds. But then, in order to end the narrative, Ken recognized that he must stop the car at about 8 seconds. He proceeded with another repair bending down the curve on Graph 6g.
- *Phase 4.* "And finished there." (line 9). The beginning and ending points of a narrative are always significant. The narrator has to decide on them so that everything that is relevant is included. Ken chose as the start of the narrative the time interval (2 seconds) of stillness, whereas the end was a single point in time (at about 10 seconds, according to Graph 6g) in which the car "stops." Diagram 1 shows Ken's narrative construction in a schematic manner:

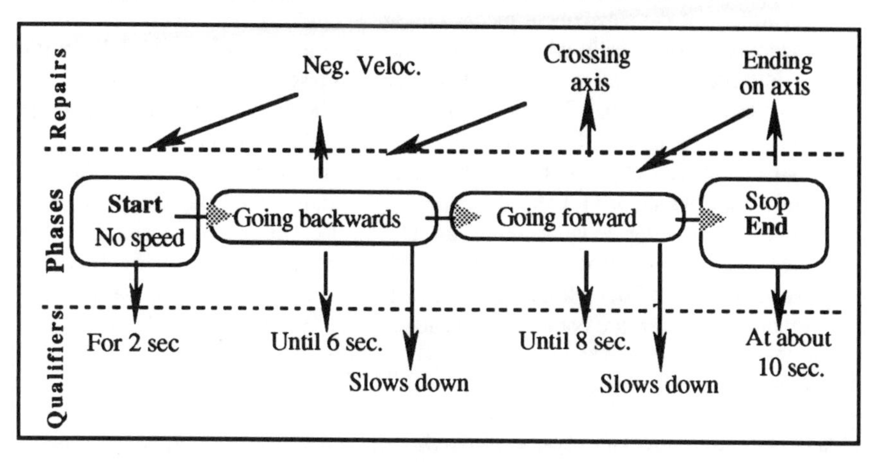

Diagram 1.

Ken split the phases based on motion directionality: backwards/forward/still. But besides these critical features, Ken also highlighted other qualifiers of the narrated motion of the car: temporal markers were a class of qualifier important to Ken. Note the different forms of the temporal markers: "for two seconds" (length of a time interval), "until six seconds" (upper limit of a time interval), "until about eight" (imprecise temporal limit). Another relevant qualifier was the condition of "slowing down," which emerged through repair. Ken was driven to slow down the car at the end of the second and third phases in order to get ready to begin the next phase.

The linguistic connectives that Ken used to link the successive phases were "and" and "then" (line 3, phase 1->phase 2; line 7, phase 2 -> phase 3; line 9, phase 3 -> phase 4). "And" and "then" as well as their combination "and then" are connectives frequently used in English narratives for indicating just chronological sequence ("so," "because," and "but" have causal or contrastive implications; see Peterson & McCabe, 1991). However, Ken's narrative is not just a chronology of phases; it reflected the continuity of the velocity versus time graph. The phase of "going forward" was not just what came after "going backwards." These two phases were also connected by a continuous curve that had to cross the horizontal axis on the velocity versus time graph, which imposed the need to insert a "slowing down" before changing direction. The continuity of the velocity curve, and how it traverses the graphical space, brought up crucial issues of connectedness that moved Ken to repair his narrative.

As a summary of this analysis, I will highlight the following general ideas:

- Ken constructed his narrative by splitting the motion of the car into phases demarcated by changes of directionality. He developed each phase following the chronology of its occurrence, as he noticed them while analyzing Graph 5, but, as he tried to incorporate the next phase, he needed to repair the narrative. Some qualifiers (like temporal markers) appeared from the beginning, others (like "slowing down") emerged through repair. The appearance of repairs--reorganizations of the narrative as a whole--expresses the fact that narratives are *not* constructed like puzzles, in which each new piece that "fits" stays in place regardless of the subsequent pieces that are assembled; narratives, instead, are meaningful totalities in which the assimilation of a new element may demand a re-interpretation of all the previously elaborated episodes.
- The connectedness that Ken developed between the different phases is based not only on a chronology, but also on the continuity of a curve in a graphical space that demarcates regions, such as positive/negative, before/after, above/below, and so on, which are traversed by the graph along certain trajectories such as up/down, slowing down/speeding up, and so on. The phases or episodes that Ken articulated introduced elements of discontinuity (like going forward, backwards), without violating his awareness that the graph, as well as the motion of the car, are continuous events.

3.2. *Laura*

This episode introduces the notion of playing out a narrative. It supports the thesis that playing out a graph, that is, performing imaginary or real actions according to one's understanding of a graph, is, above all, enacting one's narrative. Contrary to the assumption that the use of a tool like the motion detector converts the production of graphs into a process of sensory-motor feedback, this episode shows that the construction of a graph through kinesthetic actions is mediated by the narrative with which one makes sense of the graph. It also illustrates how a mathematical narrative may express an attempt to account for global features of a graph. The problem that Laura tried to solve was the production of a graph of distance versus time by moving a toy car with her hand.

3.2.1. *Description*

14 *Ricardo: You have a straight line and then another line with different slope.*
15 *[draws Graph 8] . . .*

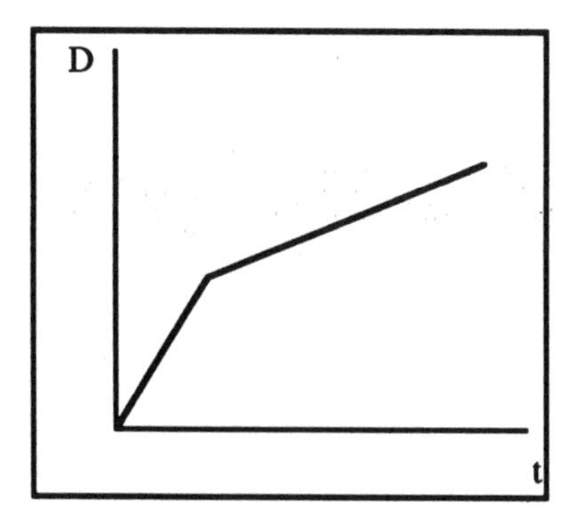

Graph 8.

Laura started to analyze the motion depicted in Graph 8.

16 *Laura: Alright, ahm, if you start it [the car] forward [points to the initial*
17 *position on the table, close to the motion detector], ... you go back*
18 *[gestures moving away from the detector] it's positive [the*
19 *velocity]. I think that's what we said. ... and then right here [when*
20 *the slope changes, Graph 8], it [the car] goes farther away so I*
21 *assume that with that bend [the change of slope in Graph 8], maybe*
22 *it [the car] would stop ... I think you would stop for a quick second*
23 *and then go back some more to get that, that edge or maybe you*
24 *would go off to the side more [gesturing the car going to the side,*
25 *see Diagram 2] ... can you, there's a [computer] record [when the*
26 *car goes to the side] or just on the straight line?*
27 *Ricardo: Just on the straight line.*
28 *Laura: OK, so it [the car] just does a straight line.*

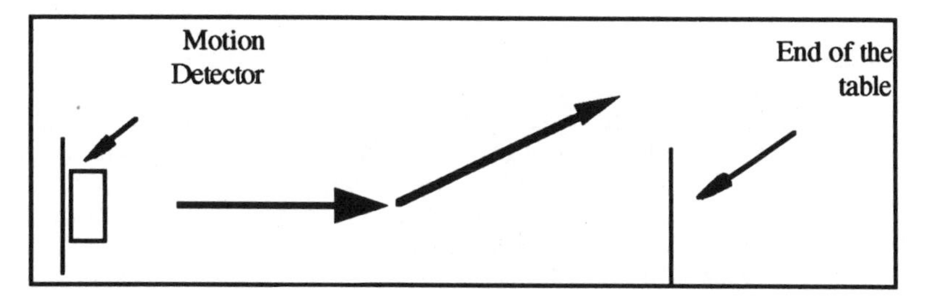

Diagram 2.

29 *Laura: So maybe you stop and then it goes back further [away from the*
30 *motion detector].*

Then Laura moved the car trying to generate Graph 8. She moved the car according to Diagram 3. Note that she moved the car with approximately the same speed before and after the stopping interval; the distance graph that appeared on the computer screen is similar to that of Graph 9.

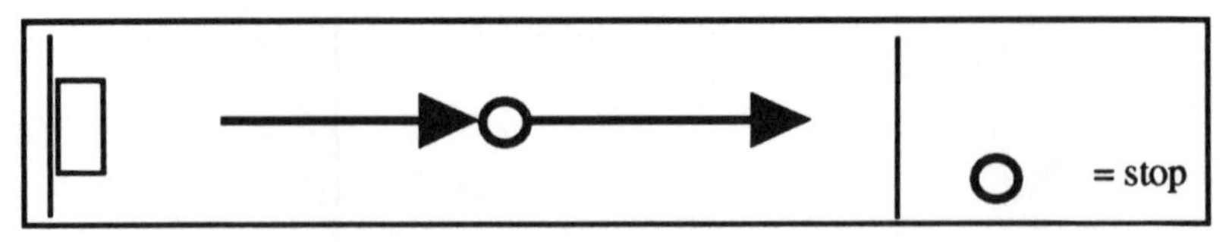

Diagram 3.

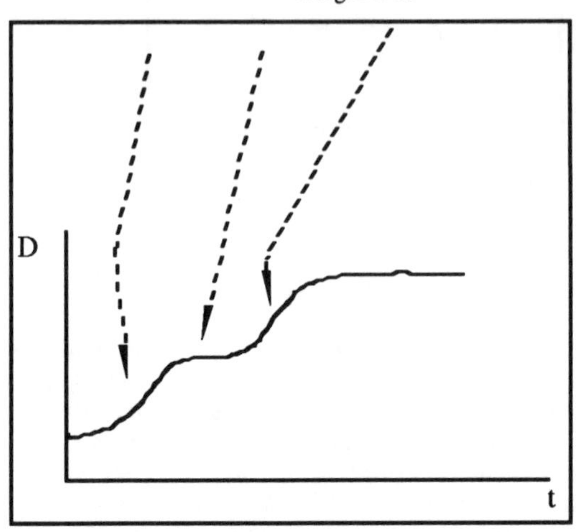

Graph 9. (on the computer screen)

31 *Laura: Hmm, I don't know, no, that [Graph 9] ain't nothing, that's not*
32 *anything like it [Graph 8]. I can't figure out how to get that one*
33 *point [change of slope, Graph 8].*

Laura made six other attempts with similar results. After the second attempt I asked her:

34 *Ricardo: What would happen if you stop here at this point? [when the slope*
35 *changes on Graph 5]*
36 *Laura: The computer would just go straight [gesturing horizontal]. I,*
37 *yeah, it [the distance graph] would just go straight [horizontal], I*
38 *would think 'cause you're not moving, so we can't measure any*
39 *distance . . . but [the car has to stop] just, just for a, a quick second*
40 *just to get it [the distance graph] to go that way [with the edge, as in*
41 *Graph 8]. I think. Maybe, it's an idea. Let me try it again.*

After her sixth measurement Laura said:

42 *Laura: I can't seem to get the edge, that point.*
43 *Ricardo: Let me try. [Ricardo moves the car generating a computer graph*
44 *similar to Graph 10.]*
45 *Laura: That's closer. I can't seem to get that edge, I don't know why.*
46 *Ricardo: What am I doing differently than what you did?*
47 *Laura: Well, at the beginning you waited a second and then you just, I don't*
48 *know, you, you didn't stop like I did. I stopped in the middle to try*
49 *to get the edge right here.*

50 *Ricardo: Yeah, you did something like this [Ricardo acted out the car*
51 *motion stopping in the middle, like in Diagram 4]. Okay, let's try,*
52 *try it once more.*

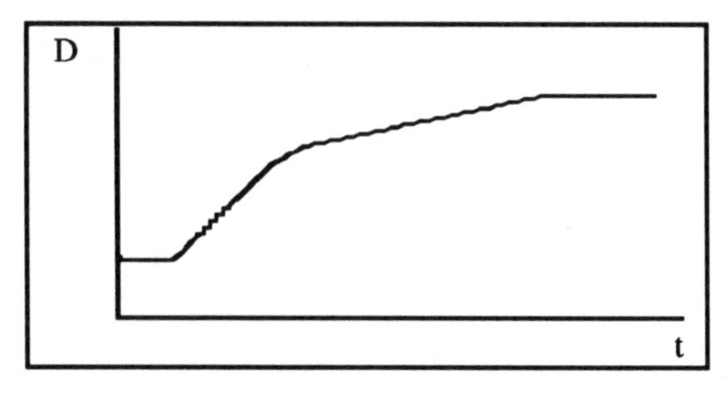

Graph 10. (on the computer screen)

Now Laura moved the car without stopping in the middle but without getting the edge either. (See Graph 11.)

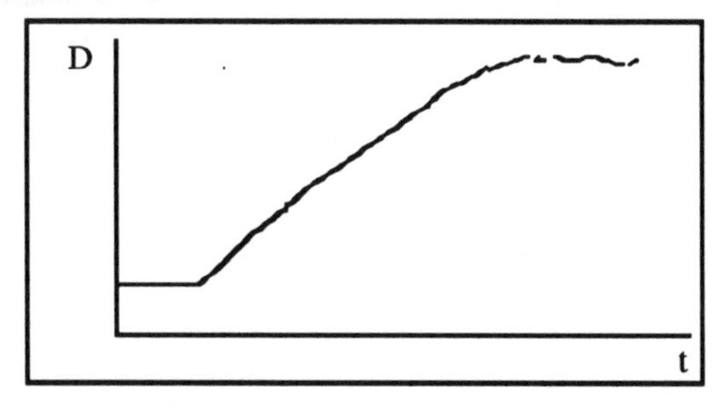

Graph 11. (on the computer screen)

53 *Laura: But if you don't stop then you don't get, it [the distance graph] just*
54 *keeps slanting, you know what I mean? That's why I was stopping*
55 *to try to get the slant. I don't seem to get a slant if I don't stop.*

3.2.2. *Analysis*

Laura first interpreted Graph 8 by constructing a narrative (lines 16-22), which is schematically described by Diagram 4:

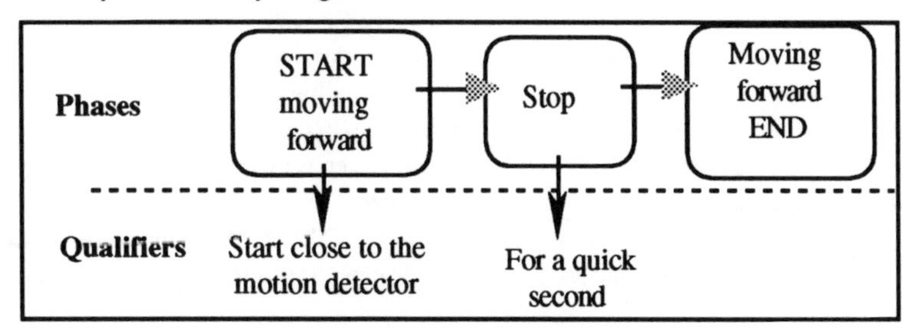

Diagram 4.

Then she tried to elaborate an alternative narrative based on the idea of the car "going to the side" (lines 22-26), an idea that she dismissed after I asserted that the motion detector measures only on a straight line (lines 28-30). This was an attempt to interpret the graph as a picture of the car's motion. The tendency to interpret graphs as a picture of the represented events (Janvier, 1978) seldom occurs with high school students using the motion detector, precisely because of the constraint of the measurement on a straight line.

After articulating the narrative represented in Diagram 4 (lines 31-33), Laura tried several times to produce a distance graph similar to Graph 8. Her repeated failure to produce Graph 8 shows that creating a graph with kinesthetic actions is not a matter of sensory-motor feedback, that is, of adjusting one's movements so that the graph appearing on the screen looks like the expected one. Producing a graph with kinesthetic actions is, above all, playing out one's narrative.

Furthermore, one is "inside" one's own narrative. For example, even if one is shown how others' playing out produces the "right" graph, such as when I showed Laura the production of Graph 10 (lines 45-52), that does not necessarily make evident the implicit narrative underlying the other's enactment. Laura noticed that I had not stopped, but the puzzle remained: "But if you don't stop then you don't get it" (line 53). What *else* could produce the slant? Laura knew that a still car would produce a horizontal distance graph (see lines 36-39), "but [the car has to stop] just, just for a, a quick second just to get it [the distance graph] to go that way" (lines 39-40).

The overall intention in Laura's narrative is to make sense of what she called "the edge" of Graph 8. This is a reminder that a narrative, beyond a chronology of phases, aims to illuminate global aspects underlying the narrated phenomena. The edge of Graph 8 is not an isolated event in time. The narrative reflects the structure of Graph 8 as a whole. It must account for what is similar and different between the previous and the ensuing graphical pieces. For Laura, the two pieces were the same because she was focusing on the directionality of the motion, and both pieces corresponded to moving away from the motion detector. So she attempted to introduce a third element in between them, one that had to be very quick. At that point during the session, Laura was framing all her stories of motion by highlighting the quality of directionality (forward/backwards/still), and because a phase of backwards was not possible (Graph 8 does not go down), she resorted to stillness.

3.3. *Steve*

The following episode exemplifies the process of idealization, namely, the use of a narrative to assess whether two graphs, in this case a hand-drawn graph and another generated with the motion detector, are equivalent--whether they tell the same "story." Idealization is the use of a narrative as a ground to recognize sameness. Steve deems that the two graphs, which look radically different from a purely visual point of view, are comparable on the basis of the narrative with which he made sense of the situation. The episode also shows Steve's awareness that empirical results indicate possibilities but not logical necessities.

3.3.1. *Description*

I posed to Steve the problem of two cars that start at the same position and move according to Graph 12:

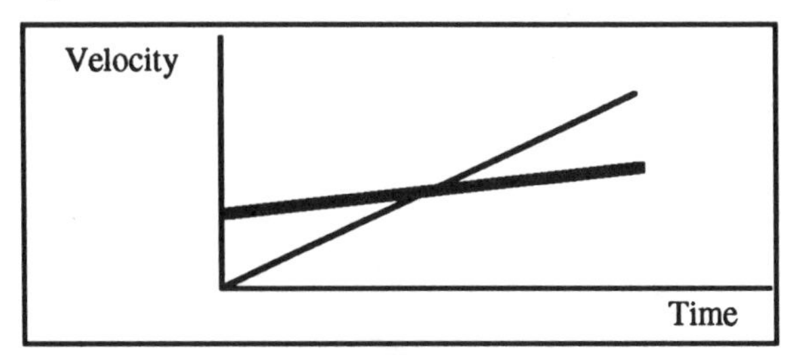

Graph 12.

56 *Ricardo: The question is about where they would meet.*
57 *Steve: I would say since the one that ends up going faster, larger velocity,*
58 *starts at a slower speed than the one, well, they would end up meeting*
59 *there [at the crossing point, in Graph 12]. That's where they would*
60 *end up being side by side if they started at two separate places and*
61 *ended up side by, they would end up side by side at that point where*
62 *the lines intersect.*

Then I asked Steve to produce Graph 12 on the computer screen by moving the toy car according to one of the lines and then according to the other. Steve measured five times trying to produce straight lines on the velocity graph. The last two measurements looked on the computer screen like Graph 13:

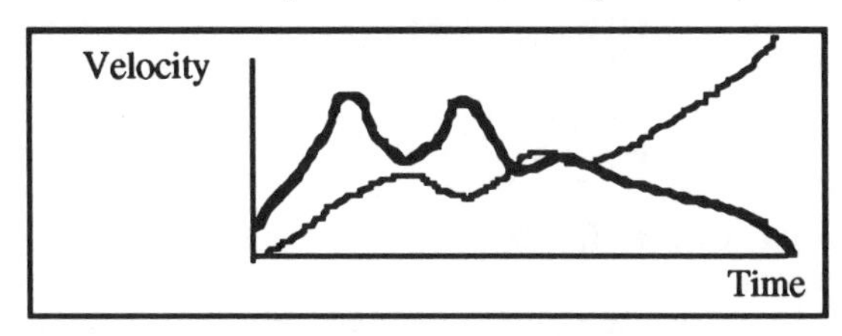

Graph 13. (on the computer screen)

63 *Steve: Hm. It's hard to work with velocity I'm finding.*
64 *Ricardo: Then the problem is how you get something.*
65 *Steve: ... How would I get a straight line. That's what I'm having problems*
66 *with, a steady hand, getting increasing speed, increasing velocity.*
67 *[But] more or less. This [black curve, Graph 13] started at a greater*
68 *speed and ends up crossing [the gray curve, Graph 13] at a point; so*
69 *this one [gray curve, Graph 13] starts at zero as this one [gray line,*
70 *Graph 12] did. This one [black curve Graph 13] starts at a slightly*
71 *higher speed and increases in velocity more or less as this one [gray*
72 *curve, Graph 13] is, but just not as great, overall it's [black curve,*
73 *Graph 13] not increasing in velocity as much as the first one [gray*
74 *curve, Graph 13] I did. So we can see that . . .they should meet there*
75 *[pointing to the first crossing between the two curves on Graph 13].*

Then Steve asked for the corresponding distance graph. Both graphs, velocity- and distance-time, appeared on the screen as in Graph 14:

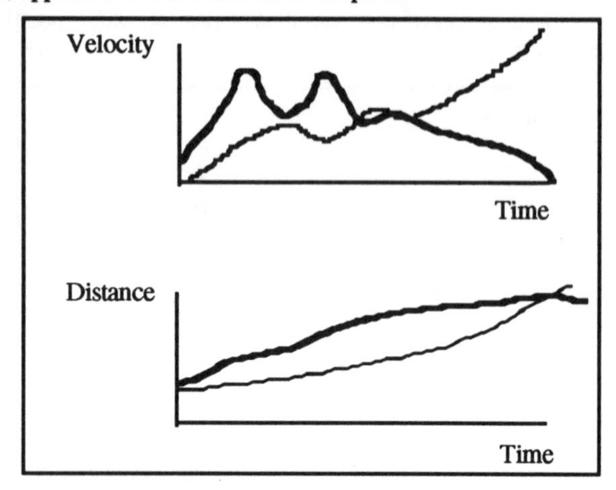

Graph 14. (on the computer screen)

76 *Steve: Oh. Hm. That's odd [observing that the distance curves do not cross at*
77 *the same time as the velocity curves].*
78 *Ricardo: They met here [pointing to the crossing point between the distance*
79 *curves].*
80 *Steve: Yeah, they met way up there. That's. Oh, well. Oh!, I get it. Yeah.*
81 *The reason they don't meet there [crossing velocities point, Graph 14]*
82 *is all of that means when they intersect is that they're going the same*
83 *speed. So that doesn't mean anything. That just means they're going*
84 *the same speed. I could be a hundred miles away from you and still be*
85 *going the same speed as you. I guess the reason they intersected up*
86 *here [crossing position curves, Graph 14] was because this one [black*
87 *velocity curve, Graph 14] stopped . . . And this one [gray velocity*
88 *curve, Graph 14] didn't stop or kept on going. So that seems to be the*
89 *reason.*
90 *Ricardo: . . . Now you don't think that they meet here? [At the crossing*
91 *velocities point, Graph 12]*
92 *Steve: Well, it's possible I just made a mistake. I mean I could have just*
93 *messed that [the computer graphs] up enough so that it didn't give*
94 *accurate results but no, I don't, they shouldn't meet there [at crossing*
95 *velocities time, Graph 12] ... No, they should meet there . . . [long*
96 *pause].*
97 *Ricardo: How do you imagine the two cars? [Steve takes two objects that*
98 *will stand for the cars and puts them side by side]*
99 *Steve: This one [Black car] would sort of take off. It would be going already*
100 *at a certain speed. . . . So it would be going like this and increasing*
101 *the speed slowly [acting out only Black car]. Whereas this [Gray car]*
102 *would start out slow and then start going real fast [acting out only*
103 *Gray car]. And at some point they'll both be going the same speed*
104 *[putting the two cars side by side];however, this one [the Gray car]*
105 *will be going around here [behind the Black car; when they are going*

106 *the same speed].And eventually this [gray car] will catch up [the Black*
107 *car] but I don't know if I really know when they would meet. I guess,*
108 *I think it would involve drawing, for me drawing a diagram of*
109 *distance, I guess. Just thinking about it in my head I can't really*
110 *picture it when they would.*

3.3.2. *Analysis*

Steve's prediction that the two cars would meet at the crossing velocities time of Graph 12 was based on his identification of equal velocity with being side by side: "That's where they would end up being side by side if they started at two separate places and ended up side by, they would end up side by side at that point where the lines intersect" (lines 59-62). This is a generalization of a very common experience. The most immediate way to recognize that two objects, or oneself with respect to something else, are moving with the same velocity is the perception of moving side by side.

Steve's words in lines 57-62 and his attempts to produce Graph 12 with hand motion expressed a narrative articulating four phases represented in Diagram 5:

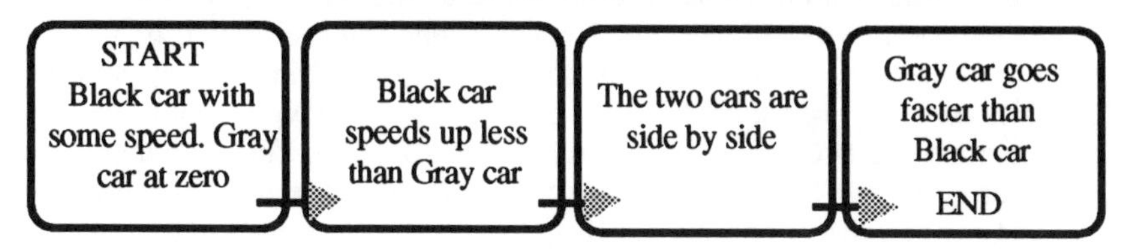

Diagram 5.

After facing difficulties in trying to produce uniformly accelerated motion, Steve pondered whether Graph 13 could be taken as an appropriate instance of Graph 12. Graphs 12 and 13 had very different meanings. Graph 12 meant two crossing straight lines indicating constant acceleration. Graph 13, on the other hand, meant the actual movement of Steve's hand. What is the basis on which it can be decided whether his hand motion represented the same motion prescribed by Graph 12? One is tempted to think that it is a matter of judging closeness or resemblance. But resemblance is not enough. There is no limit to how similar a kinesthetic action can be to an ideal target. How does one decide whether it is *close enough*? The point that I am trying to make is that in many circumstances, such as Steve's, judgments of "close enough" are based on the critical qualities highlighted by a narrative. The kinesthetic action is viewed as valid according to a criterion of relevance structured by the narrative. For Steve, Graph 13 is a befitting instance of Graph 12 because of the following.

- The Black curve "started at a greater speed" (lines 67-68); the Gray curve "starts at zero" (line 69).
- The Black curve "increases in velocity ... but overall not increasing as much as" the Gray one (lines 71-73).
- The Black "curve ends up crossing" the Gray curve (line 68).

I define idealization as the use of a narrative to reckon whether different representations are equivalent, in other words, as the process of envisioning an "ideal type" of graph by recognizing that all of them tell the same story. Steve idealized

Graph 13 as being equivalent to Graph 12 because, he deemed, both reflect essentially the same narrative.

After observing Graph 14, Steve noticed with perplexity: "That's odd" (line 76) that the cars did not meet when the velocities intersected. The almost immediate insight that "I could be a hundred miles away from you and still be going the same speed as you" (lines 84-85) allowed Steve to make sense of the computer graphs, but reopened the issue of when the cars meet. Steve thought that the cars met at the particular point indicated in Graph 14 because the Black car stopped, whereas the Gray one "kept on going" (lines 86-88). However, this explanation did not help him to repair his narrative because, in the motion prescribed by Graph 12, the Black car never stops.

In lines 92-96, Steve expressed his struggle to repair his understanding. Through his experimentation with hand motion, he became aware that it is possible not to be side by side in the condition of equal velocity and yet, it can happen: "It's possible ... I mean I could have just messed that [the computer graphs] up enough so that it didn't give accurate results" (lines 92-94). Experimenting with kinesthetic actions had an important role in Steve's envisioning of new possibilities and yet, at the same time, Steve deemed that the empirical result was not a logical necessity.

At the end of the episode, Steve began acting out the motion of the cars sequentially. First, he moved the Black car (lines 99-101), then the Gray car (lines 101-103); after this, he said: "And at some point they'll both be going the same speed [putting the two cars side by side]" (lines 103-104). This first impulse to reenact the going side-by-side condition was immediately corrected: "However, this one [the Gray car] will be going around here [behind the Black car]" (lines 104-105). Steve's new awareness, that when the cars' speeds are equal the Gray car is behind, moved him to complete the repair of his narrative: "And eventually this [Gray car] will catch up [the Black car]" (lines 106-107); that is, the cars meet after the time in which their velocities are equal. However, Steve does not feel yet that this has to happen in general: "But I don't know if I really know" (line 107). He then suggested drawing the distance graphs.

3.4. *Repair, Playing Out, and Idealization*

The three episodes with Ken, Laura, and Steve were about students who constructed narratives that united the motion of a toy car with symbolic expressions in the form of graphs. By repairing, playing out, and idealizing experimental curves, they strove to get a sense of how well their narratives could account for physical and symbolic events, that is, for the motion of the car and the related graphical patterns. As shown in the episodes, both the playing out and the idealizing are mediated by a narrative and, at the same time, they may create the need to repair the narrative itself. Repair, playing out, and idealization open up a space for sharing expectations and puzzlement between the narrator and, in this case, the interviewer; such processes thereby create important conditions for the co-construction of narratives.

Constructing mathematical narratives can foster very diverse experiences. For Ken, the successive repairs led him to a general confirmation of his expectations. In the case of Laura, it made evident to her that her narrative was not "working," prompting her and me to shift the problem of producing Graph 8 to the background and to start work on the meaning of steepness on a distance time graph. Steve's episode led him to revise distinctions like "having the same speed" and "going side by side." The centrality of mathematical narratives--their role in students' use of symbolic expressions--emerges throughout this diversity.

The construction of mathematical narratives has several implications for students' learning:

- It represents a shift from the usual focus on internal relationships between variables (e.g., time and distance) toward the connections between narratives and traits of symbolic expressions (e.g., a change in the direction of movement and a vertex on a distance vs. time graph). The examples with Ken, Laura, and Steve illustrate this phenomenon. Their struggle did not center on pairing time and distance, but on figuring out how narratives and graphical shapes correspond to each other. Ken wanted to change the velocity sign to reflect a change of direction in the car's motion, Laura faced difficulties in constructing a narrative that could account for "the edge," and Steve repaired his narrative to reflect a new meaning for the graphical crossing of lines.
- Variables in a narrative are *not* experienced as isolated and generic measures; they are situated in real or imaginary actions that combine elements of change and permanence. For example, to interpret Graph 8, Laura did not isolate "distance to sensor," and "time." What she saw in the graph was a movement away from the sensor; this action involved change (e.g., getting farther from the sensor) and permanence (e.g., the direction of motion is always away from the sensor). Similarly, Steve made sense of the changes in velocity portrayed in Graph 12 by imagining the motion of two cars, losing ground, catching up or, at times, being side by side.
- There are always many possible narratives to account for the same situation depending on what aspects are shifted to the foreground. Any narrative is a version, a possibility among others. But some narratives are more appropriate and compelling than others. For instance, in Laura's episode, she articulated narratives with phases of moving-away/still/moving-closer; the slow/fast distinction was not, at that time, an aspect that she incorporated into her narratives. She could account for some graphical attributes, such us increasing/constant/decreasing, but not for the change of slope. Later, when Laura enriched her narratives with comparisons of speed, she experienced a sense of learning and growth. In another example, the continuity of the velocity versus time graph made compelling to Ken the addition of phases of slowing-down/speeding-up.
- Even when the explicit layer of a narrative--the utterances or displays--seems to depict a linear succession of changes along a single dimension, a narrative accounts for phenomena that interrelate many dimensions; it reflects a complex web of changes and permanencies. For instance, I created Graph 8 by moving my hand back and forth. Graph 8 expressed changes over time of the distance between my hand and the motion detector among many other aspects. The movement back and forth of my hand encompassed changes in velocity, acceleration, directionality, average distance, and many other quantities. Understanding a narrative involves recognizing the unspoken. Ken's Graph 6 of

velocity versus time revealed implicit events portrayed by Graph 4, such as the slowing down before changing direction.

4. MATHEMATICAL NARRATIVES AND EARLY ALGEBRA

Children construct and interpret mathematical narratives from an early age. We have designed and implemented classroom activities for 9- and 10-year-old children (Tierney et al., 1994). They created and discussed representations for the motion of a car, the growth of a plant, the number of people in their homes over time, and so forth. In one of the activities, called "Mystery Graphs," each pair of children create a qualitative graph to show how the number of people at a familiar place, such as the school or a supermarket, changes over 24 hours. The graphs are displayed and the class discuss what is the place represented in each graph. Through the class discussions the students interpret many aspects of the graphs, such as maxima and minima, abruptness of increases or decreases (e.g., the number of people at the school increases more sharply in the morning than it decreases in the afternoon), or the distinction between zero and few people.

We noticed that, initially, children focused on the overall shape of the graph or the number patterns, disregarding issues of scale and regularity of the observations. For instance, after they had collected measurements of the height of a plant every weekday, they created graphs that excluded the weekends; consequently the graphs seemed to show a steep growth from Friday to Monday. In graphing the number of people in his house, a child decided to exclude "1" from the axis indicating number of people because "there is never one person at home." Some children were reluctant to mark the axis of time according to regular intervals because information would be lost (e.g., because the child left her house at 7:55, she wanted to make explicit this information, and not use 8:00 on her axis). The students wanted to be accurate and true to the particulars of their stories. Issues of systematicity and scale became explicit themes of classroom discussion when students were confronted with interpreting each other's representations, with merging several of them into one, and with incorporating new data.

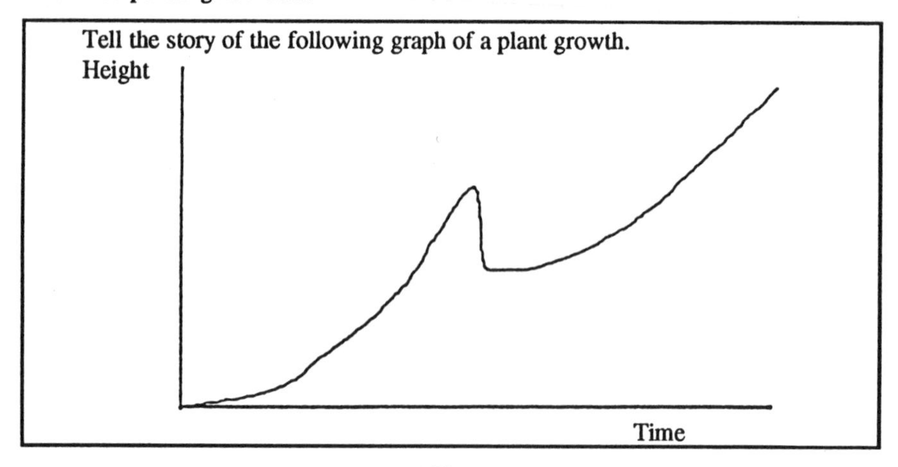

Figure 1.

In these activities, students played out narratives by imagining the events described by a graph or a list of numbers. For example, given the graph of the height of a plant versus time (see Figure 1), a student imagined the following story: "The plant started slow, the [sic] went fast, the [sic] all of a sudden it dropped (it most likely fell off) then it started up slow then it went very fast."

Students' use of narratives to idealize--to assess whether different representations tell the same story--took many forms: from the sketching of an overall shape that accounted for the discrete series of measurements of plant growth, to debates on whether different variations of the "Mystery Graphs" are equivalent versions of the same events. Repair often emerged as a response to difficulties and misunderstandings in the sharing of narratives among peers.

Other educators have also reported and discussed activities in which students interrelated stories with graphical and numerical representations, creating diverse ways of combining cultural conventions with their idiosyncratic inventions (Chazan & Bethell, 1994; diSessa et al., 1991; Krabbendam, 1982). How does this strand of experiences relate to algebra? If we frame our understanding of the nature of algebra by the dialectic between mathematical generalization and specialization (Mason, Part II of this volume), it becomes clear that early algebra must have its roots in the experiential domains that children bring to the meaningful encounter with mathematical generalization. It is through the construction of mathematical narratives--our interpretation of "modeling"--that children's experiences with change, that is, with the different ways in which change occurs, become the subject of mathematical generalization.

By elaborating on their experience with different situations of change, children may encounter rich opportunities to develop their sense of variable as a manner of continuous variation. For example, a number series such as "1, 5, 8, 10, 11, 11, ...," standing for number of blocks in a bag, number of people in a room, successive distances from a point, or any other phenomenon, could motivate descriptions like "it started increasing a lot, then increased less and less, until it remained constant." In this way the former number series might be felt as equivalent to "9, 15, 19, 21, 22, 22, ..." and different from "1, 4, 7, 10, 13, 16, ..." This kind of activity can nurture a sense of variable that is not confined to letters with unknown values, but one that opens up a field of mutually related modes of variation.

How can one nurture children's conversations on the generalization of change? What symbolic forms may become expressive to children in talking about the different modes of change? One approach that we, as well as Yerushalmy and Schwartz, have pursued is based on the idea of a "grammar" of graphical shapes (Nemirovsky, 1992). For example, the distinction between "growing steadily" and "growing but slowing down" may be expressed with these two iconic shapes:

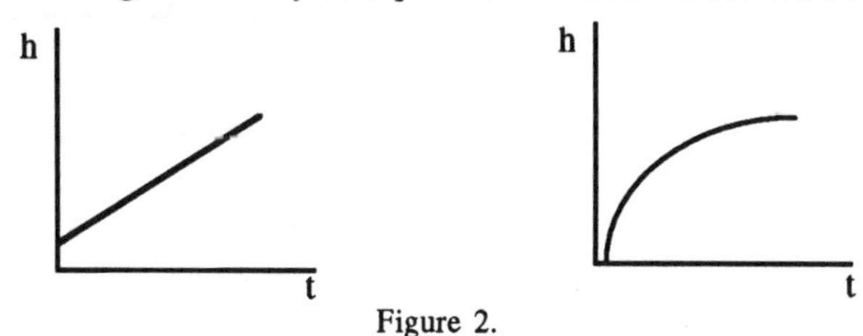

Figure 2.

The following are two examples (see Figures 3 and 4) of children's use of the graphical shapes in a posttest that we applied after the activities on plant growth.

John, Susie, and Emma measured their plants over five consecutive days and graphed the heights. They got graphs that looked like these:

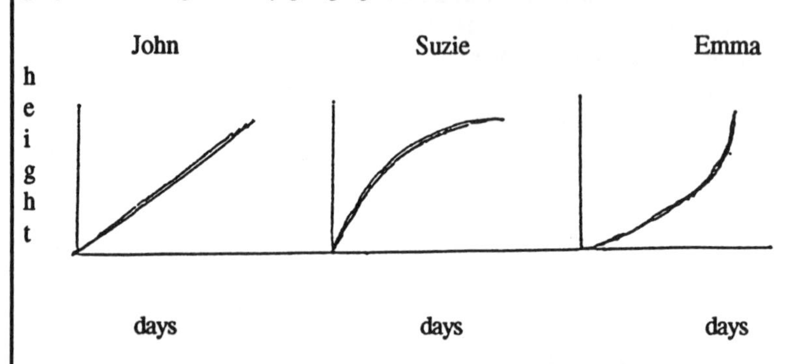

Can you guess the heights of their plants over the five days? It is not important to be accurate, just think of heights that could follow these graphs.

John		Susan		Emma	
day	height	day	height	day	height
1	3	1	2	1	1
2	4	2	4	2	3
3	51/2	3	51/2	3	31/2
4	71/2	4	6	4	51/2
5	9	5	6	5	6
It grew steady		It wasn't steady. First it grew kind of tall, then, it started slowing down, then it stopped.		It started growing slow in the beginning. Then it started growing kind of steady. It started growing fast up here (It was steady from 3 to 3 1/2). A straight line is steady.	

Figure 3.

Plants

Zachary measured the height of his plant over five consecutive days. He got the following measurements.

Day	Height
1	15cm
2	18cm
3	20cm
4	21cm
5	21 1/2cm

Before graphing his data, Zach wants to know the overall shape of the graph. Which one of the following will be most similar to his graph?

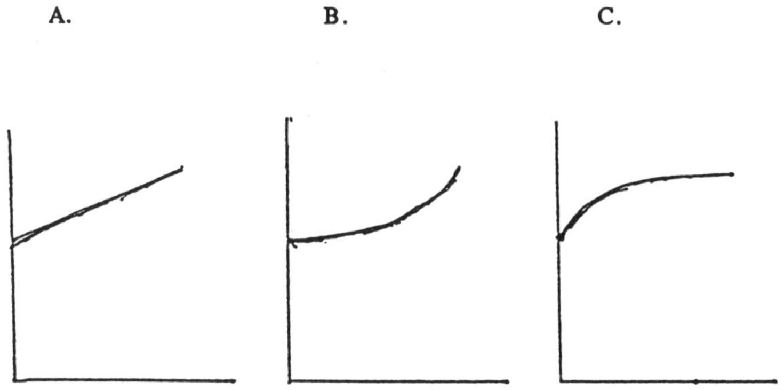

A. B. C.

C, because it started really **booming** and then it slowed done. It jumped 3cm (a really big jump), then 2cm, then 1, then 1/2. It is levelling off.
(Could it be A?)
For A it's growing steadily at the same pace, like 1cm per day.

Figure 4.

At this point an important issue that calls for reflection is the role of the equation-type of notation, of the expressions with x's and y's, that have traditionally been identified with algebra. Within the repertoire of symbolic forms that can be used to create and interpret mathematical narratives, textual notations are clearly distinct from graphical shapes or number sequences. With graphs or tables, the process of narrative construction involves a reading act in the sense that new phases appear as one traces the graphical shape, or the number table, along a certain dimension (e.g., left to right, up-down, clockwise, etc.). Textual notations do not admit literal readings in terms of stories. Some of the best documented mistakes in students' use of equations stems from the tendency to "read" the equation from left to right (Clement, Lochhead, & Monk, 1981; Kaput & Sims-Knight, 1983).[4]

In contrast to the construction of narratives emerging from the reading of a graphical shape or a number sequence along a certain dimension, equations conceal their stories in powerful but intricate ways. And yet, it can be argued that an important aspect of understanding an expression, such as $1/(x-3)$, is to recognize that a crucial symbolic event happens at $x = 3$, delimiting two modes of change that take place "before" and "after" $x = 3$. This analysis suggests the importance of graphical or numeric representations in making comprehensible the narratives articulated with textual expressions; accordingly, students' fluency with graphical and tabular representations seems to be necessary for their use in exploring the meanings of textual notations.

In this chapter, I have explored ways in which modeling--interpreted as the construction and interrelation of mathematical narratives--may become one of the avenues for the introduction to algebra. A mathematical narrative fuses events and situations with properties of symbols and notations. It is possible to argue that combining the introduction of algebra with the experimentation on phenomenologies of change, such as motion, growth, or number of people, makes the learning process more complicated because the students would have to learn about them *in addition* to learning about functions, graphs, equations, and so on. However, the point is that students have rich and personal experiences with motion, growth, number of people, and other phenomenologies. This experiential background, instead of being an additional subject, may provide unique contributions to students' understanding of algebra. Mathematical generalization does not occur just because one uses a generic "x" or "y." It has to also be grounded in the exploration of multiple contexts about which students have expertise, expectations, and ways of talking.

NOTES

1 The research reported in this paper was supported by the National Science Foundation grants MDR-8855644 and MDR-9155746. Opinions expressed are those of the author and not necessarily those of the Foundation. The author wishes to thank David Carraher, Dan Chazan, Cliff Konold, Steve Monk, Analucia Schliemann, and Cornelia Tierney for their valuable feedback based on previous versions of this paper.

2 A reference to the students'use of narratives in the interpretation of graphs is included in Nemirovsky and Rubin (1991).

3 An extended analysis of the teaching experiments with Ken and Laura can be found in Nemirovsky (1993). The episode with Steve is briefly referred to in Nemirovsky and Rubin (1991).

4 In this regard, software programming is more like graphs or tables of numbers because programs can be read as a sequence of instructions and decisions. The reading of a computer program may follow the order in which the instructions and decisions are executed.

CHAPTER 16

REFLECTIONS ON MATHEMATICAL MODELING AND THE REDEFINITION OF ALGEBRAIC THINKING

M. KATHLEEN HEID

As one reflects on the role of mathematical modeling in beginning algebra, an enriched image of algebraic thinking emerges with deeply rooted interactions between the ways in which students think about the real world and the ways in which they think about mathematical concepts. This commentary chapter, which draws on the contributions of the previous Nemirovsky chapter, raises several questions on the role of narratives in disclosing student sense-making.

In Nemirovsky's chapter, *Mathematical Narratives, Modeling, and Algebra,* a narrative is defined as "one method of recapitulating past experiences by matching a verbal sequence of clauses to the sequence of events which (it is inferred) actually occurred." A mathematical narrative is, according to Nemirovsky, "a narrative articulated with mathematical symbols," having both episodic and configurational dimensions. The central feature of mathematical narratives seems to be sense-making.

1. WHAT CAN WE LEARN ABOUT STUDENTS' THINKING FROM MATHEMATICAL NARRATIVES?

One example is offered by Nemirovsky to illustrate the misunderstandings that we might have about a student's thinking if we do not examine it closely enough. To illustrate this point, he relates an example of a fourth grader's response to the question of which of two plants represented in the following graph grew more:

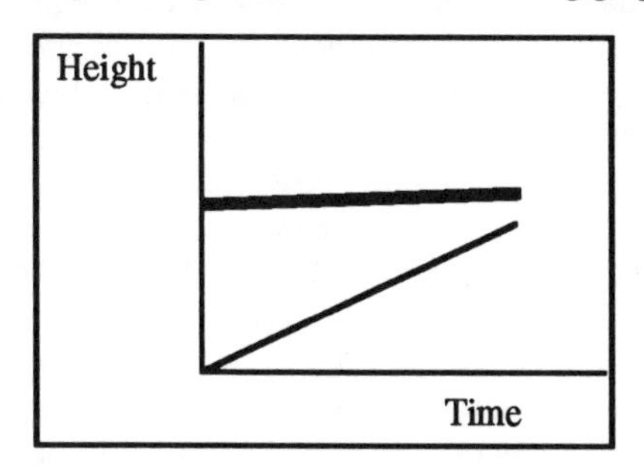

The child said the plant represented by the top line segment grew faster. The author points out that the interviewer could conclude that the child was thinking that the higher the graph the greater the growth, but then the child augmented the original drawing to explain his response.

N. Bednarz et al. (eds.), Approaches to Algebra, 221-223.

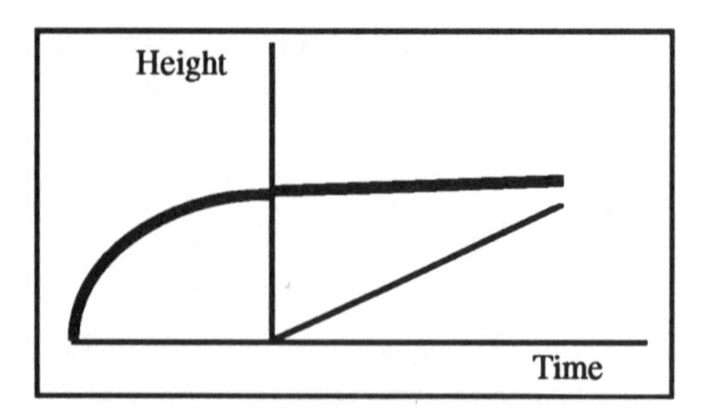

This answer is consistent with the oft-reported interpretation of the question as to which plant has a greater absolute growth, rather than which plant has a faster rate of growth. As is seen here, children may be operationalizing the knowledge that plants grow from seeds (from zero height). Without more information, however, we cannot learn more about the child's understanding or intentions. We have no evidence that the child responded as he did because he expected continuity and completeness. We can only theorize and relate a story about the child's answer that illustrates a point. Testing such a theory requires a deeper investigation.

2. HOW DO STUDENTS ADJUST THEIR NARRATIVES WHEN THEY NOTE CONTRADICTIONS? HOW DO THEY USE KNOWLEDGE OF OVERALL TRENDS?

An important point about this question is that students seem to enact a local rather than a global fix for imperfections in their models. They seem not necessarily to return to the more global aspects of the problem situation. When students have "repaired" their narratives, how (if at all) do they think about how the resulting representation fits the situation? Do they ever re-examine the whole picture? It may be that students' narratives aim "to illuminate global aspects underlying the narrated phenomena," as Nemirovsky claims, and so enhance one's ability to focus on more global issues. It seems that the narratives shared in this paper illustrate students using only a small selected part of the global aspects. Research should be conducted to determine the extent to which narratives can enable students to focus on more global features of a mathematical model.

It would be interesting to examine differences in the role of local and global fixes for students who have studied algebra through complete courses that emphasize mathematical modeling. Recent research (Heid & Zbiek, 1993) using such curricula has suggested that student responses in the face of contradictions may focus on examining and analyzing other representations with which they may be more comfortable. Under what circumstances do students look at the global picture to help them settle contradictions? Do they generally worry about apparently contradictory information or, as a student we once interviewed expressed it: "I don't wonder if nobody asks."

3. HOW DO STUDENTS USE MATHEMATICAL KNOWLEDGE IN CONSTRUCTING THEIR NARRATIVES?

How do students make mathematical sense of kinesthetic phenomena? Perhaps students' choices of what to feature in their narratives reflect their internalized understanding of mathematical relationships and concepts. Students with deeper understanding of rate of change, for example, might be more likely to focus on this concept as an important feature of the narrative.

It seems then that there is a new kind of mathematical knowledge that emerges with the matching of kinesthetic motions to graphs. To what extent are students using mathematical ideas in their attempts at matching the graphs? Does kinesthetic feedback play any part in their work? Narratives allow us to witness a struggle for dominance of mathematical ideas and have exciting potential for enhancing our understanding of mathematical modeling and non routine problem solving.

4. TO WHAT EXTENT DOES THE USE OF A COMPUTER CONFINE STUDENTS' MATHEMATICAL CONSTRUCTIONS?

In Laura's interview (Nemirovsky, this volume), she suggested an alternative representation of the motion of the car--motion at an angle. The program and the interviewer steered her away from that representation because the car had to move along a single line. It might have been interesting to engage Laura in a conversation about the consequences of her construction.

This introduction suggests that two function rules are equivalent if they describe exactly the same story of events. Is the converse also true? What does it take for two stories to be the same?

In summary, reflections on mathematical narratives have raised a number of interesting questions--all related to the centrally important issue of what does it mean to understand functions. This last aspect "has to also be grounded in the exploration of multiple contexts about which students have expertise, expectations, and ways of talking." We find here some interesting new ways of thinking about understanding--ones that may inform our understanding of the interface between everyday thinking and mathematical thinking.

CHAPTER 17

MODELING AND THE INITIATION INTO ALGEBRA

CLAUDE JANVIER

This chapter will not be concerned with the teaching of algebra per se but mostly with students' mental processes when they work on school algebra exercises. Firstly, it will attempt to describe the nature of some basic algebraic reasoning processes. Secondly, this analysis will allow a careful examination of the role of modeling in algebraic reasoning. Finally, the consequences of my considerations on modeling and algebra will be applied to examine issues that are raised by functional approaches to algebra.

1. ALGEBRA AND THE USE OF LETTERS

The discussions of the different authors in this book have clearly confirmed the view that there is no agreement on what algebra is. Moreover, they have shown that the teaching of algebra is the object of hot debate involving words or expressions whose meanings are loosely defined. The following remarks are intended to clarify the meaning of some of these.

1.1. *Letters: Unknown or Not?*

Most researchers in the domain share the view that doing algebra is not equivalent to handling letters. Indeed, letters to be considered as algebraic entities by beginners must, first of all, stand for numbers. Children can manipulate letters referring to geometric figures, points, or objects, without any algebraic mental processes being involved. This is the case when we ask a child: "What is the height of the triangle ABC?"

However, reasoning algebraically does not start when letters standing for numbers are used by students. No algebraic thinking is involved in: *Find the height h of the triangle ABC.* Actually, a letter can stand for a number without any "unknownness" being involved. I contend that this distinction is major and that it should particularly be made when formulas are handled by students. Too often, teachers and textbook authors overlook the fact that students can use the formula $A = \pi r^2$ in two very different ways. First of all, the "area formula for a circle" is used for finding the area of any circle when its radius is given. When it is utilized for such a purpose, students are in no way expected to imagine or consider the radius of the circle as an unknown. Quite the contrary, the radius may take as many specific values as provided by the exercise, but it is never expected to be unknown. In that sense, the letter r does not have, in the mind of students, the status of an unknown number. Similarly, for the letter A in the formula, students are not expected to imagine or assume that the value of A is unknown. As a matter of fact, as soon as a value is given to r, a value for A automatically exists. Researchers should agree that a letter is to be considered as an unknown only if the student shows that it is used as an

N. Bednarz et al. (eds.), Approaches to Algebra, 225-236.

unknown. This assertion is simply a corollary of a more fundamental principle in any theory of representations, namely that a symbolic representation has the meaning the interpreter who is engaged in the solution of a problem or an exercise assigns to it. Observing the solution processes is consequently crucial.

On the other hand, using the formula "the other way around" is an altogether different task. If the area is provided, the length of the corresponding radius cannot readily be found. Some calculations have to be carried out. But, those calculations are not ordinary ones because they involve using letters as general numbers having no specific values. The ability to carry out such calculations should be one of the main and basic features of algebraic reasoning. I shall not try to show to what extent it differs from the arithmetic ability involved in carrying out the same operations with numbers. I assume with Collis (1972), and Filloy and Rojano (1989), that both are substantially different. In other words, this chapter assumes that an algebraic (mental) use of letters corresponds minimally to the ability (and willingness) on the part of the student to imagine that a letter stands for a number that has to be discovered and to the competence of the student to perform the required "arithmetical" calculations on those numbers. Collis has, in fact, judiciously pointed out that the use of algebraic operations (arithmetic operations on letters as unknowns) requires that the solver carry out his operations even though the results obtained after each intermediary operation do not appear in closed form or as a single symbol or symbolic entity. For example, to find r, a student must accept that A/π stands for a number and must identify r as its square root. Collis suggests calling this ability an Acceptance of Lack of Closure (ALC). This clarification helps to distinguish a formula from an equation, a distinction without which no progress can be made in research in the domain of algebra teaching. On the basis of these preliminary clarifications about what is meant by "unknownness," I put forward now that formulas can be defined as two symbolic expressions linked with an "=" sign, which only require direct calculations on known numbers; while an equation, even though it can appear syntactically identical to a formula, must involve calculations on unknown numbers. Consequently, expressions such as $y = 3x + 2$ or $T = 9/5\ C + 32$ can be regarded as equations or formulas, depending upon the type of activity in which they are involved, because, let me recall, a symbolic expression becomes an equation (and is treated algebraically) only on the basis of the actual mental processes of the solver. If students are asked to find y (or T) when x (or C) is provided, then neither T nor y are really unknown for the students, and the expressions are not equations but formulas. However, on the other hand, if a "point" or an ordered pair (x, y) is to "belong" to the line $y = 3x + 2$ or to satisfy $y = 3x + 2$, then the symbolic expression can be regarded as an equation, namely an equation in two unknowns.

These preliminary considerations may turn out to be futile if it can be shown that years spent in replacing letters in formulas finally guarantees the development of those basic abilities involved in using letters as unknowns, such as ALC (Acceptance of a Lack of Closure). However, my experience as a teacher, as a teacher trainer, and as a researcher all concur to confirm the contrary. The distinction between formula-use and equation-use of an equality involving letters has to be considered as fundamental, but must be completed by other considerations.

1.2. *The Use of the Letter: Unknown or Variable?*

Even if students can accept that a letter stands for an unknown number and can perform calculations on that basis, they might be unable to imagine concurrently that a letter can also stand for a number that changes. As Freudenthal (1983), Schoenfeld and Arcavi (1988), and Janvier (1993) have pointed out, using a letter as a variable does not depend on the symbolic representation used, but is rather a question of a particular interpretation of the "solver." For example, one can use for a long time and successfully (as equations or as formulas) expressions, such as $A = \pi r^2$ or $d = v \times t$ or $T = 9/5\ C + 32$, without considering for a single moment that r, t, or C vary and consequently A or d or T vary too. This clarification is crucial and makes possible an additional relevant distinction.

Actually, a letter can be (mentally) used without being interpreted either as an unknown (in an equation) or as a variable (such as in $\varepsilon \to 0$). Indeed, as Freudenthal pointed out, letters are often used as polyvalent names or place-holders. The use of a in the following identity is typical of such a utilization: $(a + 1)^2 = a^2 + 2a + 1$. The letter a stands for any number and is neither unknown nor variable to the solver.

To summarize, for a student, the letter a can be interpreted in at least four different ways and each case, I contend, calls for a different mental process. Firstly, it can be used as an indeterminate value in a formula such as in P(erimeter) $= 4a$; secondly, it can be considered as an unknown, such as in $2a + 3 = 7$; thirdly, as a variable in $A = \pi r^2$ (if r varies); and finally, it can be interpreted as a polyvalent name as in the identity stated above. It could be possible to extend this preliminary manifesto with a systematic description of the differences in the processes involved. However, the basic distinctions established so far are sufficient for dealing with the issue at stake: modeling and the introduction to algebra.

2. MODELING AND ALGEBRA

2.1. *The Usual Definition of Modeling*

How is modeling related to using letters or manipulating symbolic expressions? To answer this question, let us examine how modeling as a process is defined in the literature. According to Burkhardt (1981), de Lange (1987), and others, modeling involves a formulation phase that is completed by a validation phase. During the formulation phase, a phenomenon or a situation is examined in order to establish some key relationships between the variables involved. Those relationships originate from observations or measurements, or simply come from clever guesses made on the situation under investigation. For instance, the rate of change of the height of water in a barrel over time may be assumed to be constant. In other words, one might accept the hypothesis that $\Delta h / \Delta t$ is constant. After this initial assumption (that can be considered as a sub-phase), the formulation phase comprises a more or less complex series of mathematical transformations or operations that ultimately lead to a model expressed as a symbolic expression. According to tradition, this expression is understood to be a formula expressing one variable in terms of the others. More recently, it has become common to consider also as a model the graph or the table of numbers produced as the result of computer simulation. Actually, the formulation phase requires a deduction such as: if the phenomenon displays such a feature, then it is bound, according to calculations or mathematical reasoning, to obey the following rule or law. The rule that is deduced from the assumption is precisely the model obtained at the end of this first phase of the process. In the above example, from the assumption that $\Delta h/\Delta t$ is constant, it can be deduced mathematically that $h = k \times t + h_0$.

The second phase is the validation phase. It consists of testing the validity of the model by going back to the reality that it is supposed to represent. Actually, verifications through measurements or otherwise should display the limits of the model. It may turn out that the domain for which $\Delta h/\Delta t$ is constant is different from the one expected, that over a certain domain $\Delta h/\Delta t$ changes with time, and so on. At this point, it is possible to enter another formulation phase and to improve the initial assumption. It may then be assumed that $\Delta h/\Delta t = \alpha \times t$ (α is a constant: the rate of change is proportional to t). Then a second formulation phase will yield, thanks to mathematical deductions, a quadratic model of the kind $h = at^2 + bt + c$. If for some reason, $\Delta h/\Delta t = \beta \times h$ (β is a constant: the rate of change is proportional to h), the second formulation phase will yield an exponential function.

Modeling is defined as this double process that involves (a) creating or devising a model on the basis of assumptions and (b) checking it out during a validation phase. Within this framework, the model has a double status. On the one hand, it is stated in mathematical terms and stays independent of the reality from which it emerges. This abstract status is confirmed by the "blind" application of rules involving symbol manipulation, providing ultimately a mathematical relation between the "entities" of the model. On the other hand, it is meant to represent concrete objects or relations that can be measured. Then, the elements of the model (the parameters of the formula, the global features of the graph, the characteristics of the table) that previously had no meaning must be given contextual significance in the situation under investigation. This gives the model a concrete status. Consequently, a model can be considered at the same time abstract and concrete. I shall come back to this point in the next section. Let us continue our analysis of modeling as a process.

2.2. *Modeling and Its Formulation Phase in School Algebra*

According to the description just given, modeling is certainly a rare activity in school algebra. However, modeling sub-phases can easily be recognized in students' activities. Indeed, algebra has traditionally been associated with either solving equations or solving algebra problems. According to this same tradition, learning to solve equations should come first and, as a consequence, the lessons on word problems later. More recent views (see Bednarz & Janvier, this volume; Boileau & Garançon, 1987) stress the importance of starting with the solution of easy problems, even if the symbolic representations of the relationship involved may at first be formulated in only a clumsy manner. Such approaches to algebra emphasize the formulation phase. In fact, right at the beginning of their introduction to algebra, students are asked to come up with symbolic descriptions of the relations inherent in some text or experimental setting. Their initial symbolic expressions do not necessarily involve letters for representing numbers. Often, they consist of words more or less written in an abbreviated form with arithmetic signs linking them. Progressively, it is understood, the use of letters will become more systematic.

Traditionally, students after months of lessons on the rules of correct algebraic manipulations are introduced to the art of solving word problems. They are then expected to write down the correct and unique equations fitting the relations described in the statement of the problem.

In both approaches (traditional and recent), we are very far from the genuine modeling process. Performing the entire loop, namely creating a model on the basis of extracting features and testing it, is certainly an unusual event. However, some students' activities in algebra involve some of the formulation phase included in modeling. In the well known situation in which students are asked to find the number of matches required to make a "10, 15, n-triangle frieze," the solutions they propose belong to two different categories. On the one hand, a student may take several examples, build a table, examine the numbers, and guess that the number of matches needed is one plus two times the number of desired triangles. On the other hand, using a table of numbers or otherwise, the fact that adding a triangle to the frieze requires two more matches can be made explicit and stated. In that case, the basic relationship (a rate of change of two matches for each triangle) has been singled out, and the writing of the symbolic relation is made in conjunction with the formulation of this basic relation. In the first case, it is difficult to state that the algebraic expressions obtained witness the presence of any model, while in the second one the expressions obtained at the end of the process contain potentially a family of relationships characterizing similar frieze patterns.

To acknowledge the presence of a model at the end of a formulation phase, I contend that the symbolic expressions derived from the situation must contain, in the way they are constructed, the elements through which they are made to belong to a family of relationships between variables. Then, to find the solution of similar problems, students can later recognize that a particular situation belongs to a family of relationships. This family of relationships can precisely and fruitfully be regarded as a model. In more recent approaches to algebra, it can be said that the formulas written down by students or the graphs drawn by them share many features of a model because of their "generalizability," and the process through which they are obtained resembles the formulation sub-stage of a classical modeling process. Indeed, an adequate use of modeling exercises in any approach should aim at helping students work inside a model by adjusting its parameters or generalize it by extending the range of its parameters; such use can also aim at helping them select an appropriate model from a list of previously known families of models.

As a matter of fact, the agenda of reform promoters who put forward an approach to algebra that stresses the importance of modeling seems to favor a wider range of activities than the genuine modeling process that has been granted an official status by the literature. However, the formulation phase is not always given proper and careful attention. My pedagogical view on the matter is that improving the formulation phase is not a question of adding chapters in textbooks, but more one of finding the adequate classroom approach or devising the appropriate interactive piece of computer software (e.g., Boileau & Garançon, 1987).

2.3. *Modeling and Magnitudes*

I have just shown that the crucial point of approaches advocating some form of modeling is the nature of their formulation phase. Let me now turn to the question of examining, firstly, what makes a model concrete and, secondly, if the use of models simplifies the learning of algebra.

Finding the price of objects when the unit price is given or finding the distance traveled by a moving object at a constant speed are two examples of situations within which students can work. Can we affirm that both situations belong to the same model because of the similarity of the mental processes supporting the operations to be performed? Actually, in each case, a multiplication involving a rate has to be carried out: "price per unit multiplied by number of units" or "speed multiplied by time." Now, when students are presented with price or speed problems, it is assumed that they are familiar with the basic relationships involved and, consequently, that this familiarity can support the reasoning or the solving process. For instance, students can find a price because they know that "the number of units times the price per unit" will give the price. We can assume here that the expression "so many per" triggers a conviction to use a product or at least a repeated addition.

However, teachers all know that familiarity with the "price" linear model does not transfer to the "speed" linear model. What are the objects to which the students are expected to resort when they successfully use a model? The examples just presented reveal that the objects mentally manipulated are "real" numbers measuring a feature of an object or of a relation. Measuring numbers have for a long time been distinguished from non-measuring numbers and have been called *magnitudes* or in French *grandeur* (magnus: big). A longer discussion on the subject can be found in Rouche (1992) or in Dolmans (1982). I shall make use of the term *number* to denote a "pure" number and of the term *magnitude* to refer to a measuring number. Actually, a magnitude can be imagined as a number accompanied by a unit of measure. For example, "10 km per hour" is a magnitude, "10" is a number, and "km per hour" is the measuring unit.

So, magnitudes refer to "concrete" numbers, to numbers for which can be associated (in the mind) mental images showing the features of the measured objects or relations. Actually, a magnitude can be handled as a number. In that case, no references whatsoever are made to the reality from which it takes its origin. However, when dealing with magnitudes, anyone is bound to resort to concrete features of the objects being measured, as I shall show with the notion of "length."

Line segments (such as rods) can be associated with numbers according to the number of units they contain. Each arithmetic operation can be associated with particular actions on the rods. So, when an operation cannot be automatically processed, it is always possible to resort to the inspiration supplied by the rod images. But, are the rods or the cardinal numbers the models? Actually, this example is an ideal one for exemplifying the double status of what is called model. On the one hand, arithmetic with its addition and multiplication tables deals with numbers and is the abstract guise of the model. On the other hand, the various configurations of rods is the realm of magnitudes and is the concrete guise of the model. All pedagogues agree (although probably not exactly in these terms) that the familiarity gained by the students within the magnitude model can enable them to control their action in solving exercises in the mathematical model, that students can find inspiration in the heuristic cues provided by the magnitude model as to how to apply rules, as to which rules to resort to, as to what quantities are involved, and so on.

2.3.1. *Magnitudes and their difficulties*

One kind of difficulty, when models and magnitudes are introduced, is certainly this continual switching between the concrete and abstract meanings of any magnitude. For instance, once a student arrives at an expression and has to manipulate it, the decision taken in the series of operations may be triggered by the numerical content of the magnitude or by the concrete content. It is easy to envisage the conflict resulting from the use of magnitudes in establishing formulas in algebra.

Let us mention one difficulty more related to the use of symbolism. The symbolic designations of magnitude units are often letters. and this may provoke critical misunderstanding. For instance, if 2 boys and 5 girls are abbreviated by 2b and 5g the same way 5 grams and 7 meters are often abridged to "5g" and "7m," then introducing a letter to denote a variable or unknown number clashes with the use of letters to stand for a unit or an individual. For instance, the same *g* may represent the constant used in $E = mgh$, an abbreviated form for gram, or a variable such as a particular gain (that stands for a varying magnitude expressed in dollars). I shall not carry on the analysis of this phenomenon because the usual difficulties it provokes have been reported extensively in the literature (see Clement, Lochhead, & Monk, 1981).

Magnitude units can also be classified basically as simple or composite (see Rouche, 1992). Some, such as cm, m^2, are related to visible features of objects, and others are more abstract, such as Newton or kilowatt-hour. However, composite magnitude units, such as Newton/m^2, km/s^2, belong to an even more abstract class.

Apart from controlling the double status of magnitude and handling the symbols correctly, there remains, for the students, the problem of knowing when to transfer methods that work with one magnitude to another. In other words, even when a student works on a linear model with two different magnitudes, the similarity of structure can remain unrecognized. Why is this so?

2.3.2. *Extensive and intensive magnitudes*

In Rouche's (1992) thorough study of the notion of magnitude (grandeur), he not only provides a systematic presentation of what magnitudes are, but also proposes a classification based on the idea that some magnitudes, such as price, can be isomorphically associated with the basic systems of line segments; in other words, one feature of the reality to be represented can be associated with the line segment system.

The sort of classification made by Rouche explains why some magnitudes are more difficult to apprehend and consequently that the linear function describing the relationship between two variable magnitudes does not suffice to describe per se the category of mental processes required by students to solve all linear function problems.

To understand how modeling can be used to introduce algebra to students, I contend at this point that more research should examine the mental processes involved when students make use of magnitudes. It seems to me that extending Rouche's basic categorization of magnitudes would allow us to have a better grasp of students' difficulties and consequently help us to better design a coherent set of situations that would encompass a more relevant diversity of cases. Moreover, such research could respond to some of the issues raised in the Bednarz/Janvier chapter (this volume). Indeed, the different patterns of reasoning they have identified with their diagrams--and labeled arithmetic or algebraic--might be further explained through a judicious classification of magnitudes.

2.3.3. *Models, magnitudes, and approximations*

A model is generally considered to be an ideal version of a phenomenon that has to be validated, but that will never exactly fit what is sought to be described. For instance, stretching a spring will *grosso modo* fit a linear model. The use of the theoretical framework exposed above enables us now to describe the situation as follows: The *magnitude model* of the spring is not expected to correspond to the *mathematical model* of the linear relation between quantities. This means that a mathematical model is only a fair approximation, and the discrepancy is likely to be analyzed in mathematical terms. However, linking mathematics with reality does not always involve this idea of an approximation to be measured; in other words, the difference between the mathematical model and the magnitude model is often not supposed to be considered. Indeed, in the mathematics classroom, either an experiment will exactly fit its mathematical formulation or the differences are neglected. The former is the case for the arithmetic rod model described above, or when we establish a price, or apply a linear rate to find a fee. No discrepancies are expected to arise. In those cases, there cannot be any validation phase. In the latter case, approximation belongs per se to the situation in which the differences are expected to be overlooked. Once again, I see in these distinctions a source of difficulty since textbooks or teachers' intentions are not always clearly stated.

2.4. *Modeling not Always a Panacea!*

I have attempted thus far to make clear that the key feature of introducing algebra in a modeling perspective is an emphasis on the importance of the formulation phase to make sure that the relationships between the variables that are introduced both become generalizable and lead to versatile models. I have also pointed out that the simultaneous abstract and concrete status of models entails a coordination of representations that may often pose difficulties to students. Some readers may believe that this distinction between number and magnitude is solely an intellectual fad. Recent research on contextualized reasoning (Carraher, Carraher, & Schliemann, 1985) does prove them wrong. Indeed, this research tends to substantiate the fact that students can sometimes succeed at tasks within a certain context, while they are unable to do so with conceptually similar tasks expressed in a different context or taken out of context. Actually, it has been shown that different contexts trigger different problem-solving strategies, foster different mental images and, above all, reveal astonishingly different success rates.

The issues raised by the use of different contexts (often in a modeling perspective) are far from being easy to tackle. However, they appear fundamental because research results could shed some light on a wide variety of questions, ranging from the benefits of using manipulatives or computer simulations to the likely outcomes of doing laboratory work and getting involved in real problem solving.

3. MODELING AND A FUNCTIONAL APPROACH TO ALGEBRA

3.1. *A Functional Approach to Algebra*

Several authors in this volume (Heid and Kieran/Boileau/Garançon) have proposed to introduce algebra from the start with a functional perspective on algebraic letters. One advantage of this approach is that it can be framed inside a more comprehensive theoretical setting (such as in Kieran, 1994). In addition, these approaches (despite their particular features) respect the position we expressed above, namely that variables must be distinguished from unknowns because they correspond to two different ways of interpreting letters. Now I examine some implications of a brief but relevant analysis of the notion(s!) of function.

3.2. *The Notion(s) of Function and Its (or Their) Representations*

The notion of function has recently been the object of several reflections (see Harel & Dubinsky, 1992; van Barneveld & Krabbendam, 1982). However, despite the many interesting essays and research reports contained in those books, not much can be found on the fact that the notion of function conceals a wide range of concepts (so much so that one should more correctly speak of the notions (plural!) of function). Moreover, one can also note the widespread absence of careful analyses of the notational problems that are raised by the notions of function.

My colleague, Louis Charbonneau, and I have explored the historic evolution of the idea of function (see Fikrat, 1994; René de Cotret, 1986). Our main conclusion is that the notion of function is not a single concept and that to consider it as such is both misleading and counterproductive. Indeed, if *function* refers to different entities that are processed differently in the mind, it becomes helpful and essential to use different words to refer to different concepts or, at least, to make the relevant distinctions when required. Other reflections in this direction can be found in Freudenthal (1983) and Menger (1979). For the time being and for the purpose of this chapter, I shall insist on only one major distinction, one that will show a difference between a *function-of* and a *function.*

The term function has been used for a long time to denote the relation between two variable magnitudes or numerical quantities. Frequently today, one still reads or hears that y depends on x and that y is a function of x. From Euler's time until very recently, it was common to find in mathematics textbooks: "let $y = f(x)$" read as "let y be a function of x." But, at the end of the last century, it became productive and necessary to distinguish between the variable y (the function of x) and the way in which this variable y was related to the other variable x. A function was then differently defined as a rule of correspondence by which a variable y is dependent upon another one (x, for instance). It is not yet clear to me when the rule of correspondence was attributed a proper notation, as it is common now to denote a function by a letter f, g, h, ϕ, or ψ. No historians whom I have consulted have been able to track down exactly when this started. Actually, even in the prestigious *Leçons d'analyse fonctionnelle* by Riesz and Nagy (1968), functions are defined by the expression: Let $f(x)$ be a function such as ..., even though some occasional uses of a single letter are made to denote a function, as τ or σ being used, for example, to denote the measure function of a Lebesgue-Stieltjes integral. This rule of correspondence idea was later formalized within set theory, either as a set of ordered pairs or, in very sophisticated textbooks, as a triple of sets (A, B, C)--A and B being the initial and the final sets, C being a subset of ordered pairs taken from A × B defining the correspondence.

An early introduction of the notion of function calls for an expression such as $y = f(x)$ to be presented early. Students are then expected to come across a great diversity of symbolic expressions. What could have been only $y = x^2$ can appear as $y = f(x) = x^2$ or simply as $f(x) = x^2$ or f: x --> x^2. As those examples show, designating a rule of correspondence with a particular letter (f, for instance) has given this letter a double status. On the one hand, the letter f can be used with (x) and then $f(x)$ is another name for y since $y = f(x)$. On the other hand, f may not refer to y but to a rule. Then, f can be used in expressions containing no "y", as in the two examples of the list of three given above. In those particular cases, the use of any variable becomes possible. Actually, $f(x) = x^2$ and $f(t) = t^2$ are two identical rules of correspondence, but they may define different variables if one writes $y = x^2$ and $s = t^2$. Similar notation intricacies can be met with expressions of the following kind where the main variable y is absent: f: x --> x^2.

The first preliminary comment which the above considerations suggest is that the functional approach advocated by many researchers should more adequately be called a variable approach to algebra or a y-function-of approach to algebra. Indeed, these approaches, as I understand them, do not aim at promoting an algebra of rules of correspondence. In no way have I noticed any intention of emphasizing the use of any particular "rule of correspondence" notions or concepts.

The second comment concerns the vigilance required from curriculum designers and teachers to avoid, within the framework of a variable approach to algebra, the numerous pitfalls hidden here and there in the usual exercises of standard work on functions.

3.3. *Magnitude and Functional Approaches*

Since it seems clear that the notion of variable is central to the approaches called functional, it must be remembered that our reflections about the dual status of magnitudes have tremendous consequences for the design of learning units built in this perspective. Indeed, what is called variable in any particular functional approach can be either a varying numerical quantity or a varying magnitude. All the difficulties described above, as well as their causes, will be found in y-function-of approaches that are not carefully planned and designed.

3.4. *F(x) or T(x)?*

As we already pointed out, a variable approach to algebra is bound to introduce early the notation $y = f(x) = x^2$ or simply $f(x) = x^2$. Within the context of starting from situations (with a more or less involved phase of formulation), the notation $f(x)$ is likely to take a very special turn. Indeed, when a formula is found that displays the relationship between two variables, such as $T = 9/5\ C + 32$, there is a great temptation to note that T depends on C and to suggest to students that they write $T(C) = 9/5\ C + 32$. However, if not handled with care, this practice will certainly create difficulties. Indeed, the letter T represents a variable (more precisely a varying magnitude), and students will be inclined to write (as engineers and physicists often do) $T = T(C) = 9/5\ C + 32$. The notation then becomes utterly misleading: The letter T simultaneously refers to a magnitude and to a rule of correspondence because it is involved in an expression of the kind $f(x)$. Finding the inverse function, for instance, will present the dilemma of deciding whether the notation T^{-1} is acceptable because the notation to be used for a rule of correspondence should be independent of the variables utilized to define it.

My pedagogical view on this question is that, if the notation $f(x)$ is introduced early, it can be introduced as in physics, namely that the letter used in $f(x)$-expressions could and should stand for a variable. However, the idea that the letter in front of (x) stands for a rule of correspondence should be postponed for at least a year, thus giving students time to gain some familiarity with this usage. But, in the long run, $f(x)$ will have to be given a double meaning because the letter f in front of (x) often refers implicitly to a particular way of algebraically combining the "x"s even though f is not written independently of $f(x)$. For instance, $f(x) = x^2 + 3x$ refers to squaring x and adding its triple, while $g(x) = \sqrt{x} + 2x$ means taking the square root of x and adding its double. This usage does not emphasize the role of the variable x and actually, it becomes important that students become familiar with the Eulerian idea that $f(\bullet)$ refers to a special algebraic way of combining "•". Indeed, they must learn that $f(x+h) = (x+h)^2 + 3(x+h)$ and that $g(\sin t) = \sqrt{(\sin t)} + 2(\sin t)$.

4. CONCLUDING REMARKS

In this chapter, I have tried to point out several components of school algebra. First of all, I have examined three kinds of interpretation of letters: unknown, variable, and place-holder--each giving rise to different reasoning patterns in identical simple symbolic expressions. Secondly, I have shown that algebraic units are sometimes numbers and sometimes magnitudes. At this point, it is possible to display six particular ways of looking at the processing of algebraic expressions (see Figure 1).

	QUANTITY	MAGNITUDE
UNKNOWN	X	X
VARIABLE	X	X
PLACE-HOLDER	X	X

Figure 1.

Thirdly, I have shown that the formulation phase in the modeling process is crucial for giving meaning to basic formulas, so as to make them into generalizable models. But, by the same token, I have insisted on the need to explore more carefully the notion of magnitudes, as it seems the vital link between modeling and a variable approach to algebra. Fourthly, I have stressed the importance of distinguishing between the "rule of correspondence" content of the idea of function and its "variable" content in the development of approaches based on function. Finally, I have concluded with an analysis of traditional usage of the notation $f(x)$, showing notational intricacies that ought to be taken into account.

I believe that the original aim of the chapter has been reached: to spell out some fundamental notions and principles that should allow the research community and curriculum designers to think differently about what an adequate approach to school algebra might mean.

In this field of research, I consider that too many manipulations of symbols are taken for granted, and consequently, the complexity of the many underlying reasoning processes are overlooked and not often fully appreciated for their diversity. Therefore, not only do I believe that approaches to algebra should be based on theoretical frameworks that really take into account those reasoning processes, but I also contend that, because the multiplicity of issues raised in this chapter has revealed that school algebra is far from being monolithic, it would be unwise to promote approaches that are not basically eclectic.

PART V

A FUNCTIONAL PERSPECTIVE ON THE INTRODUCTION OF ALGEBRA

CHAPTER 18

A TECHNOLOGY-INTENSIVE FUNCTIONAL APPROACH TO THE EMERGENCE OF ALGEBRAIC THINKING

M. KATHLEEN HEID

The functional approach to the emergence of algebraic thinking described here suggests a study of algebra that centers on developing experiences with functions and families of functions through encounters with real world situations whose quantitative relationships can be described by those models. Computer-Intensive Algebra is a beginning algebra curriculum that introduces the concepts of functions, variables, equations, inequalities, systems, and equivalence in the context of mathematical modeling. Our research has studied the development of mathematical understandings in this setting as well as the pedagogical and environmental contexts in which this learning occurs.

Central to algebraic thinking is the concept of variable with all of its possible connotations, uses, and connections (Usiskin, 1988). Contemporary approaches to the emergence and development of algebraic thought differ in the extent to which they emphasize particular aspects of the variable. They differ in their emphasis on the dynamic nature of the variable, on the ways in which variables represent quantities in real world settings, and on the use of variables in building formal and well defined mathematical structures. During the past decade computing technologies available to the schools have provided unprecedented access to representations of variables as quantities with changing values. What is interesting and important about the dynamic nature of the variable, however, is not solely the change in the values of variables, but rather the effects of those changes on the values of other variables. That is, what makes the study of variables interesting is the study of functions of those variables.

There are a variety of ways in which one might fashion a functional approach to the emergence and development of algebraic thought. In each such approach, there are essential perspectives regarding what mathematical content is taught and what it means for students to understand that content.

The mathematical content central to any functional approach to algebra involves the fundamental concepts of variable and function. In the functional approach to algebra described in this chapter, the mathematical ideas that are targeted for study are ones that help explain quantitative relationships in real world settings. Variables are used to describe real world quantities whose values change, and functions are used to describe the relationships among those quantities. Families of functions, like the linear, exponential, and rational families, are studied because they are reasonable models for relationships among real world quantities. Students study these families of functions and their properties as they relate to real world situations, analyzing the meaning of various rates of change, zeroes, maximum and minimum values, and asymptotic behavior in their contextual settings.

In this approach to algebra, understanding the concept of function requires being able to solve problems using multiple representations and to reason within and

N. Bednarz et al. (eds.), Approaches to Algebra, 239-255.

among graphical, numerical, and symbolic representations of function. It requires understanding properties of basic families of functions and how those properties can be seen in each of the different representations. It requires the ability to recognize function families and their properties within real world settings, as well as the ability to answer questions about those real world settings using function families and their properties.

Computer-Intensive Algebra(CIA)[1] (Fey, Heid, Good, Sheets, Blume, & Zbiek, 1991) is a beginning algebra curriculum that exemplifies the functional approach just described. Characteristic of *CIA* is that it introduces students to the concepts of functions, variables, equations, inequalities, and equivalence, in the context of mathematical modeling. The curriculum provides students with opportunities to explore functions in real world situations through the development and evaluation of mathematical models. Students are asked to develop mathematical representations of functions, explore questions inherent in real world settings, and formulate and defend predictions about realistic situations. While reasoning about important applied problems and investigating realistic questions, students use fundamental mathematical ideas and methods and learn properties of variables, functions, relations, and systems. Computing tools (with function graphers, curve fitters, table generators, and symbolic manipulators) facilitate explorations of algebra by providing students with continual access to numerical, graphical, and symbolic representations of functions, as well as to technology-intensive procedures for reasoning about algebraic expressions. *Computer-Intensive Algebra* can be characterized through its approach to mathematical content as well as through its pedagogical assumptions.

1. THE CURRICULAR APPROACH OF *COMPUTER-INTENSIVE ALGEBRA* TO THE DEVELOPMENT OF MATHEMATICAL IDEAS

The *Computer-Intensive Algebra* curriculum focuses on the development of understandings about a few fundamental mathematical concepts through providing students with opportunities to explore those concepts in realistic settings. The concept of function is central to the organization as well as the content of *Computer-Intensive Algebra*. The development of this concept can be seen in the organization of the chapters of the *Computer-Intensive Algebra* text--Chapter 1: Variables and Functions; Chapter 2: Calculators, Computers, and Functions; Chapter 3: Properties and Applications of Linear Functions; Chapter 4: Quadratic Functions; Chapter 5: Exponential Functions; Chapter 6: Rational Functions; Chapter 7: Algebraic Systems: Systems of Functions and Equations; Chapter 8: Symbolic Reasoning: Equivalent Expressions; Chapter 9: Symbolic Reasoning: Equations and Inequalities. As the *Computer-Intensive Algebra* curriculum involves students with the concept of function, it also develops the concepts of family of functions, equation, inequality, systems, and equivalence. The next few pages describe that development.

The first chapter introduces students to the concept of variable and function by engaging them at the outset with a *Talent Show* computer simulation. In this simulation, students control several input values: whether or not to hire a master of ceremonies, the price of tickets and concession-stand prices, and the quantities of soda and candy to buy for the concession stand. By running the simulation for different combinations of input values, students see the direct effect on the output values of sales, revenue, and profit. The rest of the chapter gives students a variety of experiences with graphical, numerical, and symbolic representations of single-variable functions in the context of applied settings. By the end of the first chapter, students will have been introduced to graphical, numerical, and symbolic representations of single-variable functions as well as to the connections among them. This exposure is needed early in the course because the use of function graphers and spreadsheets to explore functions through their representations depends on an understanding of the relationship between a function rule and its graph or tabular representation.

The second chapter introduces students to the use of the computer to generate tables and graphs from function rules. With this capability, students begin to answer fundamental questions of interest associated with the application of function rules to applied settings. These questions are:

- What are the function's output values or range of values for specified input values or range of input values?
- For which input value(s) or range of values does the function attain specified output values or range of output values?
- For which input value(s) does the function attain an output value of 0?
- What are the trends in output values?
- For which input values does the function attain a maximum or minimum?

Figure 1 shows the *Mason-Dixon Inn* situation and related questions, an example that appears at the end of Chapter 2 in *Computer-Intensive Algebra*.

SITUATION: The management at the Mason-Dixon Inn is planning to offer a special Winter Weekend at its resort hotel in the mountains. There will be special meals, entertainment, and outdoor recreation activities for the whole family, with all activities included for a fixed price per person. The problem is what price to charge!

Market surveys suggest that the *number of customers* is a function of the *price charged* with rule:

$$C(p) = 450 - 2.5\ p$$

After itemizing the expected costs, the management estimates that the *profit* also depends on the *price charged* with rule:

$$F(p) = -2.5\ p^2 + 600\ p - 27000$$

Use these function rules to answer the following questions in the same way you worked in the "moon ball" situation.

12. What number of customers is predicted if the price is set at $100?
13. What profit is predicted if the price is set at $100?
14. If the Inn wants 300 customers, what price should be charged?
15. What price(s) will give at least 255 customers?
16. At what price(s) does the Inn break even; that is, have profit of $0?
17. For what price(s) does the Inn have a profit of $5000?
18. At what price(s) is the profit for the Inn at least $1000?
19. What price means that no customers will come?
20. What price gives the Inn maximum profit? What is that profit?

Figure 1. A situation and related questions appearing at the end of Chapter 2 (p. 2-59) in *Computer-Intensive Algebra* (Fey et al., 1991)

In that the *Mason-Dixon Inn* situation occurs in Chapter 2, students encounter both linear and quadratic functions early in their *Computer-Intensive Algebra* experience. The intention is early exposure rather than in-depth coverage of the properties of these functions.

Chapter 3 introduces students to the family of linear functions. Particular attention is paid to a fundamental property of linear functions, the constant rate of change. Students explore graphical and numerical representations that highlight rates of change for different linear function rules. They compare the constant rate of change in linear rules to the rates of change in non-linear rules. Students associate the slope and intercept from the symbolic function rule with graphical and numerical features. The entire mathematical development occurs in the context of extended explorations of mathematical models for relationships in realistic situations.

Chapter 4 involves students in explorations of quadratic functions. Students complete a variety of explorations in which they use quadratic function rules to answer questions about applied situations. They encounter situations in which they are given a function rule as well as situations in which they develop their own function rules through collecting data and fitting curves. In some of the only *Computer-Intensive Algebra* explorations not set in an applied context, students explore the effects of changes in parameters of quadratic functions (the effects of changing a, b, and c in $f(x) = ax^2 + bx + c$) on related graphs and tables. In the last section of Chapter 4, students extend this parameter exploration to polynomial functions of degree higher than 2.

Chapter 5 continues with the exploration of an exponential function family ($E(x) = Ca^x$). In their work with these growth and decay functions in a variety of applied settings (e.g., to provide initial models for bacteria growth, depreciation, spread of an epidemic), students encounter contextual meaning for negative and rational exponents and develop and use explicitly defined, as well as recursively defined, function rules. They develop properties of exponential functions through consideration of their meaning in an applied setting. For example, the property $a^{-n} = \frac{1}{a^n}$ is developed through examples like the following (*Computer-Intensive Algebra*, p. 5-22):

Example 1. In a pool the bacteria population was 1500 per cubic centimeter at 8 a.m. on Monday and doubling every day. This led to the model:

$$N(d) = 1500 \times 2^d$$

It seems reasonable that the population density was 750 one day *before* that Monday [N(-1) = 750], 375 two days *before* that Monday [N(–2) = 375)], about 188 three days *before* that Monday [N(–3) ≈ 188], and so on. Thus:

(1) $\quad = 750$

(2) $1500 \times 2^{-1} = 750$

(3) $2^{-1} = \frac{750}{1500}$

(4) $2^{-1} = \frac{1}{2}$

(5) $2^{-1} = \frac{1}{2^1}$

As they use exponential functions in a variety of contexts, students are asked to compare rates of change and other properties with those of linear and quadratic functions. They conduct parameter explorations, examining the effects of changing the values of C and a on graphical and numerical representations.

Chapter 6, the final chapter centered on a specific family of functions, focuses students' attention on rational functions of the form $R(x) = \frac{a}{x^n}$. As in previous chapters, the emphasis is on developing student understanding of: (a) the typical patterns in graphs, tables of values, and symbolic rules for this family of functions; and (b) the kinds of relations among variables that are best modeled by the functions being studied. Throughout their explorations of situations modeled by rational functions, students compare the patterns they see with those of other familiar families of functions. The culminating activity in Chapter 6 engages students in a modeling activity in which they must gather data, generate function rules, and compare and evaluate how well functions from different families model the real situation.

Although the first six chapters concentrate on functions of a single variable, students encounter and explore situations in Chapter 7 that are best described through systems of single-variable functions or through multi-variable functions. For example, Figure 2 describes one situation (the *Dapper Dan* game--*Computer-Intensive Algebra*, pp. 7-9 to 7-18) that involves students in analyzing the interplay among sales, cost, revenue, and profit functions for a high school all-star game.

In answering an array of questions related to the *Dapper Dan* situation about the relationship between attendance and ticket price, operating cost, revenue, and profit, students explore the situation through a variety of other lenses. They consider questions that require solving systems of equations and inequalities: For which values of $f(x)$ is $f(x) = g(x)$; for which values of $f(x)$ is $f(x) < g(x)$; and for which values of $f(x)$ is $f(x) > g(x)$? They revisit revenue, cost, and profit as functions of the ticket price; and they look at ticket sales as a function of several variables (advertising expenses and ticket price). In other applied settings (e.g., protein as a function of the amount of meat, cheese, and grain in a diet), students encounter and interpret graphical and numerical representations of functions of two variables; and they symbolize and explore functions of three and four variables. They get a variety of experience with the notion that a system of equations and inequalities may be satisfied by one, many, or no combinations of input values. They then get some practice with finding graphical solutions to systems of equations like ($f(x,y) = c$ and $g(x,y) = d$).

Unlike more traditional approaches to algebra, *Computer-Intensive Algebra* does not focus on (or include) by-hand symbolic manipulation as a formal part of the curriculum. Instead, students use a symbolic manipulation program as a tool for dealing with symbolic rules. If symbolic manipulation programs are to be used as a major exploration tool in the absence of exposure to by-hand symbolic manipulation, however, students will need to develop an appropriate and useful "symbol sense" in alternative ways.

Situation 2.1. Every year at the end of the high school basketball season there are many all-star games in major cities around the country. One of the most famous is the Dapper Dan game in Pittsburgh, PA. It has nationwide attraction for players, coaches, news media, and college recruiters.

In planning for this game the organizers must consider many variables and make many decisions. It looks like a complex system to model. But most variables and decisions can be sorted into two major categories—those that affect *operating expenses or costs* of the game and those that affect *income or revenue* from the game.

Operating expenses or costs include items like rent for the stadium, travel expenses for players, pay for officials, advertising of the game, printing of tickets and programs, and so on. Sources of income or revenue include sale of tickets, programs, and concessions as well as possible television broadcast rights.

The critical variable in both cost and revenue functions is *attendance* at the game, and a critical factor influencing attendance is *ticket price* charged for admission. You have seen in earlier chapters how these variables can be used as inputs for cost and revenue functions. For the Dapper Dan game those functions might be like the following examples:

(1) The fixed and variable *operating costs* might combine to give a function with rule $\mathbf{C(x) = 3x + 30000}$, where x is the number of people who attend the game.

(2) The sources of *revenue* might combine to give a function with rule $\mathbf{R(x) = -0.005x^2 + 40x}$.

(3) The *ticket price* for the game is related to the expected attendance. The function rule might be $\mathbf{T(x) = -0.005x + 40}$.

(4) Finally, *profit* for the game is the difference between revenue and operating costs, $\mathbf{P(x) = R(x) - C(x)}$.

Figure 2. The *Dapper Dan* situation from p. 7-9 of *Computer-Intensive Algebra* (Fey et al., 1991)

Chapters 8 and 9 focus on the development of symbol sense. In Chapter 8, students encounter equivalent expressions as they construct alternative representations for expressing the relationships among two variables. They develop graphical and numerical techniques for providing evidence that supports or contradicts the equivalence of several algebraic expressions, and as a result they begin to recognize the need to establish generalized rules for producing equivalent expressions. Thus, rather than giving students extended practice in applying symbolic manipulation rules to produce equivalent algebraic forms, the *Computer-Intensive Algebra* curriculum is designed to develop in students a solid understanding of why such rules are needed and of graphical and numerical meanings of equivalence of expressions. The goals of Chapter 8 center on:

- developing graphical, numerical, contextual, and symbolic meaning for the concept of equivalent expressions;

- transforming function rules into more familiar form;
- recognizing expressions that are not in familiar form; and
- getting different information from different equivalent forms.

Each of these goals is approached through investigation of applied contexts. Chapter 9 extends this development of symbol sense to equations and inequalities as it introduces students to graphical and numerical meanings for the addition and multiplication properties of equality and inequality.

The entire *Computer-Intensive Algebra* curriculum is carried out within the context of the development, use, and interpretation, of mathematical models for the functional relationships in realistic situations. The curriculum focuses on three modes of representation in the exploration of these functions: graphical, numerical, and symbolic. Through the exploration of situations that are reasonably modeled by linear, quadratic, exponential, and rational functions, students develop techniques generalizable to the exploration of the properties and applications of families of functions. Although most of the curriculum is focused on single-variable functions, systems of equations are viewed as represented by multi-variable functions. Symbolic representations are used throughout the *Computer-Intensive Algebra* curriculum, and particular attention is paid to the development of contextual, graphical, and numerical meaning for those representations as well as for the algebraic rules that govern the production of alternative symbolic forms for these representations.

2. THE PEDAGOGICAL ASSUMPTIONS OF *COMPUTER-INTENSIVE ALGEBRA*

The *Computer-Intensive Algebra* curriculum makes a number of assumptions about the teaching and learning that should accompany use of the curriculum. There are environmental and organizational assumptions, assumptions about learning, and related assumptions about teaching.

2.1. *Environmental and Organizational Assumptions*

The *Computer-Intensive Algebra* curriculum is written premised on the assumption that students will have constant classroom access to computing tools, including a function grapher, a curve fitter, a table generator, and a symbolic manipulator. Although it is not absolutely necessary, it is desirable that many of these capacities be contained in a single computer program. At present, the most desirable software configuration is a computer algebra program in which symbolic outputs can be captured and used to construct graphs or tables without requiring the user to reenter the symbolic rules. In addition, these programs commonly allow for easy graphing of table values. This capability facilitates the translation from symbols to tables and graphs, as well as from tables of values to graphs. The translation from tables to symbols is facilitated by the curve fitter, which gives the user the ability to generate a variety of different symbolic rules which fit the data in different ways. The translation from graphs to numerical values is facilitated by the point-by-point reading and recording of output values.

In addition to the ability to translate to and from contextual situations, what is missing from today's computer algebra systems is the ability to translate from graphs to symbolic rules (see Figure 3).

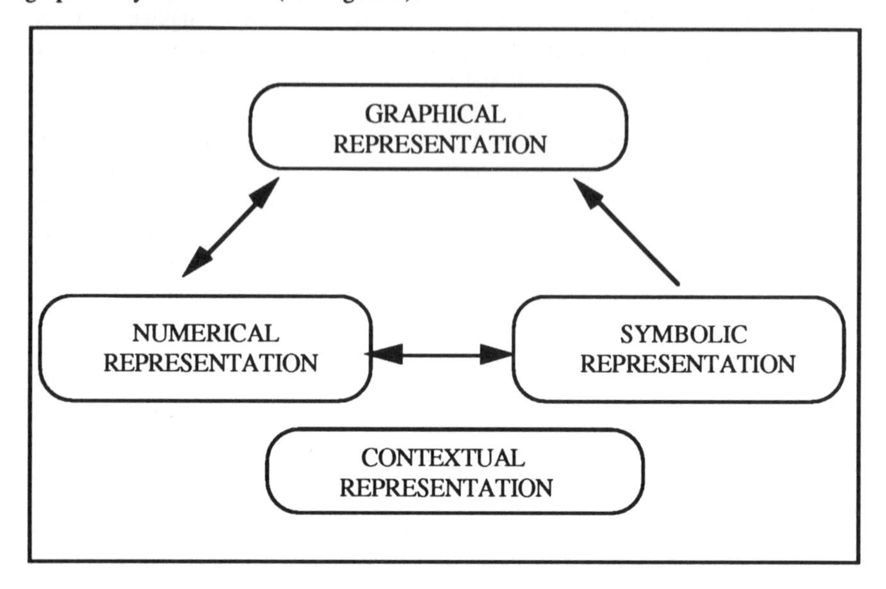

Figure 3. Current capabilities of computing tools to produce translations among representations. The arrows represent the capabilities of computing tools to assist substantially in producing the indicated translation. A double-headed arrow suggests that the computing tool can translate in either direction.

Although it is conceivable that computing tools will improve greatly in their abilities to translate from graph to symbolic rule, current tools which attempt such translations are fairly limited in their capabilities. Confrey's *Function Probe* (1992b), for example, will allow the user to see immediately the effect on the function rule when he or she acts: (a) to grab and drag a graph horizontally or vertically; (b) to reflect a graph over a horizontal, vertical, or diagonal ("$y = x$") line; or (c) to stretch a graph by some fixed factor in either the horizontal or vertical direction. These capabilities are likely to expand in the near future. Future tools could, for example, allow composition of functions when only the graphs (and not the symbolic rules) for the functions are known.

The *Computer-Intensive Algebra* approach could be extended as new technology (videodisk, sensory-controlled direct data input like TI's *Computer-Based Laboratory*, user-friendly and inexpensive handheld computer algebra systems, etc.) becomes available. With such availability, the *Computer-Intensive Algebra* approach could easily extend to assuming the constant classroom and home access to a powerful tool kit of computing tools.

Not only is the *Computer-Intensive Algebra* curriculum premised on particular software and hardware assumptions but it also depends on a classroom organization that centers on the use of collaborative pairs working together at the computer. The explorations are intended to be conducted and discussed in groups rather than individually. This requirement (which crosses over into *Pedagogical Assumptions*) draws on the belief that conceptual understanding grows more rapidly when students are communicating orally with each other about mathematics.

2.2. *Assumptions about Learning*

The *Computer-Intensive Algebra* curriculum is designed to develop concepts gradually and through examples rather than through definitions and applications of those definitions. Students are not expected to master each concept and procedure when they first encounter it, but rather to develop continually their mathematical understandings through encounters with mathematical models of realistic situations. Through these encounters, it is the intention of the curriculum that students develop their ability to reason within and among graphical, numerical, and symbolic representations for functions. These assumptions are consistent with a view of concept learning that suggests the need for a variety of examples, with a view of practice that suggests that it is best if distributed over a longer period of time, and with a view of knowledge that suggests that it is constructed by individuals through their interactions with their environment.

The *Computer-Intensive Algebra* approach, then, is built on the belief that students learn mathematical concepts best when they are using those concepts in applied settings and when they are exploring various representations of those concepts. It depends on students building their own robust concepts of the mathematical ideas under study. Rather than allowing students to roam freely in problem-solving explorations, however, the *Computer-Intensive Algebra* text sequences suggested experiences to enhance student development of particular concepts. Careful examination (including contrasts and comparisons) of series of situations whose relationships can be modeled by linear, then quadratic, then exponential, then rational functions will presumably result in growth, for example, of student understanding of the concepts of function, linearity, and family of functions.

2.3. *Related Assumptions about Teaching*

There are a range of assumptions the *Computer-Intensive Algebra* approach apparently makes about teaching, many of which are consistent with current NCTM recommendations as well as with current understandings about teaching. Generally, it seems that teaching *Computer-Intensive Algebra* naturally leads to changes in teaching and learning roles and responsibilities. These changes are discussed in more detail elsewhere (Heid, Sheets, & Matras, 1990), but are based on carefully controlled documentation of what happened as teachers in diverse settings implemented the *Computer-Intensive Algebra* curriculum.

3. RESEARCH ON THE *COMPUTER-INTENSIVE ALGEBRA* CURRICULUM

Analysis of the *Computer-Intensive Algebra* curriculum was based on a multi-level concept of curriculum. In short, different parts of the analysis centered on different parts of the curriculum. Those phases were: the written curriculum (what appears in the text); the taught curriculum (what actually happened in *Computer-Intensive Algebra* classes); and the learned curriculum (what *Computer-Intensive Algebra* students seemed to understand and be able to do as a result of the *Computer-Intensive Algebra* curriculum). Figure 4 shows how different parts of the analysis fit this tri-part conception of curriculum. The following pages will reflect on some of the results from the analyses of the *Computer-Intensive Algebra* curriculum.

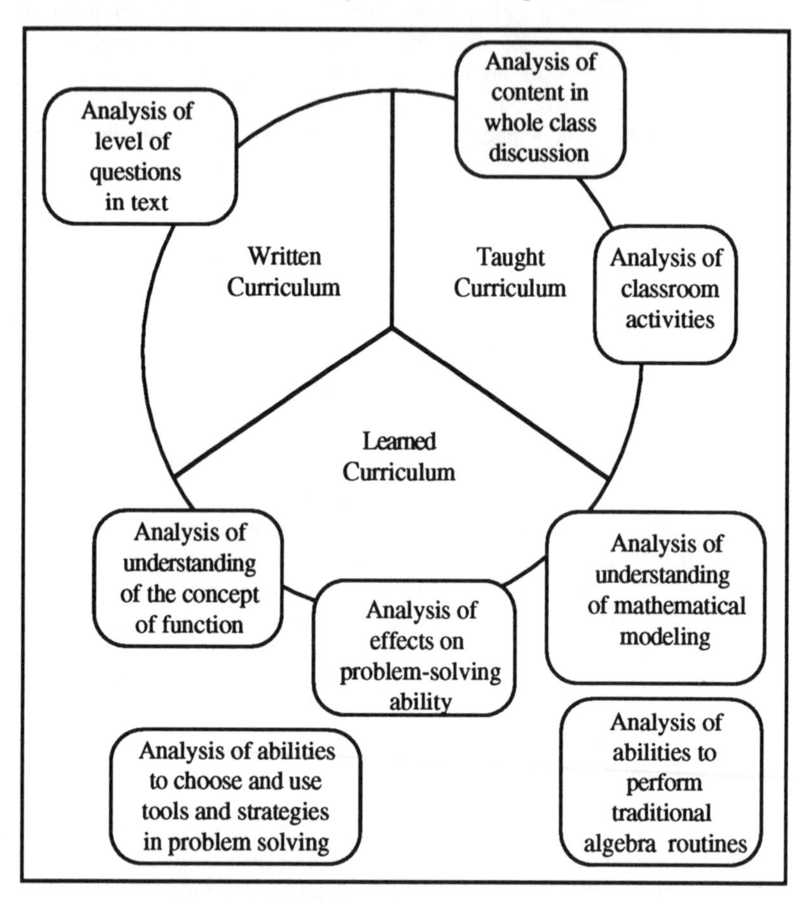

Figure 4. Diagram of studies analyzing the *Computer-Intensive Algebra* curriculum

3.1. *Analysis of Level of Questions of the Computer-Intensive Algebra Text*

A comparison study (Engelder, 1991) analyzed the cognitive levels of questions appearing in the *Computer-Intensive Algebra* text and being asked in *Computer-Intensive Algebra* classes. In the textbook analysis, the study compared questions and items in linear function and quadratic function chapters from a popular 1960s beginning algebra text, from the 1980's beginning algebra text being used at a *CIA* pilot school, and from the *Computer-Intensive Algebra* text. The *Computer-Intensive Algebra* text asked more higher level questions than the other two texts (the difference was greater between the *Computer-Intensive Algebra* text and the 1980s text). This difference seemed attributable to the preponderance of questions in the *CIA* text that asked for explanations and justifications. The study also compared the levels of questions asked by the same teacher teaching algebra to two classes, one using the 1980s targeted text and one using the *CIA* text. The study found that the same teachers asked higher level questions in their *CIA* classes than in their traditional classes.

3.2. *Analysis of the Taught Curriculum*

Several types of data were gathered in an attempt to analyze the taught curriculum. Over 130 hours of classroom and computer-lab observations, along with the coding of over 120 hours of whole-class discussions, gave a picture of classroom activity in *Computer-Intensive Algebra* classrooms. The analysis of hours of classroom and computer-lab transcripts and observers' field notes suggests (Heid, Sheets, & Matras, 1990) that implementation of the *Computer-Intensive Algebra* curriculum seems to lead not only to different classroom activities but also to a different set of roles, responsibilities, and challenges for teachers (e.g., facilitator, technical assistant, catalyst) and students (e.g., new responsibilities, new goals). Teachers attend to formulating questions that encourage group exploration, to finding new ways to assess student understanding as they work on lab explorations, and to dealing with the difficulty of drawing closure in a class whose members are moving at widely varying rates. Students struggle with finding ways to work with lab partners, to take class notes, and to study for tests.

A different type of data arose as a result of analyzing the coding of classroom discussion in *Computer-Intensive Algebra* classes along with that of traditional algebra classes. When whole-class discussion in *Computer-Intensive Algebra* classes is compared with that of traditional algebra classes, *Computer-Intensive Algebra* classes seem to spend their discussion time differently. The data collected during the second year of trials of *Computer-Intensive Algebra* materials was the percent of the 30-second observation periods during which selected types of mathematical content were discussed in *Computer-Intensive Algebra* classes at one site as well as in traditional algebra classes at that same site (Heid, Sheets, Matras, & Menasian, 1988). *Computer-Intensive Algebra* classes seemed to spend a greater percentage of their class discussion time on problems that are represented graphically or in applied settings; they spent less time discussing problems whose representation is symbolic. Similarly, they seemed to spend much less time discussing symbolic routines and somewhat more time on routines about the Cartesian coordinate system. With respect to time spent on problem interpretation, *Computer-Intensive Algebra* classes seemed to spend more time than traditional classes on interpreting answers. With respect to time spent on problem formulation, *Computer-Intensive Algebra* classes seemed to spend more time than traditional classes on conceptualizing problems and on planning solutions.

When category definitions were refined and expanded to allow for a greater range of traditional content to be classified under each of the categories, some of these differences reoccurred in comparisons made for *Computer-Intensive Algebra* classes at a demographically different school three years later (Heid, 1992). However, the sizes of these differences, fairly large in the first school with teachers who were in their second year of teaching *Computer-Intensive Algebra*, seemed smaller for teachers in the second school who were in their second and third years of teaching *Computer-Intensive Algebra*. Traditional classes spent significantly more time on symbolic representations and on discussing additional strategies. *Computer-Intensive Algebra* classes spent significantly more time on applied problems, on diagrammatic representations, on using different representations for the same problem, and on comparing representations. Finally, traditional classes spent more time on step-by-step procedures (symbolic, numeric, and Cartesian graph).

Thus, *Computer-Intensive Algebra* seems to result in different content for whole class discussion with more talk about applied problems, with more movement between and comparison of different representations, and with less time spent discussing step-by-step procedures. It is in the context of knowing both about the written and the delivered curriculum that differences in learning can be understood.

3.3. *Analysis of the Learned Curriculum*

Analysis of what students learned as they used the *Computer-Intensive Algebra* curriculum took a variety of forms. Researchers have gathered data comparing students in *Computer-Intensive Algebra* classes with students in traditional classes on problem-solving abilities (Matras, 1988/1989), on understanding of the concept of function (Sheets, 1993), on the development of abilities to use computers as tools in solving realistic algebra problems (Heid, Sheets, Matras, & Menasian, 1988), on understanding of the concept of variable (Boers-van Oosterum, 1990), on facility with beginning mathematical modeling activities (Heid & Zbiek, 1993), on ability to explore unfamiliar families of functions, and on traditional algebra skills (Heid, 1992).

3.3.1. *Problem-solving abilities*

After one or two semesters of work with *Computer-Intensive Algebra* (approximately the first three or four chapters), *Computer-Intensive Algebra* students were compared with their counterparts in traditional algebra classes on two measures of problem-solving (Matras, 1988/1989). The Problem Solving Test consisted of ten typical beginning algebra problems that students were asked to solve in any way known to them; the *Triads* task consisted of 24 groups of three problems each, and students were asked to identify which pair of each triad would be solved in the same manner. The tests were conducted with students at two demographically different sites; Site One used the *Computer-Intensive Algebra* materials for one semester and Site Two used the *Computer-Intensive Algebra* materials for all except the last month of the school year. With *California Achievement Test* (CAT) total reading, mathematical concepts, and mathematical computation scores as covariates, the *Computer-Intensive Algebra* students at both test sites performed significantly better than students in traditional classes on the Problem Solving Test. A similar analysis of covariance was conducted on the *Triads* task scores with no significant difference in adjusted means at Site One and with the *CIA* students outperforming their counterparts in the traditional classes at Site Two. This study gave evidence that *CIA* had a positive effect on the abilities of students to solve typical algebra problems as well as on their ability to perceive similar structures in word problems.

3.3.2. *Understanding of the concepts of function and variable*

Studies have been conducted analyzing the understandings of fundamental algebraic concepts by *Computer-Intensive Algebra* students. One study (Zbiek, Hess, & Rodriguez, 1989) compared the understanding of the concept of variable held by traditional algebra students and *Computer-Intensive Algebra* students with comparable prior mathematics experience. The study measured understanding through a written posttest completed by students in three *Computer-Intensive Algebra* classes and two traditional classes and two sets of interviews with a subset of students in the two classes. Whereas the traditional class interviewees developed only a concept of variable as unknown number, a majority of the *Computer-Intensive Algebra* interviewees developed a concept of variable as generalizer as well as variable as missing number. *Computer-Intensive Algebra* interviewees had a more flexible approach to problem solving than their counterparts in the traditional algebra classes. Finally, on both the written posttest and the interview measures, *Computer-Intensive Algebra* students were better than traditional algebra students at using variables to model problem situations, at reading and interpreting graphs and tables of values, and at relating graphs to their equations.

A second interview study (Sheets, 1993) analyzed understandings of the concept of mathematical function held by ninth grade *Computer-Intensive Algebra* students and by students who were enrolled in a class that studied functions, but which had the successful completion of Algebra I and Geometry as prerequisites. After conducting a series of carefully designed interviews with each student, the researcher concluded that the *CIA* students demonstrated greater flexibility in reasoning with multiple embodiments of mathematical functions than did the comparison students who had had at least two years of additional experience in traditional school mathematics programs.

3.3.3. *Tool-assisted strategies for solving problems*

An interview study of students who had begun *Computer-Intensive Algebra* in January of their eighth grade year (Zbiek & Heid, 1989) characterized the strategies that *CIA* students used in solving problems. Students in this study had access to a variety of tools (computers with function-graphing, table-generating, and symbolic manipulation software; scientific calculators; and paper and pencil) and a variety of representations (numerical, graphical, and symbolic). The study found that students in *Computer-Intensive Algebra* classes chose and used computers (with function-graphing, table-generating, and symbolic manipulation software) with significantly greater frequency than scientific calculators and paper-and-pencil. They chose and used symbolic representations (with symbolic manipulation software) three times as often as they chose and used tables and graphs. They showed that they were able to use different strategies with a single tool and representation. The data also suggested that students remain aware of the realistic context of applied algebra problems during the solution process. This study gave evidence that students could learn to use a variety of tools and representations and that they could use them with a certain degree of flexibility.

3.3.4. *Facility with beginning mathematical modeling activities*

Several studies have been conducted on *CIA* students' mathematical modeling abilities. In one study (Heid, Sheets, Matras, & Menasian, 1988), students in the *Computer-Intensive Algebra* class outperformed students from traditional algebra classes during task-based interviews in every area of mathematical modeling that was tested: constructing representations; solving equations or evaluating functions; interpreting results; linking representations; explaining the modeling process; and outlining the solution procedure. A second study (Heid & Zbiek, 1993) analyzed the use of mathematical modeling concepts by students who had experienced a full year of *Computer-Intensive Algebra*. Results from the second study suggest that students develop some competence, but vary widely in the nature of their competence in the following areas: making sense of a mathematical model, making sense of their answers to specific questions about a realistic situation, seeking consistencies in their observations about the model and their work with representations of it, and making sense of contradictory observations.

3.3.5. *Traditional algebra skills*

Finally, in many of the *Computer-Intensive Algebra* trials, after students studied the *Computer-Intensive Algebra* curriculum, they studied traditional algebra topics for a period of time (ranging from 6 weeks to one semester). In each of the three cases that were studied in depth, student performance on tests of traditional algebra skills was compared with the performance of students who had studied traditional algebra for the entire year. In the first study, when a semester of *Computer-Intensive Algebra* was followed by a semester of traditional algebra for average-ability ninth grade students, there was no significant difference between students in the two types of classes on a test of traditional skills administered at the end of the year (Heid, 1987). In a second study, when ninth grade students studied *Computer-Intensive Algebra* from the beginning of September until May 6 (six weeks before the end of the school year) and began their work with traditional algebra on May 6, there was a significant difference in traditional algebra final examination scores (Heid, Sheets, Matras, & Menasian, 1988) favoring the traditional algebra classes. In this study, however, the differences (68% for the traditional algebra classes versus 59.3%, 56.8%, and 54% for the *Computer-Intensive Algebra* classes) were not as large as one might expect given the large difference in time spent on traditional algebra skills. Finally, in a third study, when above-average students began their study of *Computer-Intensive Algebra* in January of their eighth grade year and continued through March of their ninth grade year (with the remaining 8 weeks spent on traditional algebra skills), their scores on the common items of the traditional algebra final examination were compared with scores of a high ability eighth grade algebra class and an average ability ninth grade algebra class (Heid, 1992). Using a sixth grade *CAT* total mathematics score as a covariate, an analysis of covariance was computed for the three groups on the common set of items. There was no significant difference in the scores of the three groups on this set of items.

3.4. *Current perspectives on Computer-Intensive Algebra research*

Several themes emerge from the research on *Computer-Intensive Algebra*. First, it is now clear that in a computer-intensive environment, students can develop an understanding of fundamental concepts of algebra (functions, systems, equations, inequalities, equivalence) without the concurrent or prior development of by-hand symbolic manipulation skills. When the routine procedures are left to computing tools, students in an introductory algebra course can learn to use families of functions in describing and analyzing realistic situations.[2] Also, this development does not deter the learning of traditional algebra skills later, and such learning often occurs in a shorter period of time. When the computer is used as a graphing, numerical, and symbolic manipulation tool, students can learn important algebraic ideas that would not have been encountered previously prior to late high school or early college. Our research has shed some light on the relationship between the acquisition of symbolic manipulation algebra skills and the development of symbol sense in a computing-intensive environment.

As we think about research on the emergence of algebraic thought, there are several related issues of importance. First, we need to continue our examination of the relations between by-hand symbolic manipulation skills and the understanding of algebraic concepts. Second, we need to continue our investigation into the development of "symbol sense." With the imminent widespread availability of calculators and computers equipped with computer algebra systems, students will need to understand symbols in more sophisticated ways. Students with differing levels of symbol sense will perform differently in interpreting the meaning of changes in parameters involved in symbolic representations and in interpreting the meaning of symbolic expressions as they represent quantitative relationships. In addition, when symbolic expressions represent real world situations, students with well developed symbol sense will be able to obtain different information about situations from different forms of its symbolic representations. Choosing and using appropriate technological tools, such students will be able to transform symbolic expressions into more familiar and useful forms, and they will be able to recognize members of families of functions even when they do not appear in their familiar form. With the growing availability of computer algebra systems, the goals of school algebra will shift from refinement of by-hand symbolic manipulation to development of adequate symbol sense. More research will be needed to understand the nature of this development.

4. HOW DOES *COMPUTER-INTENSIVE ALGEBRA* RELATE TO THE OTHER THREE APPROACHES?

Computer-Intensive Algebra is a functional approach to beginning algebra. It differs from existing traditional approaches to algebra for several major reasons. First, unlike traditional algebra courses, *Computer-Intensive Algebra* does not include the study of by-hand equation-solving or production of equivalent forms for algebraic expressions. Instead, students have access to computing tools including a symbol manipulation program during every class. Second, unlike traditional algebra classes, *Computer-Intensive Algebra* focuses on the concept of function and on families of functions.

Computer-Intensive Algebra has some of the characteristics of other approaches (modeling, generalization, and problem solving) discussed in this volume. First, the functional approach used in *Computer-Intensive Algebra* is like the *modeling* approach in that its content centers on the creation, evaluation, use, and interpretation of mathematical models for relationships in realistic settings. Although *Computer-Intensive Algebra* frequently proposes function rules to describe real world situations, like curricula using a modeling approach it also involves students in gathering data and deciding on function rules that might describe the relationships in that data. It is probably unlike a true modeling approach in that its content is chosen and sequenced to exemplify particular families of functions, whereas a true modeling approach might be more open-ended in its choice of problems and models and fuzzier in its fit of data to particular families.

Second, there are ways in which the *Computer-Intensive Algebra* approach addresses the role of *generalization* in algebra. Regularly, *Computer-Intensive Algebra* involves the generalization of patterns. For example, in Chapter 8 on symbolic reasoning, one way in which students generate mathematical models is to generalize patterns from geometric figures. Other experiences with generalizing (although not resulting in a symbolic rule) occur when students engage in explorations of how changes in parameters affect changes in graphs. In fact, *Computer-Intensive Algebra* may enhance students' abilities to notice patterns and to make generalizations. Research cited earlier showed that *Computer-Intensive Algebra* developed an understanding of variable as generalizer, and teachers using the *Computer-Intensive Algebra* curriculum have provided anecdotal observations of the enhanced ability of their *CIA* students to notice patterns both in symbolic procedures and in data.

Finally, the *Computer-Intensive Algebra* curriculum addresses the role of *problem solving* in beginning algebra curricula. *Computer-Intensive Algebra* involves some aspects of problem solving to a greater extent than others. Its problem-solving focus is on planning solutions and looking back. Some attention is given to formulating problems, but for the most part students are given the mathematical model with which they will work. The *CIA* focus on problem solving can perhaps best be seen through some of the research studies discussed earlier. Those studies have documented that whole class discussions in *CIA* classes are more likely than traditional classes to focus on problem formulation and interpretation (Heid, Sheets, Matras, & Menasian, 1988), and that students in *CIA* classes often outperform their counterparts in traditional classes in solving typical algebra word problems and in recognizing the mathematical structure of algebra word problems (Matras, 1988/1989).

5. CONCLUSIONS

As definitions of algebraic thinking change, and understanding of algebra relies more substantially on deep and broad understanding of the concept of function, new challenges await curriculum writers, teachers, and students. Especially in functional approaches that capitalize on technology-intensive environments, students will need to develop abilities to work smoothly between and among representations and strategies (tool-assisted and other). They will need to develop deeper understanding of the meaning of various representations, and they will need to develop the ability to reason about families of functions through multiple representations. Fortunately, research studies have shown that much of this is quite possible. In the process of completing curricula that de-emphasize by-hand manipulative skills, however, students may not be as adept at complicated by-hand symbol manipulation. The challenge of the future is to learn to capitalize on the new skills and understandings that will arise in technology-intensive functional approaches to algebra.

NOTES

1 The *Computer-Intensive Algebra* text is being distributed by Janson Publications (Dedham, MA) as *Concepts in Algebra: A Technological Approach.*

2 Teachers in at least 20 different states are now successfully implementing *Computer-Intensive Algebra* in a wide variety of classroom settings (e.g., inner-city Miami, Philadelphia, Gary, Knoxville, Washington, D.C., and San Jose; rural Wisconsin and Virginia; small-town Pennsylvania and Iowa; suburban New Jersey) with students ranging from average seventh graders, accelerated eighth graders, average ninth graders, and "at-risk" eleventh graders. This implementation is being funded in part by the National Science Foundation, TPE# 9155313, "Empowering Mathematics Teachers in Computer-Intensive Environments" (Project Directors: M. Kathleen Heid and Glen Blume). Any opinions, findings, conclusions, or recommendations expressed herein are those of the author and do not necessarily reflect the views of the National Science Foundation.

CHAPTER 19

INTRODUCING ALGEBRA BY MEANS OF A TECHNOLOGY-SUPPORTED, FUNCTIONAL APPROACH

CAROLYN KIERAN, ANDRÉ BOILEAU, MAURICE GARANÇON

This chapter describes the main highlights of a seven-year program[1] *that was devoted to researching various aspects of introducing algebra to different groups of students by means of a technology-supported, functional approach. The software that was developed as part of the research program was CARAPACE, a functional problem-solving environment that includes three representations: algorithmic, tabular, and graphic. The six studies that are synthesized were carried out with students in the 12- to 15-year age range and focused on the cognitive processes engaged in while the students worked with each of the three functional representations. The chapter concludes with a discussion of the contributions of this functional approach to the initial learning of algebra.*

For some mathematics educators, functions are considered the core of algebra. For example, Schwartz and Yerushalmy (1992) have proposed that function is the "primitive algebraic object." The aim of this chapter is to describe what we mean by a functional approach to algebra and to report some of our empirical findings of studies designed to inquire into the feasibility of different aspects of this approach.

1. WHAT IS A FUNCTIONAL APPROACH?

A functional approach to algebra does not necessarily mean the study of functions. It does, however, entail the use of letters as variables--as opposed to unknowns. For example, the expression $3x + 5$ can be viewed as a function, that is, a mapping that translates every number x into another. As such, x is interpreted as a variable because it can take on a range of values. In contrast, when two functions such as $3x + 5$ and 23 are equated (i.e., $3x + 5 = 23$), x is clearly to be interpreted as an unknown--the value for which the two functions are equal. But there is more to a functional approach than seeing letters as variables; it also comprises, for example, viewing a function from the perspective of the relationship between the x-values and the corresponding functional values, that is, from the perspective of how a change in x produces a particular variation in the values of the function.

The historical event that moved algebra in this direction was an ingenious contribution on the part of Viète in the late 16th century: the introduction of letters in two different roles--as unknowns and as givens. His use of letters in this double role was crucial to the development of, what we might call, functional algebra; it now became possible to express general, non-numerical solutions to problems, something that had not been imagined earlier. In other words, the letter could now be viewed as a variable. Over the course of time, the parameterized expressions that were conceptualized by Viète came to be viewed in a functional way. Sfard and Linchevski (1994) have pointed out that "after the new invention was transferred (mainly by

N. Bednarz et al. (eds.), Approaches to Algebra, 257-293.

Descartes and Fermat) to geometry to serve as an alternative to the standard graphic representations, and then applied in science (by Galileo, Newton, and Leibniz, among others) to represent natural phenomena, algebra was ultimately transformed from a science of constant quantities into a science of changing magnitudes" (p. 200). However, further developments, notably by Dirichlet in the 1830s, led to a modification in the view of functions from a dependency relation to an arbitrary correspondence between real numbers; this latter was generalized even further a hundred years later by Bourbaki who defined function structurally as a relation between two sets.

Most current textbooks, in their introductions to functions and variables, reflect the structural advances of the last 150 years. This has been lamented by many in mathematics education; for example, Shuard and Neill (1977) have sadly pointed out that the idea of functional dependence has been totally eliminated from the current definition of function, and Freudenthal (1982) has advocated a return to notions of dependency in the teaching of functional ideas to elementary and secondary students. Students too have been found to resist structural approaches to instruction on functions and variables and to form more operational conceptions that involve viewing a function as a process for computing one value by means of another (e.g., Sfard, 1987, 1991; Soloway, Lochhead, & Clement, 1982). Thus, in the functional approach we describe in this chapter, not only are letters viewed as variables, but also the representations used for expressing functional relations are presented in an operational, process-oriented way. The representation that is favored at the outset is the computer algorithm, which involves the use of input and output variables, and which in turn provides the computational foundation for tables of values and Cartesian graphs.

The context in which these algorithmic processes and pairs of input-output variables are elaborated is one of problem solving. Typically, algebra problem solving involves generating an equation with an unknown. A functional approach to problem solving, on the other hand, entails the setting up of a more general functional relation among the problem givens and viewing the input-output letter names as variables. Students attempt to establish the functional relation by means of trying out the operations of the problem situation with a few trial numerical values. As well, in functional approaches to problem solving, there is usually a greater variety (linear, quadratic, and exponential types of functional situations) than one tends to find in standard algebra curricula. To obtain a sense of the problem-solving contexts that are used in our functional approach (for a fuller description, see the appendix at the end of this chapter), let us look at the following situation:

Karen works part-time after school selling magazine subscriptions. She receives \$20 as a base salary per week, plus \$4 for each subscription she sells. How much can she earn in a week? How many subscriptions must she sell to earn at least \$50?

The general functional relation of this situation is: Karen's base salary of \$20 plus \$4 per subscription yields the total that she earns in a week. Thus, a response to the first question is: "It depends on how many subscriptions she sells; for example, if she sells 5 subscriptions, she would earn \$40; if she sells 10 subscriptions, she would earn \$60, and so on." Thus we see the variable aspect of the problem situation. The second question, which looks more like a typical algebra word problem question, might be responded to by trying a few values in the general relational statement and finding that 8 would be an appropriate answer. It is emphasized that, in a functional approach to the introduction of algebra, questions that could be interpreted as involving the use of the letter as an "unknown" are handled--in passing--through the functional relation by providing some numerical value to the function.

A side issue, but one that is of relevance to the goals of the book to which this chapter is addressed, concerns the role of generalization in a functional approach. Setting up the general functional relation in the form of an algorithm involves a certain amount of generalizing for beginning algebra students. As noted above, they often attempt to make sense of a problem statement by trying a few numbers. They read and reread and carry out a few operations by hand. When they believe that they have figured out the sense of the problem and how to compute the functional values of the situation, they then translate their numerical computations into a general formulation. Thus, they have generalized their localized numerical work into a more global algorithm. In this sense, a functional approach intersects with a generalization approach to introducing algebra.

Returning now to the functional situation presented earlier, we show its computer-algorithm representation (see Figure 1) in the *CARAPACE* environment (see the appendix for details of *CARAPACE*). The entering into the computer of the functional relation in the form of a computable algorithm allows the computer to carry out the student's numerical tries. Figure 2 illustrates the detailed-calculation display that can be accessed when a student enters a trial value. The reader will observe that the variable names used in these figures are not single letters; significant variable names are introduced and used for an extended period of time before moving on to single letter names.

It is to be noted here that a functional approach does not necessarily demand the use of a computer. Rubio (1990), for example, used a similar functional approach in a problem-solving study that did not involve computers. However, such non-computer-aided work bears with it a heavier cognitive load with respect to the student's carrying out all of the numerical trials by hand and, as well, does not permit the easy access to the other functional representations that are afforded by computer environments.

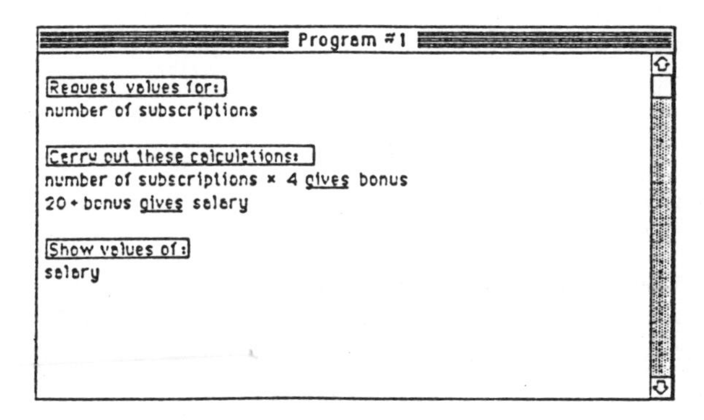

Figure 1.

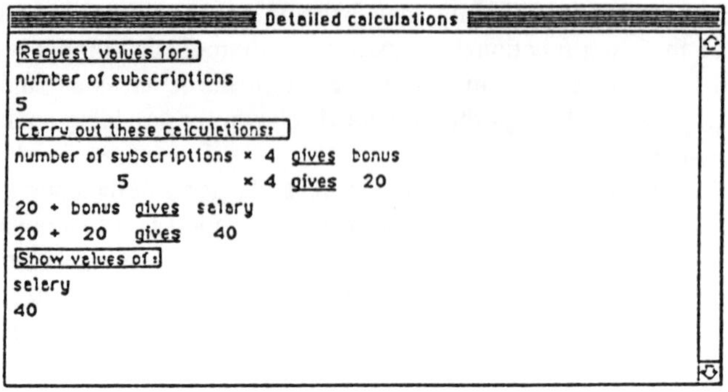

Detailed calculations

Request values for:
number of subscriptions
5
Carry out these calculations:
number of subscriptions × 4 gives bonus
5 × 4 gives 20
20 + bonus gives salary
20 + 20 gives 40
Show values of:
salary
40

Figure 2.

The use of several representations is an integral part of a functional approach to algebra problem solving. Up to now, we have only emphasized the procedural functional algorithm and its input/output variables. Equally important are the tabular and graphical representations as problem-solving tools. In the *CARAPACE* environment, trial values for the input variable can also be entered while in either one of these two other representation modes. These trial values along with the results of the calculations that are carried out by the computer are dynamically displayed as pairs of numbers in the table of values and as discrete points on the Cartesian graph. See Figures 3 and 4 for an illustration based on the subscription problem above.

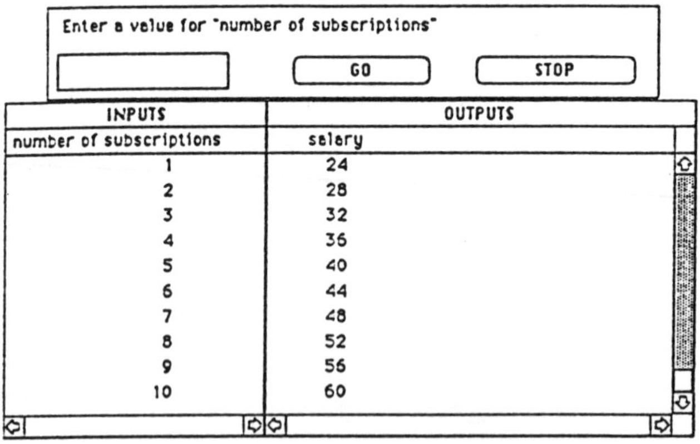

Enter a value for "number of subscriptions"

GO STOP

INPUTS	OUTPUTS
number of subscriptions	salary
1	24
2	28
3	32
4	36
5	40
6	44
7	48
8	52
9	56
10	60

Figure 3.

It is important to point out that, because the procedural algorithm entered by the student serves as a basis for calculating the functional values represented in both the table and graph, the interpretation given to these representations tends to be process-oriented.

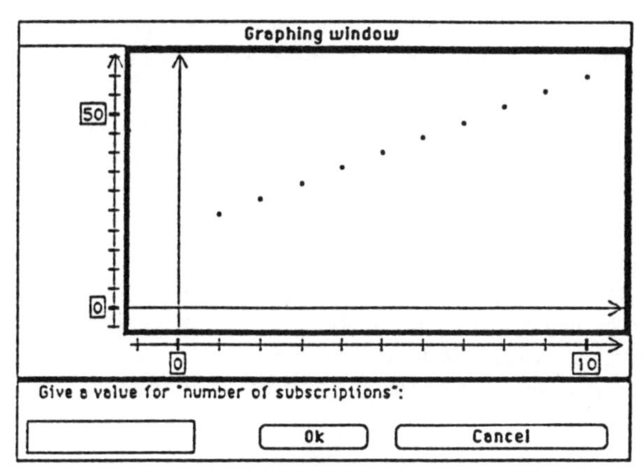

Figure 4.

This is in contrast with the object-oriented approach of Schwartz and Yerushalmy (1992) in the *Function Analyzer* environment, an environment that permits direct transformations of the graphical representation by means of operations such as translation, dilation, and contraction. In *CARAPACE* any alteration to the underlying function that is represented by a given graph requires that one first change the procedural algorithm that produces the graph. (See Sfard, 1991, and Sfard & Linchevski, 1994, for an elaboration of process-object distinctions.)

2. WHAT DOES SUCH A FUNCTIONAL APPROACH BRING TO THE INITIAL LEARNING OF ALGEBRA?

We began our research with the functional approach in 1986. After having developed the first version of the *CARAPACE* computer environment (Boileau & Garançon, 1987), we initiated a series of studies designed to examine in detail various aspects of this approach. Subsequent versions of *CARAPACE* were integrated into these studies.

2.1. *Study 1 (1987-1988): Guess-and-Test Numerical Solving Methods*

This study (Kieran, Garançon, Boileau, & Pelletier, 1988) focused on the guess-and-test numerical solving methods and strategies that are used by students when the input and output values are displayed in tabular form. Subjects were two sixth graders (12 years of age) considered by their teacher to be of average-to-high ability in their school mathematics. Each child worked individually with the researcher during a ten-week period that involved one 60-90 minute session per week. The low-intervention teaching experiment uncovered students' spontaneous approaches to guess-and-test problem solving in a computer environment.

Fey (1989) has argued that the presence of technology makes guess-and-test methods more attractive than they once were because the testing phase of the guess is now speeded up considerably. As well, such methods provide students with the opportunity to link algebra problem solving with their earlier arithmetic knowledge and experience. The substitution of numerical test values into algebraic representations of problems permits beginning algebra students to construct meaning for problem representations that may be different from those they experienced in the past, while at the same time using solving methods based on familiar arithmetic techniques.

When the students of this study began to solve each word problem on paper, doing some of their calculations by hand, their real-world knowledge of the semantics of the problem suggested: (a) suitable starting values, (b) the domain of variation, and (c) whether or not the situation was an increasing or decreasing one. As soon as they moved from their paper-and-pencil calculations to the tabular display, they tended to forget the contextual information they had previously been relying upon in their non-computer work and used strategies based solely on order relations and pattern recognition. Figure 5 shows the trial values used by one of the students for the problem:

A town is inhabited with 256 434 people. Because of pollution, many people leave the town to live in the country. Each year the population decreases by 1623 persons. In how many years will there be only 128 217 people left in this town?

That the students did not, in general, use contextual cues in deriving their guess-and-test trial values while in the tabular mode was further evidenced with the "hidden problems" (problems for which the students were merely told the value of the target output; they had to generate trial input values permitting them to reach that target, without knowing the underlying calculations). The students used the same approaches to attain the target output of the hidden problems as they used for the regular word problems.

INPUTS	OUTPUTS
Number of years	Remaining population
95	102249
98	97380
96	100626
94	103872
93	105495
89	111987
88	113610
86	116856
80	126594
79	128217

Figure 5.

Two of the main strategies that were used to derive test values in tabular form were the *increasing* and *decreasing* strategies. The *increasing* strategy, the one most commonly used, can be specified by the following rules: (a) If the result is too small (with respect to the target), then increase the trial value; (b) If the result is too large (with respect to the target), then decrease the trial value. The initial use of the increasing strategy in most problems was generally marked by a search involving fairly gross test values until a result was found that had the same number of digits and the same first digit as the target. From that point on, the students tended to try to obtain a digit-by-digit match with the goal value.

The *decreasing* strategy can be summarized by the following rules: (a) If the result is too small (with respect to the target), then decrease the trial value; (b) If the result is too large (with respect to the target), then increase the trial value. Attempts to use this strategy generally showed confusion and interference from the preferred increasing strategy.

The numerical solving approaches used by the students of this study suggest that several factors tended to thwart their attempts to be systematic in deriving test values: (a) decreasing situations (i.e., situations in which the trial values had to be decreased in order to increase the result, and vice versa), (b) numbers with more than five digits, and (c) numbers involving decimals. The strategies they used, though primitive and often inefficient, eventually yielded the targeted result. The fact that the computer handled the laborious calculations allowed them to focus on certain patterns in the numbers. Some strategies we might have expected to see, but did not, were: proportionality, interpolation, and reliance upon the given relations of the word problem.

That there were no context-effects while our subjects were generating trial values in the tabular mode was a surprising finding. It seemed that the tabular representation was very conducive to non-contextual processes based simply on order relations and number patterns, and suggests that we might expect similar behavior with other tabular forms of data, such as spreadsheets.

2.2. Study 2 *(1988-1989): Generation of Functional Algorithmic Representations*

This study (Kieran, Boileau, & Garançon, 1989) focused on the processes used in the generation of functional algorithmic representations. Subjects were 12 average-ability seventh graders (13 years old) who participated in hourly sessions twice a week for a four-month period. Some worked individually with the researcher; others worked in pairs.

Before summarizing the findings of this study, we remind the reader that the algorithmic representations of *CARAPACE* do not use traditional algebraic symbolism, rather a kind of natural language representation. One of the advantages of a representation that is close to natural language can be inferred from the historical development of algebraic symbolism. The first evolutionary stage through which algebra passed was the period before Diophantus, a period which was characterized by the use of ordinary language descriptions for solving particular types of problems and which lacked the use of symbols or special signs to represent an unknown value. The later development of a specialized symbolic language stripped away meaning from the language in which algebraic activity had been previously expressed. It is this premature stripping away of meaning that we attempt to avoid in *CARAPACE*.

The hypotheses that we had generated for Study 2 find their bases in ideas stated by Usiskin (1988) and Filloy and Rojano (1984). Usiskin has pointed out that:

In solving problems such as "When 3 is added to 5 times a certain number, the sum is 40," many students have difficulty moving from arithmetic to algebra. Whereas the arithmetic solution involves subtracting 3 and dividing by 5 [i.e., using inverse operations], the algebraic form $5x + 3 = 40$ involves multiplication by 5 and addition of 3 [i.e., using forward operations]. That is, to set up the equation, you must think precisely the opposite of the way you would solve it using arithmetic. (p. 13, parenthetical remarks added)

Furthermore, Filloy and Rojano (1984a) have emphasized that problems which can be represented by equations such as $x + a = b$, $ax = b$, and $ax + b = c$ can be easily solved by arithmetic methods. They have claimed that a "didactic cut" occurs with problems that are representable by equations of the type $ax + b = cx + d$, for students are generally not able to solve this type of problem by arithmetic methods. They must use algebraic methods that involve operating on the letter. This suggests that, in algebra, not only do students have to shift from thinking about inverse operations to thinking about forward operations, but also they must generate a representation that involves the application of algebraic methods.

We hypothesized that a functional approach to the early learning of algebraic problem solving in a computer-supported environment would be conducive to helping students think in terms of forward operations. We also conjectured that a procedural representation that has the dynamic, "operational," features, which according to Sfard (1987, 1991) and Soloway, Lochhead, and Clement (1982) allow students to be more successful problem solvers, would be more helpful than the static, equation form of representation.

In this study, the first eight sessions began with the presentation of the functional situation, without the actual problem-solving question--for example (to reuse our earlier problem): "Karen has a part-time job selling magazine subscriptions in her neighborhood; she is paid $20 per week, plus a bonus of $4 for each subscription she sells." The question, which would follow later in the same session, was: "How many subscriptions must she sell if she wants to earn $124 in a week?" One reason for separating the problem situation from the actual question was to prevent the students from attempting to solve the problem immediately by means of inverse operations. From the ninth session onwards, the students were presented with entire problems to be solved at once, that is, with both the situation and the question.

The findings of the first three sessions showed the immediate ease with which students took to a functional representation of word problems. After they had entered their program into the computer, using "forward operations"--in the sense of Usiskin--and had tried out a few values for the input variable, they were then given the question to go with the problem situation. At this point, we noticed a predominance of attempts to apply inverse operations to solve these problems, that is, to bypass the functional representation that they had entered into the computer and to use their old arithmetic solving methods involving inverse operations. From the fourth to the eighth sessions, we decided to ask the students to verbalize their own question to go with each of the problem situations. Most of the problems they generated were ones that their "function program with its forward operations" could answer. This technique appeared to be of some help in getting them away from their spontaneous use of inverse operations. When we presented our question afterward, the majority continued to rely on the functional representation that they had been using for their own particular question (or an extension of it) and to solve the problem by trial values for the input variable.

In Session 9, when the students were presented for the first time with the entire problem at once, they again tried to use inverse operations. The first problem of this session was an $ax + x = c$ type, that is, one in which the use of inverse operations involving simply c and a would not lead to success:

The price of a radio is 33 times the price of a cassette. If a radio and cassette together cost $324.70, what is the price of each?

For half of the students, the spontaneous approach was to divide 324.70 by 33, even before trying to generate a functional representation of the problem. For the next problem (a problem of the type $b - (dx + eax) = c$), we again saw the same tendency to incorrectly use inverse operations (i.e., $(c - b) / (d + e)$) when the entire problem was presented at once.

Eventually, the students became aware that they would not be successful unless they used the functional approach to problem solving that we were proposing. From Session 18 onwards, many of the problems that were presented were those that algebra experts would likely model with equations involving occurrences of the letter on both sides of the equal sign (e.g., $ax = b + cx$, $ax \pm b = cx \pm d$). There was no evidence to suggest that the students found the generating of a functional representation for these problems any more difficult than for the simpler "equation-types." In actual fact, the translation of these latter types of word problems into multi-line programs containing a single operation per line is no different from the translation of less complex problems that, were they to be modeled by equations, do not contain several occurrences of the letter on both sides of the equation. For these more complex problem-types, none of the students attempted to use inverse operations as a spontaneous first approach. This suggests they were beginning to realize that, for certain kinds of problems, their old arithmetic methods would be of little use. It also leads us to conclude that there is nothing intrinsically easier about viewing letters (or words) as unknowns rather than as variables, a phenomenon that many research reports would have us believe. The students of this study could switch effortlessly from one to the other, depending on the demands of the task at hand.

2.3. *Study 3 (1989-1990): Transition to More Standard Algebraic Representations*

This study (Garançon, Kieran, & Boileau, 1990; Luckow, 1993) examined the transition from "natural language" functional algorithmic representations to more standard algebraic representations. Subjects were eight average-ability eighth graders (14 years old) who worked in pairs for a period of 4 months, at a pace of two 50-minute sessions per week. Four of these subjects had been participants in the previous year's study. The other four worked with the researcher for an additional period of three months at the front end of this year's study in order to bring them to the same level of experience as the students who had been with us the year before.

In her analysis of the passage from operational conceptions to structural conceptions of a mathematical notion, Sfard (1991) identified the three phases of interiorization, condensation, and reification. The goal of Study 3 was to investigate the phases of interiorization and condensation as they might apply to the learning of single-letter variable names and algebraic expressions while students attempted to make the transition from the algorithmic procedural representations of *CARAPACE* to the more traditional forms of algebraic notation for equations (but without the equal sign) and to deal with the associated use of parentheses, order of operations, shortening of variable names, and the conventions for implicit multiplication. Luckow's (1993) study, which was carried out within our research program, focused in particular on the phase of condensation for the algebraic expression of functional relations. She found that the students adopted some condensation steps more readily than others. Those that were learned fairly easily were: using a single letter to replace a word phrase, using parentheses to group an expression, and eliminating the multiplication sign. One area of difficulty for all the students was the use of a variable name (or an expression representing a variable name) in two or more places on a single line. For example, in condensing the following two-line program:

price of pen × 3 gives price of watch
price of pen + price of watch gives total cost

into a one-line program:

(price of pen × 3) + price of pen gives total cost,

the students had the tendency to suggest:

(price of pen × 3) + price of watch gives total cost

in order to avoid having the same variable mentioned twice on the same line, which looked incorrect to them.

Another difficulty was the replacement of a variable name or an expression by an equivalent variable name or expression in two or more locations of the procedural representation at once. For example, the problem:

The area of Mr. O. Macdonald's farm is 80 Ha (hectares) greater than his neighbor's. Mr. Macdonald sold one-sixth of his land to his neighbor and now their two farms have the same area. What was the area of their individual farms before the sale?

led to program algorithms such as (the following sequence, which was written on paper first, shows the student's reliance on identifier labels for single-letter variable names):

$x + 80$ gives y – Mr. M's farm
$y / 6$ gives z – sold to neighbor
$x + z$ gives a – new farm of neighbor
$y - z$ gives b – Mr. M's new farm.

When asked to shorten the above program using fewer lines, the student wrote:

$(x + 80) / 6$ gives z
$x + z$ gives a
$y - z$ gives b.

Even though the student, whose work is displayed here, realized that the condensed program would no longer run because of the unidentified y value on the last line, she seemed unable to spontaneously replace both occurrences of y by $x + 80$.

A third area of difficulty identified by Luckow, and it was the most prominent one, was simplification. Most of the students did not spontaneously apply the syntactical operation of expression simplification to the contextualized procedural representations. Simplification requires a relinquishing of the contextual meaning of an expression and was found to be an operation that the students thereby did not carry out. For example, the problem:

A group of girl guides went on a 3-day bicycle trip which covered a total distance of 256 km. On the second day, they covered 23 km more than on the first day. On the third day, they did 30 km less than on the second day. How many km did they cycle on the first day?

led to functional algorithms such as:

$$x + (x + 23) + ((x + 23) - 30) \text{ gives } T.$$

When asked if they could write this in a shorter way, they removed the parentheses, but said that they could not abbreviate it any further. Yet a follow-up interview involving similar expressions in a non-contextualized setting that reminded them of the algebra they were doing in class suggested that they could simplify further, as long as the expression was not tied to a particular context.

The pedagogical implications of this study include the following. Carrying out initial numerical trials would appear to be extremely helpful for students in making sense of word problems, in representing them, and also in solving them. The use of significant names for variables can assist students in retaining the sense of a problem and in performing operations such as substitution. But, as noted above, this study uncovered the lack of any real compulsion or motivation on the part of the student problem-solvers to simplify expressions in a computer-supported, functional, problem-solving environment such as *CARAPACE.* In order for students to accept the notion of simplification, not only must they relinquish the semantic meaning of expressions, but, as well, they must see some reason for shortening the work for the computer, which is rather senseless since the computer will quickly carry out all of their calculations in any case. Thus, even though both substitution and simplification are integral components of algebraic manipulation, this study found that preparing the ground for the former seems a lot less problematical than the latter.

A final remark to be made here concerns the use of *CARAPACE* as an alternate to algebra, or at least as an alternate to a part of what we mean by school algebra. One of the aims of our study had been to use these multi-line procedural representations and significant variable names as a device for building a meaningful bridge to algebraic expressions and equations. However, as we have seen, once students became skillful in using this problem-solving environment, they felt no desire to shorten their procedural representations or to simplify them in order for them to resemble more canonical forms of algebraic expressions and equations. In effect, the computer tool *CARAPACE* had usurped the potential of algebra as a problem-solving tool. This remark is intended as a reminder that not all technology-supported roads that are intended to be algebraic lead to developing meaning for traditional algebraic representations and transformations. Some of these roads may lead to the creation of alternate problem-solving tools, which may not be unfortunate--especially if one is of the opinion that not all students need a traditional algebra course during their high school careers.

2.4. *Study 4 (1990-1991): Use of Cartesian Graphs in Solving Problems*

This study (Kieran, Boileau, & Garançon, 1992) focused on the use of graphical representations as a problem-solving tool and the role that these representations play in helping students to decide whether a problem has more than one solution. The problem situations that were used most often were non-linear ones. Subjects were four ninth graders (15 years old) who were at that time in their third year of participation in the research project. They worked in pairs for fourteen sessions, usually two sessions per week. They had not yet begun to work with graphical representations in class, although the pretest we administered at the beginning of this study indicated that all of them had learned to plot points at some time in the past.

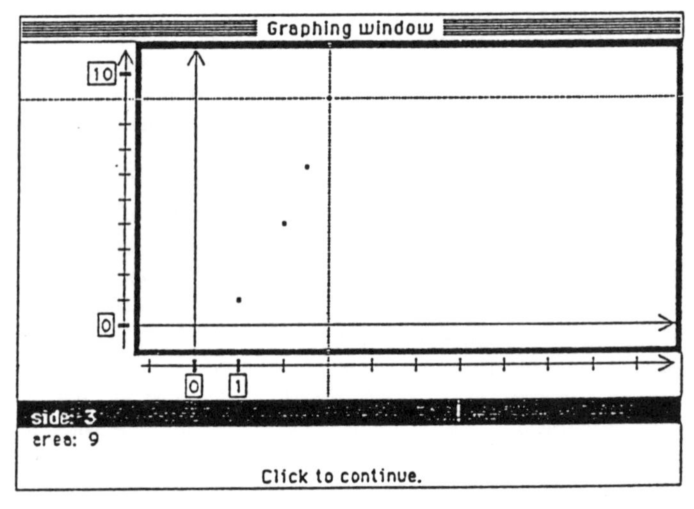

Figure 6.

The graphical representation in *CARAPACE* has been designed to permit one to teach elements of problem solving as well as technical aspects of graphing, such as scale modification and zooming, to novice algebra students. A typical functional situation that was explored graphically during the second graphing session of this study was the following: "Write a program that will calculate the area of a square, based on the length of one side." After students wrote a *CARAPACE* program, they then went to the graphing window (recall that all graphs in this environment are based on the student-provided computer algorithm, which includes input and output variables). Figure 6 illustrates the first four points plotted by a pair of students after they had first decided on an appropriate scale. Note that all the input-output values represented in the graphing window are also recorded in the table of values, which can be accessed at any time. As soon as further points began not to be visible in the graphing window, we showed the students how to change scales dynamically: by clicking on one of the numerical labels of the axes and moving it closer or farther away from the other label (Figures 7 & 8). As the scale is changed, *the already-plotted points simultaneously move.* This movement, which takes place before the students' eyes, leads to many pedagogically interesting questions (see the appendix).

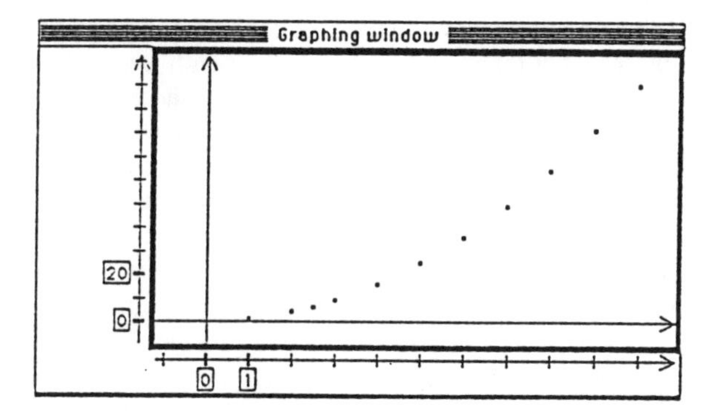

Figure 7.

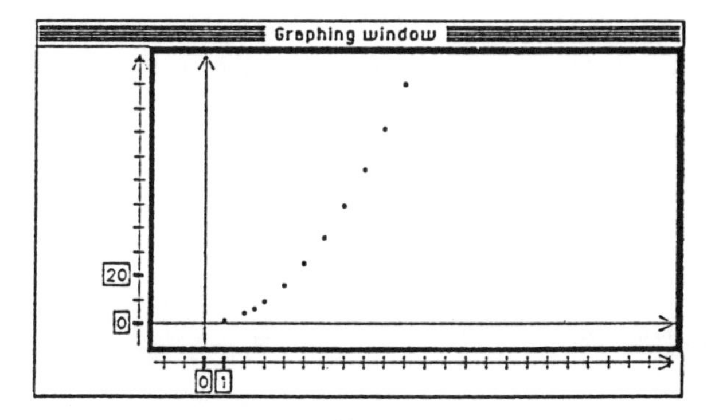

Figure 8.

The four major findings of this study were as follows: (a) The students believed that multiple solutions result from multiple processes; (b) they did not spontaneously seek more than one solution unless the wording of the problem clearly oriented them in this direction; (c) their heuristic search for multiple solutions was much more effective when they used graphical representations of non-linear functions than when they used only a table of values representation of these functions; and (d) they found the technical aspect of changing scales in the *CARAPACE* environment to be very easy to do. The latter two results merit some discussion at this point.

The finding related to the effectiveness of graphical representations over tables of values in the search for multiple solutions should not be surprising. In an environment involving functional representations of problem situations and the use of numerical methods to find problem solutions, non-linear functions can be especially difficult to work with when using a table of values representation only. The evidence from one particular quadratic problem that was presented to the students on two occasions--once before the graphing environment had been introduced and again afterward--showed very clearly that students at this level of experience could not derive an overall view of the behavior of non-linear functions from a table of values and thus could not use this tool effectively to find multiple solutions to problems. Ursini (1992) has discussed a similar result in her study of children working with functional relations in a Logo environment. In contrast, plotting several points on a Cartesian graph provided the students of our study with a global perspective that, along with the use of the graphics cursor which displays the coordinates of individual points, allowed them to focus on those areas of the graph where solutions were to be found. They soon became aware that the set of solutions to a problem could be found along the horizontal line which represented the goal or output value and where it intersected their set of graphed points.

The other finding to be discussed concerns the ease with which the students modified the default scales of the axes. The design of the graphing environment of *CARAPACE* made it seem almost effortless to change the size of the viewing subregion of the Cartesian plane in order to search for problem solutions. Not only was it easy to change scale, but the students usually had no difficulty in deciding which values to use for the modified scales--especially the vertical scale. Past studies carried out by, for example, Goldenberg (1988) have suggested that students experience perceptual illusions and even misunderstandings about the behavior of graphs when scales are changed. Such was not the case with our students. We might conjecture that the dynamic shifting of points at the same instant that the students changed the scale helped them to keep a visual track of the result of scale changes and thus avoided the difficulties reported by Goldenberg. We cannot overemphasize in this regard that our students worked with graphs of discrete points, not continuous curves, and that this aspect of our graphing environment was likely crucial in helping them see not only the movement of points but also changes in the distances between the points.

2.5. *Study 5 (1991-1992): Producing, Interpreting, and Modifying Graphs*

This study (Garançon, Kieran, & Boileau, 1993) aimed at uncovering areas of ease/difficulty experienced by four seventh graders (13 years old) in learning how to produce, interpret, and modify graphs that represent functional problem situations. The researcher met each pair for an hour twice a week over the course of 13 weeks. In the first part of the study, students learned how to write procedural algorithms to represent functional situations and to use both this representation as well as the table of values as problem-solving tools. During the 11th session, the students were introduced to the graphing environment and to its use in problem solving. A major focus of our data analysis was the ways in which the students coped with the two types of infinity they encountered in a dynamic graphing environment that plotted intervals of discrete points rather than continuous curves.

Students have traditionally encountered difficulty in interpreting graphs. Kerslake (1981), in presenting the graphing data from the large-scale longitudinal Concepts in Secondary Mathematics and Science (CSMS) study of 13- to 15-year-olds, pointed out, for example, that pupils having plotted points and joined the line on which they lay were unable in general to recognize that other points also lay on the line and that there were indeed infinitely many points both on a line and between any two points on the line. Brown (1981) found similar results for the CSMS questions related to the infinite nature of the set of numbers expressible as decimals (e.g., "How many different numbers could you write down which lie between 0.41 and 0.42?"). In a study on intuitions of infinity involving 500 11- to 15-year-olds, Fischbein, Tirosh, and Hess (1979) found that, for a question on whether the successive division of a segment into smaller and smaller parts would eventually come to an end, the majority at each age level believed that the process could not go on forever. Only 5.4% believed that the process is infinite because there is an infinite number of points. On another question of the Fischbein et al. study involving successive division of a segment in two and whether any one of the points of division would coincide precisely with a given point of the segment, the percentage of wrong answers was equally high (e.g., 91% for 12-year-olds).

The matter of student understanding of infinity/continuity is, thus, an important one that we decided to include in this study. It seemed reasonable to us that the dynamic features of computers might have a positive effect on enhancing student understanding of this conceptual domain. However, very few computer-related studies have addressed this issue (Romberg, Fennema, & Carpenter, 1993). One notable exception is the work of Goldenberg and his colleagues (Goldenberg, 1988; Goldenberg, Lewis, & O'Keefe, 1992). Goldenberg et al. have reported the positive effects of using a graphing environment that "instead of presenting the graph as a gestalt--either a fully formed static picture or the canonical smooth left-to-right sweep of graphing software--presented one point on the curve at a time" and which counteracted students' tendency to "in fact, lose track of the points when they face a continuous curve" and to "fail at tasks that require consciousness of the points" (p. 238). Goldenberg (1988) has also remarked that:

> In a pedagogy that makes regular use of translation between graphic and symbolic representations of functions, we cannot avoid dealing with issues such as these [the finite and the infinite, ...], and yet notions of infinity and the continuous nature of the real line are totally foreign to beginning algebra students. Research needs to be done to find appropriate ways of dealing with these issues. (p. 164)

We found that discrete plotting of points, along with the availability of different options for actually placing the points on the screen, not only permits students to maintain for a longer period of time a link between graphical and numerical aspects of functions but also provides for a very gradual transition from algebraic notational (procedural) representations to Cartesian graph representations. In spite of this, we observed that, in the beginning, students tended to easily lose partial awareness of the link between the two representations and that the strategies of successive approximation which were used in the numerical context were not immediately called upon in a graphing context. We also observed an initial tendency to rely on the appearance of graphs and to extrapolate from what was visible. A comparison of these findings with the results from our previous studies suggests that a prolonged period of experience with tables of values before being introduced to graphical representations in the *CARAPACE* environment does not significantly alter students' graphical, problem-solving, approaches.

It is impossible here to make comparisons with software that plots "continuous" graphs that are defined for "all the real numbers" because, if the problems that we are raising do not occur in continuous graphing environments, we do not know if it is because they do not exist or because the conditions necessary to provoke them are not present. In any case, we believe that there is a fundamental difference between reading the abscissa of a point from a ready-made graph and generating an abscissa in order to construct a point of a graph (as is done in the *CARAPACE* environment).

Nevertheless, we saw a rapid improvement in the students with respect to the following aspects: the development of successive approximation strategies in the graphing context, and the replacement of an approach favoring the search for continuity and a multitude of points by an approach favoring the search for local density of points—sufficient to solve a problem.

2.6. *Study 6 (1992-1993): Structural Extension of a Process-Oriented Conception*

This study (Kieran, Garançon, Lee, & Boileau, 1993) aimed at exploring the nature of the cognitive shift that is involved in extending a numerical, process-oriented perspective on functional problem situations and their representations to a form-oriented, structural one. By means of a single-subject case-study of a bright 14-year-old eighth grader, Kim, who had already spent 20 hours learning to solve problems in the *CARAPACE* environment the previous year, but who had not yet been introduced to traditional algebraic manipulation, we followed the subject's evolving awareness of the linear and quadratic families of functions and of the role of the parameters of these functions over the course of 12 sessions (usually two sessions per week). During the study, there was no explicit teaching; rather the interventions took the form of a structured sequence of tasks accompanied by probing questions. In each session, the interviewer presented Kim with functional problem situations, which he was to represent in the *CARAPACE* computer environment with three different representations: program algorithms, tables of values, and Cartesian graphs. The first five sessions focused on linear functions; the next seven sessions were devoted to quadratic functions (problem situations dealing with area, ticket sales, stopping distance of a car, throwing a ball, etc.) and the comparison of quadratic and linear functions in various representations.

The most salient finding of our study was the persistent development of Kim's numerical approaches for dealing with tasks involving various representations of functions. This is illustrated by the way in which he explained how he could predict whether the program algorithm he had generated for a given situation would produce a linear graph or not--by looking at its table of values and checking for the presence of a constant difference, as opposed to a changing difference, from one output value to the next.

During several sessions, Kim was asked a variety of questions that could have been answered by referring to the form of the program, but never were. His constructions were always numerically based. For example in Session 10, he was asked to generate any program that would have a line for its graph. He thought of a simple context (a helium-filled balloon rising at a rate of 2 meters/second, which he remembered from the previous session) that he expressed with the program "$m \times 2$ gives h." He was then asked to generate a program whose graph would not pass through the origin; this led to "$m \times 2 + 1$ gives h." Thirdly, he was asked to generate one which climbed faster (to use his vocabulary); he produced "$m \times 5 + 1$ gives h." When asked to explain why this would climb faster, he replied: "Because here if you put 1 (in $m \times 2$) it gives 2, and here if you put 1 (in $m \times 5$) it gives 5; it's higher."

In the next question of Session 10, Kim was asked to write a program that would produce a curve for its graph. He thought of a context and mentally substituted a few values (0, 1, 2) to see if they resulted in a nonconstant difference for successive output values. He showed no sign of being aware of and being able to rely upon specific forms, such as x^2 (or $x \times x$). When he finally generated by trial-and-error a program that yielded a curve, he was asked if he could produce one to go in the other direction. Again, he resorted to trial-and-error numerical testing.

In Session 11, he was asked to enter the program "*3(x + 5)* gives *k*" and to try to create another program without parentheses that would give the same graph. He thought for a while and then began to write down a table of values (with inputs of 0, 1, 2, 3). He calculated the differences between successive output values and thus was able to come up with the program "*15 + x × 3* gives *k*." When asked by the interviewer if he could go directly from "*3(x + 5)*" to "*15 + x × 3*," he said he could not. This example suggests quite dramatically that students who have not yet learned to carry out some basic symbol manipulation might be hampered in their explorations of alternate forms of expressions and in their attempts to generate equivalent expressions of a function.

One of the questions of the last session was designed to disclose what resources Kim might fall back on if he were not permitted to rely upon his numerical approaches. It involved a set of 11 program excerpts (e.g., *x × x - 105* gives *y*) and for each one Kim was asked to state, "without doing any calculations in his head," which ones would have a graph that forms a line and which would have a graph that forms a curve. For all of those expressions having more than one occurrence of x, Kim responded that they would produce a curve. Thus, over the course of the study, he had not become aware of the crucial differences between, say, $x \times x$ and $x + x$, with respect to their graphs. When asked to draw a rough sketch of the graphs of these programs, he graphed $x + x$ just as he graphed $x \times x$; in addition, he stated that he had no idea how to do the graph of *x × x - 105*.

In a subsequent question asking him to write a program that would calculate the area of a square when the length of the side is supplied and to predict the graph that would result, he drew a half-parabola but with the concavity inversed. After viewing the computer-generated graph of his function, he was asked to modify the program so that the graph would pass through the point (0, 10) instead of through the origin. He relied once more on numerical trial-and-error, not realizing that the simple addition of 10 to $x \times x$ was all that was required. Even though there had been several indirect attempts to focus his attention on the form of these second-degree programs, he had not abstracted the $ax^2 + c$ form nor the role of the two parameters a and c. (Note that there had been no intent on our part to include in this study any explicit focusing on the role of the parameter b in $y = ax^2 + bx + c$.)

3. CONCLUDING REMARKS

We conclude this chapter by synthesizing a few of the main aspects of students' early algebra cognition that we have learned from our research over the past seven years on the functional approach embedded in the *CARAPACE* environment. Note that in none of our studies were students introduced to the manipulation of algebraic expressions and equations. The aspects to be discussed in the remaining pages are as follows: (a) letters as unknowns versus letters as variables, (b) the roles of the various representations of functions, and (c) our approach to problem solving in the *CARAPACE* computer environment: is it algebra?

3.1. *Letters as Unknowns versus Letters as Variables*

Historically, the notion of unknown developed earlier than the notion of variable. Symbols were used to represent a single-valued unknown before they were seen to represent "givens" or a range of values. Many algebra curricula introduce letters as unknowns prior to their presentation as variables. The documented difficulties (e.g., Küchemann, 1981) that students have with conceptualizing letters as variables and the finding that "students shift their preference from rhetorical approaches to using letters as unknowns to using letters to express givens as they grow in mathematical maturity and experience" (Harper, 1987) have led some researchers to conclude, for example, that "the attainment of a variable level of conceptualization in algebra is related to the development of higher-order cognitive structures" (Booth, 1984). However, the effect of traditional algebra instruction on this unknown-to-variable sequence of conceptual development of students has never been seriously tested.

In the functional approach we proposed, letters were presented as variables. Nevertheless, some of the student work was with problems whose algorithms could be considered analogous to the algebraic models $ax + b = c$ and $ax + b = cx + d$. In the algebraic domain of equation solving, the letter x in these two equations is usually considered to be an unknown, that is, it stands for one unknown number (unless there is no solution or the solution is the set of real numbers). For algebra beginners, solving the former equation is considerably simpler than solving the latter; it can be done by undoing, according to their old arithmetic methods. The distinction between these two types of equations was further refined by our studies.

We found that, when the students were presented with the actual question for certain problem types, they tended to turn to their arithmetic methods as a means of solving these problems; these were the problems that in a traditional algebra classroom would have involved equations with a constant on the right side. In other words, for problems that we might model with equations such as $ax + b = c$ or $ax + bx = c$, the students initially attempted to solve both types by undoing, even though it was not an appropriate method for the second type. Thus, situations that, in the *CARAPACE* environment, were represented by either of the two following types of program

"3 × cost of one meal + 10 gives total cost "
and "3 × cost of one meal + 4 × cost of one meal gives total cost"

were both tackled by undoing when the final question, "What is the cost of one meal if the total cost is $70?" was posed. On the other hand, for problems that algebra experts might model with $ax + b = cx + d$, the students attempted to solve their *CARAPACE* analogs by using the tabular representations of *CARAPACE*. For this latter kind of problem, when the problem question was posed, for example, "When will the two plans cost the same amount of money?" students seemed to have a sense that their arithmetical, undoing methods would not be suitable.

These early observations led us to realize that our beginning students had to learn to distinguish between the $ax + b = c$ problem type and the $ax + bx = c$ type and, then, to become aware that, if they wanted to call upon their undoing method for answering a problem question, it would not be directly applicable to the $ax + bx = c$ type. All three problem types ($ax + b = c$, $ax + bx = c$, $ax + b = cx + d$), as well as their variants, are equally easy to represent in the functional, *CARAPACE*, computer environment; the difficulty arises when students are presented with the problem question for the second type and attempt to resort to their old arithmetical, undoing method for solving the problem--that is, starting with the given constant value and operating on it. Much of the available research literature has discussed the difficulties associated with moving from the $ax + b = c$ type of equation to the $ax + b = cx + d$ type, but few have explicitly addressed the obstacles connected with the $ax + bx = c$ type.

Returning now to our earlier references to the algebra research literature regarding the cognitive problems identified with the movement from unknown to variable interpretations of literal terms, suffice it to say that our students could switch with facility from a "variable" interpretation of the letter (in their programs) to an "unknown" interpretation when it came time to solving certain problems. This flexibility was apparent, both in the action of some students' turning aside from their variable-based programs to their undoing methods when asked to actually solve certain types of problems, as well as in others' use of their functional representations to solve problem questions by means of a search for, say, a specific input value corresponding to the given output value of a problem question. It seems that, for most of our students, the single-valued interpretation of the letter came to be nested within the multi-valued interpretation of the letter. Thus, the approach that we followed in introducing students to a multi-valued interpretation of literal terms from the outset--an approach that permitted them to view unknowns within the larger context of variables--showed that the difficulties of early algebra learning need not reside in single- as opposed to multi-valued interpretations of letters. In other words, our research findings made it patently clear that beginning algebra with a variable interpretation of the letter and then going on to include single-valued situations seems not to produce the cognitive obstacles that can be encountered when one begins with a single-valued conception of letters and attempts to move from that to a multi-valued interpretation. For a number of our students, the greater difficulty resided in learning that situations of the $ax + bx = c$ type cannot be solved by undoing.

3.2. *Roles of the Various Representations of Functions*

The procedural, algorithmic representation was the foundation for the tabular and graphical representations of functional situations in our studies; the *CARAPACE* environment was unequivocally designed around this principle. The algorithms set up by the students represented those "forward" operations they would carry out on the problem data in calculating with any trial values. The tables of values and Cartesian graphs were generated according to the underlying algorithm entered by the student into the computer. Thus, the links between each of these two other representations and the numerical, computational aspects of the algorithmic representation were intended to be quite explicit.

Nonetheless, in the beginning, the students tended to easily lose partial awareness of the link between the graphical and algorithmic representations. The strategies of successive approximation that were used in the numerical context were not immediately called upon in a graphing context. We also observed an initial tendency to rely on the appearance of graphs and to extrapolate from what was visible. Similar findings were noted in our first study that focused on students' strategies with tables of values. As soon as they moved from their paper-and-pencil calculations to the tabular display, they tended to forget the contextual information they had previously been relying upon in their non-computer work and used strategies based solely on order relations and pattern recognition. It seemed that, when the students moved from the explicit computational representation to another in which the computations were less explicit, they had a tendency initially to generate other testing mechanisms. However, as they became more used to these other representations and to their numerical underpinnings, the computational interpretation was never far from the surface. This was especially evident in our last study in which the research focus was on the movement from a numerical, process-oriented perspective of functional situations and their representations to a form-oriented, structural one. In fact, there was no spontaneous shift from the former to the latter. As we noted in our discussion of the findings, all of the questions that might have been answered by referring to the form of the program never were; the justifications were always numerical. This clearly shows that the canonical forms of the expressions (algebraic or algorithmic) for different families of functions must be taught; this knowledge does not develop spontaneously.

Another conclusion to be emphasized in this section is one regarding the relative utility of the various functional representations in problem solving. Students' heuristic search for multiple solutions was much more effective when they used graphical representations of non-linear functions than when they used only a table of values representation of these functions. However, a graphical representation alone can be restrictive in finding solutions to problems, unless the computer environment presents the potential for a good quality of "zoom-in." Once the general locations of the problems solutions were determined from the Cartesian graph and in order to arrive at a better approximation of the solutions, the students found it helpful then to either move to a table of values representation and refine the interval of input values, or remain with the graphical representation and manipulate the axis-scales or successively zoom-in. Even though they lacked the tools of algebraic manipulation and thus could usually not arrive at exact solutions, they were, nevertheless, able to produce problem solutions which were close enough for all practical purposes.

Before concluding this section, there is one other important aspect to be emphasized regarding the setting up of the procedural algorithm representation. The students of our studies seemed to find it most helpful to try out different numerical values while they were attempting to make sense of the problem situation. The standard approach to solving word problems in a traditional algebra environment is to establish which quantity is the unknown one and to move immediately towards setting up an equation. This bypasses what we have found to be a critical step in becoming aware of all of the constituents of a problem situation, that of taking some number and running it through the given operations of the problem situation. This technique would likely also be helpful in traditional algebra classrooms where the focus is on generating equations, thus once again blurring some of the distinctions between functional and algebraic-equation approaches to problem solving.

3.3. *Our Approach to Problem Solving in the CARAPACE Computer Environment: Is it Algebra?*

The functional approach to problem solving that was described in this chapter suggests the power of this approach to endow the process of problem solving with meaning, meaning that has hitherto tended to be absent from student algebra learning (see the reviews of past problem-solving research included in Grouws, 1992). This approach has opened up a new avenue whereby students have been shown to be able to develop a deeper understanding for the process of translating problem situations into notational representations without having to concern themselves with acquiring skills in equation solving. But is it algebra? It is clearly problem solving. If one defines algebra in a traditional way as the use of standard literal notation, accompanied by the transformation rules for operating with this notation, then it is difficult to call this algebra. (One can use expressions that involve standard literal notation in *CARAPACE*, but there is no equal sign.) But, certainly, the functional approach as it is embodied in *CARAPACE* calls upon some of the thinking processes that can be considered to be part of working with traditional algebra, for example, thinking in terms of forward operations when setting up a problem representation, and seeing the general in the particular. Thus, it shares with algebra some modes of thinking without participating in all of its notational features or in its manipulative approaches. So, we might then ask if it can spontaneously serve as a bridge to algebraic forms and methods.

Our findings in this regard have been mixed. Students spontaneously used shorter, but often still significant, variable names. They also attempted to combine several single-operation program lines into one; even substitution of one "expression" by another seemed reasonable to the students. But none seemed to think about the possibility of "simplifying" any parts of their procedural representations. Thus, they did not get to make any "natural" contact with the more canonical forms of algebraic expressions or with the notions of families of functions and equivalence of expressions. It appears that making such contacts will require a slightly different orientation on our part, perhaps one that is less heavily weighted toward the problem-solving dimension of algebra.

One of the ironies of introducing algebra with the strong technological support afforded by *CARAPACE*, or other similar tools such as spreadsheets, is that these computer environments can very easily remove the need of algebra as a problem-solving tool--at least at this level of problem solving. But algebra is more than a tool with which to solve problems--it is a notation, a set of transformations, a way of mathematizing situations, ... But algebra is again more than all of this, for its notations hide underlying concepts, such as unknown, variable, function, equation, and so on. Thus, introducing algebra in a function-oriented, problem-solving environment provides students with only part of the picture of algebra. Even though this part is a very important one, and its approach via *CARAPACE* is highly accessible, such an introduction to algebra would be incomplete if it were not followed up with the opening of some of algebra's other doors.

4. APPENDIX: AN INITIATION INTO ALGEBRA IN THE *CARAPACE* ENVIRONMENT[2]

In the text that follows, we describe a new approach to the introduction of algebra for students who have not previously received any instruction in algebra. Alongside this, we make explicit and justify the didactic choices that governed the conception of this approach, and, where relevant, we also make reference to some of the results obtained from our experimentation over the past seven years.

Our decision to propose an alternative approach to the introduction of algebra was motivated, on the one hand, by an awareness of the enormous difficulties students have when they are faced with algebraic activities (difficulties that persist for a long time in many cases) and, on the other hand, by what we see as a striking lack of meaning in the development of algebra that most curricula offer students.

4.1. *The Use of Functional Situations*

We have been studying children of approximately twelve to fifteen years of age who are already very familiar with arithmetic (especially that of the natural numbers), but who have not yet had any contact with algebra. Our aim has been to introduce these children to algebraic language.

Let us clarify what we mean by *algebraic language*. One can describe it as the domain of general statements of algorithmic functional relationships based on arithmetic. Thus, when we say that "the cashier got the total price by adding the prices of all the articles," we are in the presence of a *statement* (whose formal symbolism goes beyond the secondary level) of the algebraic language. On the other hand, we consider that "the cashier got the total price by adding the prices of all the articles, given the discounts and applicable taxes" is not an algebraic statement because the arithmetic relationships that are involved are not sufficiently explicit.

Note that what we call here algebraic language comprises in actual fact a diversity of languages, going from wordings expressed in one's natural language to standard symbolic expressions. In the following text, we use the expression *algebraic language* to more specifically designate a language of transition between natural language and standard symbolic expression.

Our experience has shown that it is possible to teach students of about twelve years of age to formulate the above kind of algebraic statement about functional situations provided that:

- the context (in the preceding example, the purchase of objects) is familiar to them (Note: The arithmetic context must also be familiar; for example, one must avoid percentages if this notion still poses problems for them);
- the creation of arithmetic examples is not only allowed but encouraged (in the preceding example: If one buys two objects worth \$2 and \$3 respectively, the total price will be \$5);

- the use of natural language is permitted--while maintaining the requirements of exactness (we will come back later to the evolution towards traditional algebraic language);
- students are initially confronted with pure functional situations (i.e., in the context of activities where they must establish functional relationships independently of any equation solving).

This last point merits some clarification. We observed that when confronted with problems of the type, "The price of an object after a tax of 15% is $23; what was its price before tax?" the first instinct of students who had had no experience with algebra was to perform an arithmetic operation (here, e.g., divide by 15%) and to study the researcher's reaction to see if they had made a mistake or not. However, if we preceded the problem with a question about the underlying functional situation, "If we know the price of an object before tax, describe how to calculate its price after a tax of 15%," we noticed a definite improvement in the subsequent solution attempts of the students.

We would like to make two remarks concerning the preceding paragraph. Firstly, we feel that a response to the previous question, which is of the type, "First we calculate the sales tax by multiplying the selling price by 0.15; next we add the price before tax and the sales tax, and we get the price after tax," is not only correct but entirely appropriate. The use of intermediate steps in establishing a functional relationship helps to reduce the complexity, from the student's perspective, by reducing the problem to a sequence of sub-problems. Note that the above response would have been preceded and followed by numeric examples ("if the price before tax is $10, ..."), aimed first at bringing out the pattern and then at testing it. We mention in passing that this description of the functional situation is based on arithmetic in that it calls on assignment ("this calculation *gives* this result") rather than equality ("this expression *is* this thing").

Secondly, in a traditional approach, the rapid recourse to algebraic manipulation in order to solve this type of problem seems to be an attempt to channel this drive toward the calculation we have just described. However, even when they are seen as an extension of arithmetic calculations, symbol manipulations do not have any more meaning, for the majority of students, than the arithmetic calculations they replace.

4.2. *The Contribution of the Computer*

The kind of activity that consists of producing general descriptions of functional situations in an algebraic language may seem less interesting than the traditional approach, where one establishes equations using the traditional algebraic symbolism. Effectively, the usual symbolism has the merit of being free of ambiguities[3] (even if the difficulty of the correspondence with the original problem remains) and of readily allowing for manipulations that may produce new knowledge (e.g., the number and values of solutions of an equation). By comparison, the *algebraic language* may seem sterile from the start.

Even so, one can imagine breathing some life into the algebraic language by using the computer. Although this imposes some additional constraints (the automatic recognition of the meaning of natural language is not yet sufficiently developed[4]), the computer can then operate on the text it is given, for example, by performing certain calculations and by showing the student various representations (tables of values, Cartesian graphs, etc.).

It is also possible to ask the computer to perform, automatically or on request, symbolic manipulations on the algebraic texts. But, since this is necessarily accompanied by a loss of meaning (since symbol manipulations preserve the numeric but not the conceptual meaning[5]), we think that it is first necessary to make certain that the algebraic language rests on a sound numeric base.

Note that here the computer strengthens the algorithmic aspects of the algebraic language. In this context, the student must represent a functional situation in the form of a program that tells the computer how to perform certain arithmetic calculations. This representation is by nature essentially procedural and, according to some, should facilitate access to underlying concepts.

4.3. *The CARAPACE Environment*

4.3.1. *Initial activities*

We will now examine more specifically the learning environment that we have created and have been experimenting with for seven years. It is based on a study of functional situations using a specially developed software package called *CARAPACE* (*Contexte d'Aide à la Résolution Algorithmique de Problèmes Algébriques dans un Cadre Évolutif*--or in English: an environment to help in algorithmically solving algebraic problems in an evolving framework).

Let us briefly describe the approach taken with a student of about twelve years of age who has never had any contact with algebra. The initial activities are enacted with pencil and paper, in preparation for the computer activities. At first, we present him[6] with functional-situation contexts, such as:

> Carine works part time in the neighborhood. She sells subscriptions to a magazine. She earns $20 a week, plus a bonus of $4 for each subscription sold.

We ask the student questions involving numeric calculations, such as, "If Carine sells 5 subscriptions, how much will she earn?" The questions are aimed at checking the student's understanding of the situation, and we try to minimize the complexity of the arithmetic calculations, in order not to divert his attention. We insist that he explicitly write (afterwards) all the calculations he performed. For example, here:

$$5 \times \$4 \text{ gives } \$20$$
$$\$20 + \$20 \text{ gives } \$40.$$

Then, after several similar calculations, we encourage him to organize his findings into a two-column table. In the left hand column, we put the numbers that correspond to the questions asked, and in the right hand column the corresponding answers.

Next the student is asked to find names to describe each of the columns of numbers, the results of which are illustrated in Figure 9.

number of subscriptions	total salary
2	28
5	40
7	48
8	52
10	60
13	72

Figure 9. Table of values, of the type required of the student before writing an algorithmic description of his calculations

Next we ask him to identify/write/name the numbers appearing in the preceding calculations. We obtain, for example, the work shown in Figure 10.

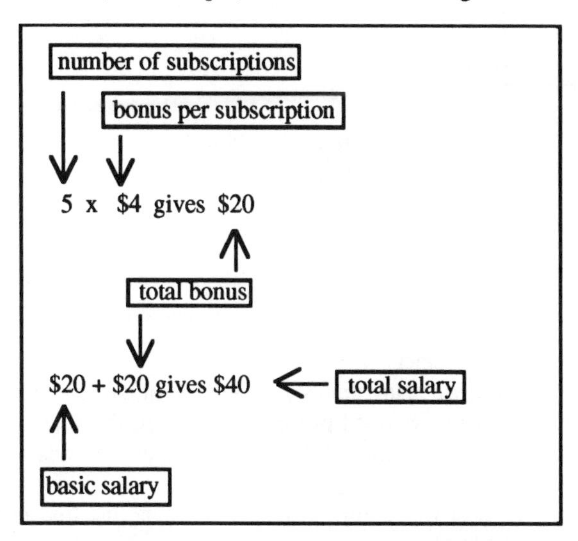

Figure 10. Labeling of a calculation, of the type required of the student beforewriting an algorithmic description of his calculations

Since all preceding calculations were labeled in this way, it is easy to get the student to identify the labels on numbers that vary from one calculation to another (here: the number of subscriptions, the total bonus, the total salary) and those labels on numbers that do not vary (here: bonus per subscription, basic salary).

The next stage is crucial. It involves directing the student to produce a general description of the calculations, which may resemble this:

number of subscriptions × $4 gives total bonus
$20 + total bonus gives total salary.

The preceding activities take place without calling on the computer. At this point, we introduce the *CARAPACE* software: The student is faced with a window like the one shown in Figure 11.

Request values for:

Carry out these calculations:

Show values of:

Figure 11. Editing window before program entry

We explain that the "form" must be "filled in" as illustrated in Figure 12.

Request values for:
number of subscriptions
Carry out these calculations:
number of subscriptions × 4 gives total bonus
20 + total bonus gives total salary
Show values of:
total salary

Figure 12. Editing window after program entry

During these first contacts with *CARAPACE*, we do not insist on explaining the "form." In fact there are many things to be learned at the same time: moving the mouse, the functioning of the word processor by which the student can enter his programs, the choice in the menus, etc. An approximate explanation is therefore sufficient at first:

- the section Request values for: contains the name we find in the left hand column of the table of values. Later we will construct a table with more than two columns--where there can be several independent variables. This section will then be the list of all the independent variables (or input variables).
- the section Carry out these calculations: contains a general description of the calculations. The student is encouraged to complete this section first since it supplies the information for the other two sections.
- the section Show values of: contains the name that appears in the right hand column of the table of values. Later we can fill in the list of all the variables (called output variables) that interest us. In mathematical terms, we can then say that *CARAPACE* corresponds to a finite number of functions (or to a vector function).

With a beginning student, we insist that calculations be done in "detailed calculation" mode. The program first asks for input variables, then goes through the general calculations in the second section step by step. Figure 13 gives an example during calculations.

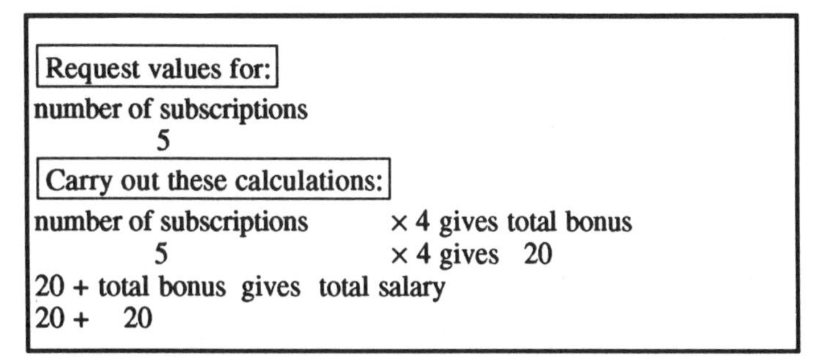
Request values for:
number of subscriptions
5
Carry out these calculations:
number of subscriptions × 4 gives total bonus
5 × 4 gives 20
20 + total bonus gives total salary
20 + 20

Figure 13. Detailed calculation window (in the process of execution)

This way of functioning is crucial because it allows the student to check whether his general description of the calculations corresponds to what he intended. It also contributes meaning to the algebraic language communicated to the computer and thus facilitates the transition between arithmetic and algebra.

CARAPACE offers a table that sums up all the preceding calculations. The student can ask for calculations in the "table" mode (intermediary calculations are then not shown); but we encourage him not to do so until he has checked his program. Figure 14 gives an example of such a table.

INPUTS	OUTPUTS
number of subscriptions	total salary
2	28
5	40
7	48
8	52
10	60
13	72

Figure 14. Table of values window

Note that the student has been encouraged to re-do with *CARAPACE* the calculations he did "by hand," and he is asked to compare the table provided by the software with the one he made himself. In this way, he again verifies that his program works correctly, which contributes to increasing the transparency of *CARAPACE*.

The beginning student is not initially given problems to solve (e.g., "How many subscriptions must Carine sell to earn $140?"). As mentioned above, this kind of problem prompts students to look for a direct response, which often involves inverse operations that do not make sense in this context and which make it more difficult to clarify the functional relationships involved.

We introduce this type of problem later however (which, in a traditional context, involves solving one or several equations). First, we have the student model the functional situation using *CARAPACE* before giving him the problem. Then, when he has acquired some experience, we formulate the problem on the spot and encourage him to find on his own the components of the "functional situation" and the "problem."

In this kind of problem-solving context, the student must resort to a numeric solution (using *CARAPACE* or by hand) by trial and error, rather than by using symbolic manipulations. We will come back to this, but let us say immediately that this kind of approach has the merit of constantly recalling the arithmetic sense of looking for solutions. Finally, we should mention that, later, the student will be able to call upon graphical methods as well.

4.3.2. *Some features of CARAPACE*

At first, the algebraic language accepted by *CARAPACE* may seem very limited. It consists of a sequence of allotments of the form:

Name of variable 1 Operation Name of variable 2 "gives" Name of variable 3

where the only possible operations are the four basic arithmetic ones, plus exponentiation. In fact, it formally has the same power of expression as the symbolism of elementary algebra if one considers the fact that a sequence of such allotments, involving intermediary variables (such as *total bonus* in Figure 12), can be used.

However, at the same time, the algebraic language is perceived by the student as being less complex than the usual algebraic symbolism. It allows for the use of variable names that are significant (with respect to the context being studied). The use of intermediary variables allows one to chop complex calculations up into steps, thereby reducing the complexity of the enterprise and identifying (with meaningful names) the results of each of the steps in the calculation.

Note that the algebraic language recognized by *CARAPACE* can be extended in a series of steps, which correspond to the habitual stumbling blocks of students. Several operations can be allowed in an assignment, provided that the order in which the calculations are carried out is indicated by complete bracketing. Incomplete parentheses can still be acceptable if we agree that the usual order of operations will determine the order of calculations. Operations of multiplication (in the case of juxtaposition) and exponentiation (in the case of powers) can be understood if the variable names are limited to a single letter. We underline the fact that in all cases the "detailed calculation" mode is there to illustrate to students the conventions used in the calculations.

At the highest syntactic level, *CARAPACE* accepts all the usual expressions of elementary algebra such as

$$ax^2 + bx + c \text{ gives } y,$$

but still doesn't use equality. In fact, equality is used implicitly in searching for solutions whenever the student attempts to get a given number in a given column or when he tries to find the same number on the same line of two given columns. For example, the problem of solving $3x - 4 = 2x + 3$ takes the following form with *CARAPACE*: "Given the program that is provided in Figure 15, find the value(s) of x that produce identical values for y and z."

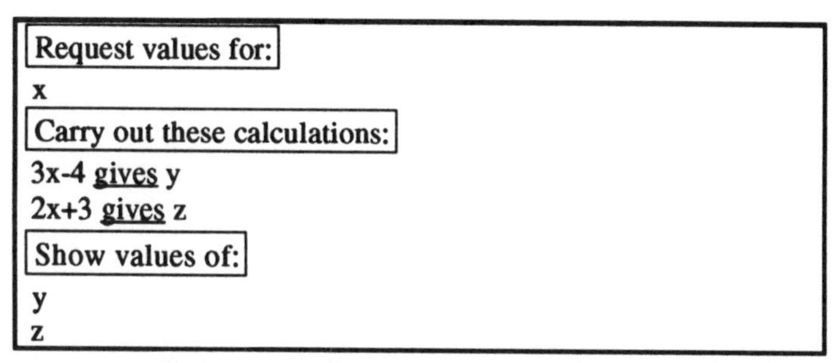

Figure 15. Program in standard symbolism in *CARAPACE*

In the table shown in Figure 16, we see a sequence of numeric attempts leading to a solution.

INPUTS	OUTPUTS	
x	y	z
3	5	9
5	11	13
9	23	21
7	17	17

Figure 16. Table of values obtained from the program in Figure 15

One might in fact prefer to work on another version of the table, whose output headings reflect the expressions in the initial problem (see Figure 17).

INPUTS	OUTPUTS	
x	3x-4	2x+3
3	5	9
5	11	13
9	23	21
7	17	17

Figure 17. Table of values obtained from the program in Figure 18

The table-output-headings of Figure 17 (displaying the calculation process rather than the variable name) were automatically generated by a special assignment mechanism in *CARAPACE*: When writing the program, one need only precede the variable name by the indicator result (a single key stroke), as in the program shown in Figure 18.

Returning to the solution of equations by numeric trials, as illustrated by the two tables provided in Figures 16 and 17, it would have been possible to avail students of a mechanism for the automatic solution of equations (such as the SOLVE key on certain calculators), but we feel that we would have lost a good opportunity to add meaning to the activity of searching for solutions.

Students who take to carrying out numerical trials generally find the experience enjoyable[7] because it calls upon both their arithmetic knowledge (which is solid) and their problem-solving strategies, which they can develop to suit themselves.

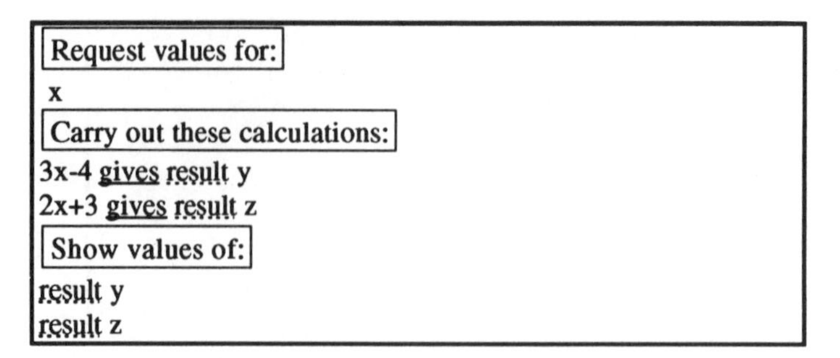

Figure 18. *CARAPACE* program using result

To illustrate, let us look at the table in Figure 16. A student can see that in going from the row where x is 5 to the row where x is 9, the respective order of the results for y and z has been switched--from which one can make the hypothesis that the solution lies between 5 and 9. Another student may notice that in looking at the first two rows, when x increases by 2, the difference between z and y decreases by 2--from which one can make the hypothesis that the solution we are looking for is 7.

It must be pointed out that this type of numeric search, characterized by inductive formulation of hypotheses that are then checked by new tries, is quite different from traditional, equation-solving methods. In the cases where the latter are applied, one obtains exact results[8] (even in the case of rational, irrational, or even complex numbers)[9] and also the assurance that all solutions have been found. The former method gives only approximate solutions, but has a much greater applicability. However, the problem of knowing if one has found all the solutions remains: We will come back to this when we discuss the graphing aspects of *CARAPACE*.

4.3.3. *The role of Cartesian graphs in the CARAPACE environment*

Let us now look at the use of Cartesian graphs in the *CARAPACE* environment. As we shall see, Cartesian graphs have a very special place in the *CARAPACE* approach because they allow us to address important situations, such as:

- A student has produced a long table of values with many numeric tries without having found a solution (perhaps because of the unfamiliar type of variation of the underlying function). He begins to wonder *if the solution he's looking for exists at all.*
- Very often the student who finds a solution does not try to find out if other solutions exist. This is quite defensible if the context of the problem assures us intuitively that *there is only one solution.* But, in general, one should attempt to develop a more critical attitude in students.

In both these cases, one can use Cartesian graphs for a rational approach to these situations. In effect, when we know how to interpret them correctly, Cartesian graphs can provide a very useful global view of the situation.

There is good reason, however, to find a more accessible approach to the introduction of Cartesian graphs than the traditional one. From a cognitive perspective, difficulties (such as infinity, the density of the real numbers) arise the moment one envisages the graph of a real function. Studies where young students are asked to count the points of a given segment, and where the results obtained are more often than not a finite number (often closely tied to the given drawing), are a case in point.

In the framework of *CARAPACE*, we decided to never pretend to show more than a finite number of points belonging to a given Cartesian graph. Moreover, the points that are shown are found automatically in the table of values and each row of the table of values corresponds to a point: The correspondence between the two representations is therefore perfect, which means we can rely on one to better understand the other.

The decision to use discrete Cartesian graphs contrasts with the usual practice, where one attempts to make discrete computer representations appear to be continuous. In addition to conforming more closely with the other representations in *CARAPACE* (calculation algorithms, discrete table of values), this discrete approach has the advantage of allowing the student to construct by himself graphs that he will use later.

In our approach, of course, we can show a sufficient number of points to give the visual impression of a continuous graph. But, at any moment, one can do a *zoom-in* on a part of the graph and dynamically see the points separate and get further apart (if, of course, the zoom is sufficiently defined).

To illustrate, here is an example (see Figure 19), where one can see that a rectangle, which will serve as the basis for the zoom-in, has been specified. Note that the graph appears to be continuous.

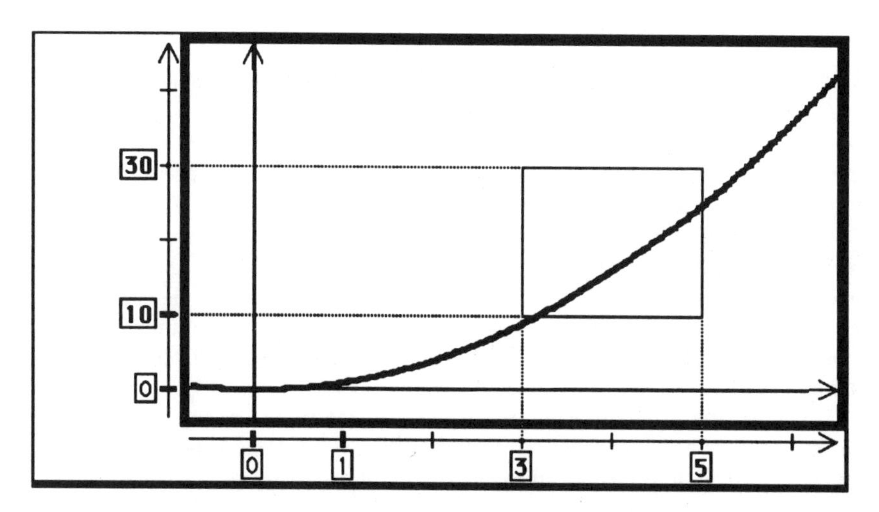

Figure 19. Cartesian graph just before a zoom-in. Note the rectangle that serves to specify the boundaries of the zoom-in, and the corresponding numeric labels (in bold type, to distinguish them from the scale labels on the axes).

Here now is the result of the zoom (see Figure 20). In fact, the user was able to see an animated sequence during which the small rectangle grew until it filled the whole region and, simultaneously, the points composing the graph got further apart.

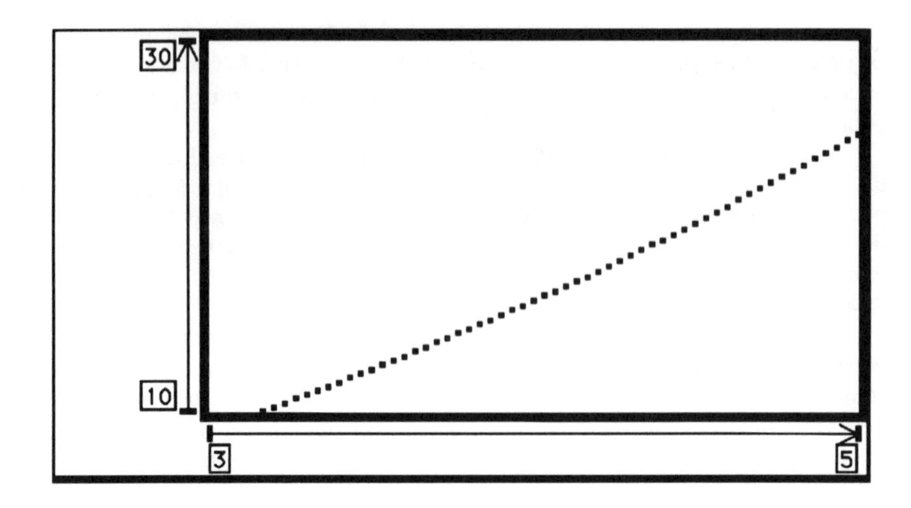

Figure 20. Cartesian graph just after a zoom-in. Note that the points on the graph have been separated by the zoom-in.

Following this, one can plot more points, thereby obtaining a graph that appears to be continuous. Thus, by a succession of alternately zooming and plotting, the student can build the concept of continuous graph, with similarities to building the notion of equality (which is absent in the software but omnipresent in the problem-solving activities).

4.3.4. *Initiation into Cartesian graphs in the CARAPACE environment*

We will now briefly describe how we introduce graphs in the *CARAPACE* environment. The students we work with have already had considerable experience in writing programs and in numeric problem solving using *CARAPACE.*

We give them a functional situation that is very simple for them, so that they can write the corresponding program easily. We then ask them to go into graphic mode. The computer asks them to choose the coordinate system to be used. During their first contact with the graphing environment, the students do not linger over the choice of axes: They just choose the default system and proceed directly to plotting points.

At first, the program asks for the value of the input variable. Then it asks students to situate this variable on the horizontal axis, helping if necessary. It then traces a vertical line passing through the point in question. Then the detailed calculation is made, allowing students to find the output variable. (This emphasizes the nature of the functional link uniting the coordinates and serving to define the point.) *CARAPACE* next asks students to find this point on the vertical axis, always helping if necessary. It draws a horizontal line passing through this last point, then an intersection point for the two lines (see Figure 21). It terminates by erasing the two lines (but not the point).

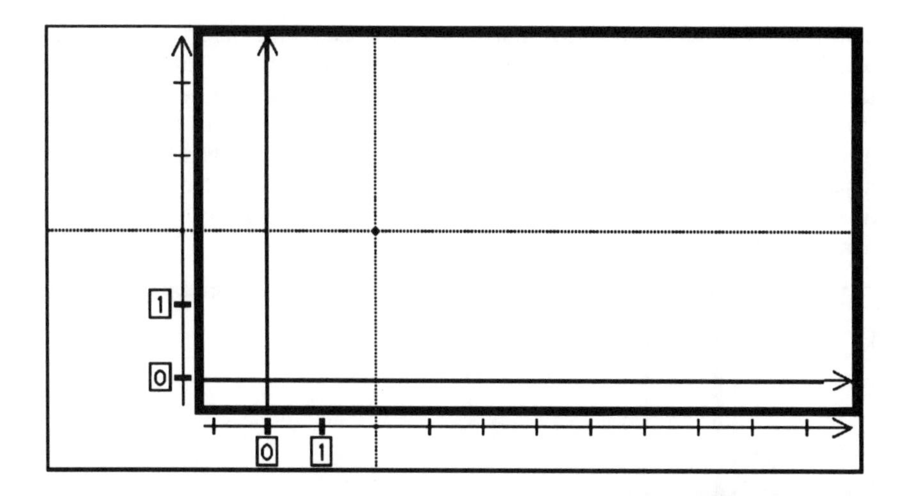

Figure 21. A data point towards the end of the process of being located, at the most elementary level of *CARAPACE*

We would like to underline, in passing, an interesting activity within this context: replotting points after changing the scale of the axes. In responding once again to the questions posed to students in *CARAPACE*, the student is led to realize, little by little, the elements that are invariant in the plotting of a point, as well as those that depend on the scale chosen. This activity may also serve to verify that the software moves the points using the same algorithm that it originally used to plot them.

We have just described the point-plotting mode that requires the most participation on the students' part. As the students gain experience, we can successively eliminate the detailed calculations, allowing them to choose the starting point with the mouse. They can draw up to four functions on the screen (the points will be different colors to distinguish each function). They can also plot several points at a time, whether by distinguishing them numerically (e.g., from 10 to 180 by jumps of 10) or via the mouse (the points corresponding to all the pixels in the designated interval will be drawn).

We have thus established that the *CARAPACE* environment allows the user to plot points belonging to graphs of functions, that it is the user who must specify the points to be plotted, and that he must more or less "assist" the computer in plotting the points in question (according to the level of graphic expertise selected in the software). Recall, furthermore, that the intended goal of the use of Cartesian graphs in *CARAPACE* is not only to help in the solving of problems, but also to allow discussion of certain questions, such as the existence and the number of solutions of a problem. These objectives have had a direct influence on the choice and the characteristics of the graphic tools in the *CARAPACE* environment. In fact, these tools belong to two main families: the "adapted" cursors, and the tools for specifying the graphic region which is to be visible on the screen.

In Figure 22, one can see a type of "adapted" cursor. It appears automatically whenever the software is not involved in executing a command or whenever the cursor/mouse (which it replaces) moves into the "Cartesian graph" portion of the window. We have chosen this geometric "set square" representation rather than that of the ordered pair (x,y) to avoid problems of inversion (use of (y,x) by our beginning students. Note that the two numbers/labels constituting the "feet of the set square" are displayed in bold type to distinguish them from the numbers/labels on the adjacent axes. This "set square" is dynamically transformed as the mouse moves along the screen,[10] and the students can observe the constant modification of the numbers/labels.

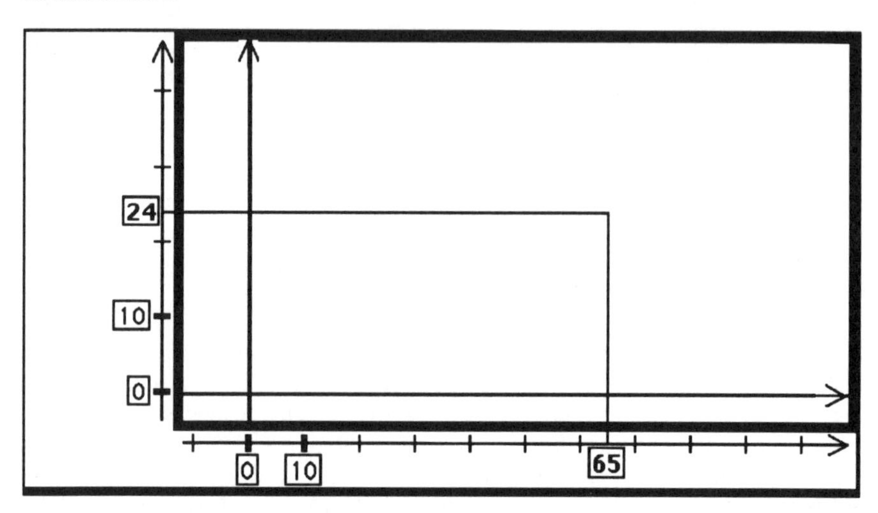

Figure 22. An "adapted" cursor in the *CARAPACE* environment

An important characteristic of "adapted" cursors is the number of decimals displayed: This number is not fixed, but is always the smallest possible. The last statement could seem a bit surprising, but one must remember that the basic building block of a computer screen, the pixel, corresponds to a square of the Cartesian plane: We are therefore free to choose to use the coordinates of one specific point of the screen square as the coordinates to be associated with the pixel. Some software chooses the middle point in the square, others the lower left point, etc.; we have chosen the point whose coordinates have the least possible number of decimals. From inception, we have made this choice in order to simplify the situation as much as possible, numerically speaking; but we have also found this choice to be a crucial one because it permits, in most of the "usual" cases, indication of a value corresponding exactly to that determined by the scale intervals, which is not necessarily the case otherwise. Finally note that the same kind of "adapted" cursor is also used in the zoom-in (see Figure 19).

There are two main ways of specifying the rectangular region to be used for Cartesian graphs: either by directly typing in the parameters, or by one of three types of zoom: zoom-in (already briefly described), zoom-out, and "zoom assisted by *CARAPACE*." The axis specifications are automatically requested the first time one enters the graphic mode, but one can also call on this feature at any later time.

The screen appears as shown in Figure 23.

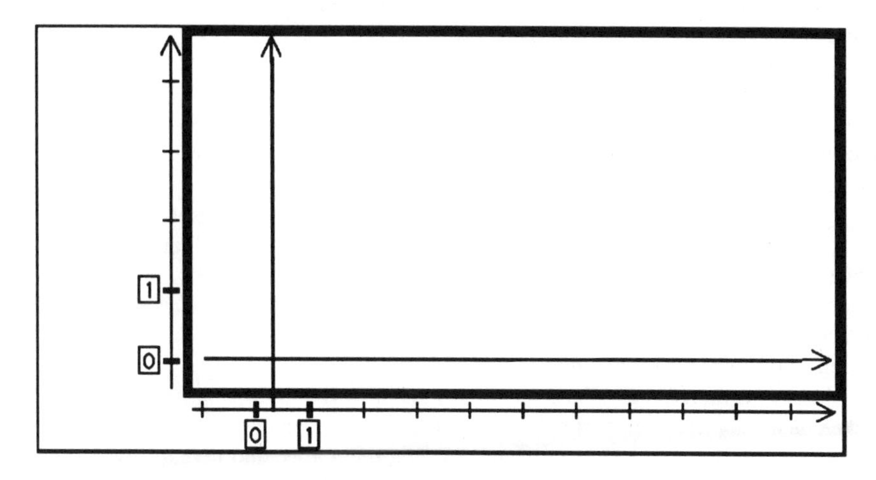

Figure 23. Direct specification of scale. One can change the position and the numeric contents of the labels, and choose the number of intervals between each.

The thick frame corresponds to the chosen portion of the real plane. Note that there are two systems of axes. The first, within the frame, appears on the screen in pale yellow (in order not to interfere too much with the plotted points) and can disappear according to the choice of the area of the plane. The second, outside the frame, is the one on which the students will work when they are more familiar with point plotting. Each axis has two labels, for which the value and position can be changed, as well as graduations, the number of which is chosen by the students.

Note the advantages of this non-standard representation of the axes: On the one hand, they are always visible (whereas the traditional set of axes isn't always) and, on the other hand, the definition of each of the axes is general (an axis is a line on which two distinct points are arbitrarily chosen and on which attributed arbitrary values have been assigned). This representation of axes, in contrast with the traditional computer representation (where the reference points correspond to the extremities of the segment), allows dynamic variation (using the mouse to displace one of the numeric labels) of the position of the defining points of the axes and observation in real time of the changes that result for the plotted points.

Using the graphic tools of *CARAPACE*, we have led students to seriously discuss difficult questions (number of solutions, maximization, etc.) about problems that are relatively complex for their age, such as:

A basketball team must participate in an international competition in Paris, and certain fans would also like to make the trip. A travel agent suggests the following deal: For a group of 20 fans, each person will pay $500 for his plane ticket; for each additional person, the price of tickets will be reduced by $2 per person. How many fans make the trip if the agent receives $36,400?

We were also able to note that these same students could construct adequate models of phenomena as complex as the density of the decimal numbers and the relations between mathematical points and computer pixels.

4.4. *In Closing*

We would like to conclude with two remarks, one on computer environments and their pedagogical use, and the other on the school algebra curriculum.

At the present time, some very sophisticated computer software exists (symbolic manipulation systems, electronic spreadsheets, solvers for systems of equations, etc.) and we tend to introduce them into teaching more and more. They use several types of representation that are familiar to mathematicians and scientists, but one need not necessarily conclude that they are easily accessible to students. Often, students don't see the same things in them that we do.

There is, therefore, a place for mathematical software of a more specifically pedagogical character, where simpler intermediary representations are proposed, which can progressively lead students to standard representations, and perhaps even help their teachers to become more aware of certain learning difficulties as well as certain strategies for dealing with them. We believe that the algebraic language, the detailed calculation mode, and the various graphic levels of *CARAPACE* are examples of these.

A large portion of students following the current algebra curriculum gets little or no benefit from it. It may perhaps be necessary to envisage a new approach, where all students would be required to acquire minimal algebraic knowledge, and where some could go further. We think that the activities involving the translation of functional situations into algebraic language, followed by numeric and graphical analysis (such as are found in the *CARAPACE* environment), can be an interesting candidate for minimal algebraic knowledge. When one thinks about it, the knowledge and abilities thus acquired constitute an acceptable base for the majority of personal and professional needs. The practice of symbolic manipulation techniques could be left to students who are more specifically interested in a scientific career, it being understood that the potential role of computer symbolic manipulation systems needs also to be taken into consideration.

This is one of the directions our future research will take. We are interested in exploring the introduction of symbolic manipulation to students who have done their basic learning in the *CARAPACE* environment.

NOTES

1 The authors acknowledge the support of the Social Sciences and Humanities Research Council of Canada, Grant #410-90-1041, and the Quebec Ministry of Education, FCAR Grant #92-ER-1207. Any opinions, findings, conclusions, or recommendations expressed herein are those of the authors and do not necessarily reflect the views of the Social Sciences and Humanities Research Council of Canada or the Quebec Ministry of Education. The authors are also extremely grateful not only to the students who participated in the studies but, as well, to all of those who helped out as interviewers, research assistants, and camera crew. Without their cooperation and assistance, these studies would not have been possible.

2 This appendix, which was written separately and authored by André Boileau, Carolyn Kieran, and Maurice Garançon, has been included here in order to help the reader better visualize the *CARAPACE* learning environment.

3 Although its interpretation by students can be surprising. Note, for example, the case of an 18-year-old student who, having noticed that the program refused "2 sin *x*," was told that multiplication must be made explicit (indicated by * for the computer). He then tried to use "2*sin**x*"; or of another student who simplified "(sin *x*)/*x*" to get "sin."

4 Moreover, from a pedagogical point of view, it seems more interesting to ask the student to be more precise than to ask the computer to try to correctly interpret what is communicated to it.

5 For example, let us quote the case of those students who, having found that the distance traveled was x the first day, $2x - 7$ the second day, and $x + 11$ the third day, refused to express the total distance traveled by $4x + 4$ because on the one hand information was lost in adding the three distances, and moreover the 4 in the "answer" $4x + 4$ conflicted with the three days in the problem. They therefore presented their answer in the form $x + (2x - 7) + (x + 11)$.

6 Even though the masculine gender is used here, it is to be understood that both boys and girls participated in our research.

7 This is not the case when the search must take place in a context that poses difficulties which are too great for a given student: insufficient mastery of the numeric domain (e.g., the decimal numbers), excessive time requirements (too many decimals), complexity of variation in the given function (e.g., decreasing functions, parabolas).

8 And these cases are less frequent than we think. Even a polynomial of degree 3 is not solvable in general by real roots. We don't notice this as much since traditional textbooks obviously only choose cases where everything works out well.

9 Though several students have developed the reflex of "finishing the calculation" with the help of their calculator, thereby obtaining a less exact result.

10 ... or if the arrow keys are pressed, which allows purely horizontal or vertical displacements, and in certain cases allows greater accuracy.

CHAPTER 20

A FUNCTIONAL APPROACH TO ALGEBRA: TWO ISSUES THAT EMERGE

RICARDO NEMIROVSKY

Although the two previous chapters presented in this book report and reflect on research in a functional approach to the introduction of algebra within a computer environment, they differ greatly in what they have chosen to focus on. This commentary chapter will tackle a central issue in each of the chapters. In my comments on the Kieran et al. chapter, I will address the authors' choice of the pointwise approach to functions and offer some suggestions for broadening their research and development program. The question of "real world problems" is central to Heid's chapter. In my comments on that chapter, I will explore issues around three implicit assumptions concerning real world problems . Examples of students' work that arose in my own research will be used to explore the issues and raise the questions that emerge in these chapters.

1. POINTWISE/VARIATIONAL APPROACHES TO FUNCTIONS

The chapter, "Introducing Algebra by Means of a Technology-Supported, Functional Approach," reflects a whole program of research and development that has been conducted over the last seven years. The importance and quality of the program is made salient by the summary of the six studies carried out with 12- to 15-year-old students. The careful analysis of the relationships between contextual knowledge and notational operations seems to me particularly valuable, for example, the observation that students abandoned contextual cues as soon as they started to work with tabular forms, or the students' reluctance to simplify expressions tied to a specific context in order to preserve recognizable links between the symbols and the situation.

I will focus my comments on some suggestions to broaden the research and development program. In order to articulate these suggestions, I need to elaborate on the distinction between pointwise and variational approaches to functions. Kieran et al. (this volume) say: "The representation that is favored at the outset is the computer algorithm, which involves the use of input and output variables, and which in turn provides the computational foundation for tables of values and Cartesian graphs. The context in which these algorithmic processes and pairs of input-output variables are elaborated is one of problem solving." This is a characterization of a pointwise approach to functions, that is, a conception in which a function is basically a relation involving binary pairs. A variational approach, on the other hand, focuses on a function as describing how a quantity varies. Let me give an example of a problem that we have explored with students:

Suppose that two cars move according to the following graph (see Figure 1) of velocity versus time.

N. Bednarz et al. (eds.), Approaches to Algebra, 295-313.

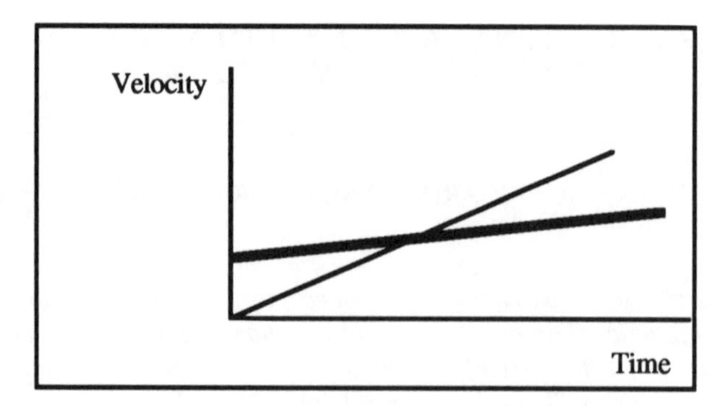

Figure 1.

How do you imagine movement of the two cars? Do the initial positions matter? How? Do they meet? When? Is one of the cars always ahead of the other?

This problem introduces the analysis of two linear functions. Each one of them involves infinite ordered pairs (time, velocity), but they are implicit. One can approach the problem without necessarily knowing the input-output pairs and the process or algorithm for computing these pairs. Figure 1 shows how the velocity of each car changed over time (i.e., increased steadily) and in relation to each other (i.e., first the thick car went faster, but then ...). Figure 1 does not tell what the cars' velocities were at any particular time, but it does indicate how the velocity of one of the cars at a certain time, *whatever it is*, relates to its velocity before and after, as well as to the velocity of the other car.

The distinction pointwise/variational is made for analytical purposes. The two aspects are profoundly related. The *CARAPACE* environment may elicit them in both the tabular and the graphical representation (e.g., by exploring differences between successive points). But the chapter impresses me with an overall strong emphasis on functions as pointwise entities. Note that the distinction pointwise/variational is not the same as process/object. To illustrate this, let me describe the following example. During a recent interview, Dina, a 9 year-old girl, was thinking about producing a "V" with the motion detector (see Figure 2).

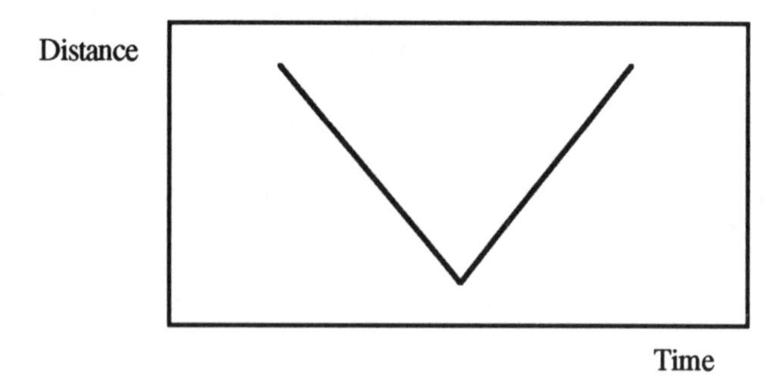

Figure 2.

Tracey Wright, the interviewer, asked:

Tracey: *What's your plan?*
Dina: *To get closer and then to get further away [from the motion detector].*

Dina was not seeing the "V" as an object--as an entity that exists "all at once." She envisioned how it represented a process over time of getting closer and then getting farther away. The process that she anticipated was not notational or algorithmic, it was kinesthetic. Her understanding of the "V" was not pointwise either; she was not basing her analysis on time-distance pairs and "the algorithm for computing these pairs." Dina's understanding was variational: The "V" described how her distance to the motion sensor had to vary.

Is a pointwise understanding of functions more basic? Do students grasp the variational aspects of a function *after* they have mastered its pointwise attributes? I believe that the answer is, "no." Dina, as well as all the other children that we have observed creating functions with sensors, did not go through a process of determining time-distance pairs to make sense of the variation described in the corresponding functional graphs. If they were able to read correctly the distance corresponding to a certain time, this was a product, more than a requisite, of their practice of creating and interpreting functional graphs.

The authors' preference for a pointwise approach to functions relates to their arguments for allowing only discrete points in a graphical representation. They introduce two types of arguments: (a) that continuous graphical representations induce students to misunderstandings and perceptual illusions (as described in the work of Goldenberg et al., 1992) and (b) that they force students to prematurely deal with complex problems on the infinite and the continuous (as described in the work of Fischbein et al., 1979).

I share the authors' sense that seeing the points moving as the scale changes may provide important clues for visualizing the effects of scale change; in particular, it helps to make salient that stretching the scale spreads apart the points, an effect that is hidden in a continuous graph. But there are also other ways to help students learn about scale issues. One of them is to develop activities to construct what might be called, in analogy to number sense, scale sense--in other words, to develop a sense for the significance of being in the different regions of a Cartesian plane for a given scale. Scale sense would be anchored in certain landmarks--notable positions on the axes such as the boiling temperature in a temperature graph, or the height of the student in a height versus time graph, and so forth--and in a qualitative metric on the basis of which one can assess how far or close a position is to a landmark.

The authors (in the appendix to their chapter) "decided to never pretend to show more than a finite number of points belonging to a given Cartesian graph." It is impossible, of course, to display more than a finite number of points with the current computer technology, but the act of showing is more than a display of marks on a screen. When one draws a line on paper and says, "this is a straight line," others can understand that this *is* a straight line, in spite of the fact that it is not. From a literal point of view, it is impossible to draw a straight line.

The authors also state in the appendix to their chapter (this volume): "From a cognitive perspective, difficulties (such as infinity, the density of the real numbers) arise the moment one envisages the graph of a real function." It is very clear that issues of infinity/continuity are complex and likely to elicit difficulties. But the same is true of all the central mathematical ideas. The length of a circumference involves irrational numbers. However, this is not a reason to postpone the use of circles until the students are "ready" to understand irrational numbers. Students have rich experiences with circles that provide a ground for their use from very early ages. The same can be said about continuity. We are surrounded by phenomena that we experience as continuous: the passage of time, the movement of our hand, the motion of the moon, and so many others. When one walks a path from A to B, one has the sense that one has to traverse *all* the intermediate positions on the path. Let me exemplify the significance of this experience.

During an interview, Eleanor, a 10-year-old girl, was experimenting with an electric train and a speed controller. Eleanor and Tracey discussed whether, in order to change direction, the train has to momentarily stop. In a previous session, she had been using a speed controller with a knob to set the speed and an independent switch to change direction; but the speed controller she used in this session had a single knob with a zero in the center and two equal scales, to the right and to the left, for each direction of motion on the track. Figure 3 shows the interface of each speed controller.

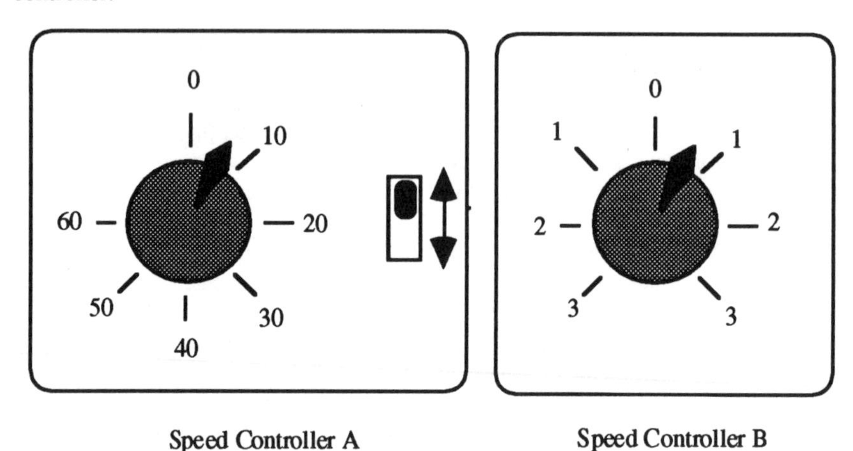

Speed Controller A Speed Controller B

Figure 3.

Eleanor's experience with changing the direction of the train was very different with each speed controller. With speed controller A, she acted out her intention of changing direction by a discrete act on the switch that was independent from the knob controlling the train speed. With speed controller B, on the other hand, to change direction required turning the knob continuously crossing positions of lower speed, zero speed, and lower speed in the opposite direction. Eleanor intuited that, as opposed to speed controller A, speed controller B forced the train to stop momentarily as it changed direction:

Eleanor: *... and it definitely comes to a kind of stop [when changing direction], because it has to. Because it has to go through the kind of stop position.*

Tracey: *Right. In order to?*
Eleanor: *In order to change [direction]. Because in the other one [speed controller A] it just. I don't think it really needed to stop.*
Tracey: *It didn't?*
Eleanor: *No. I don't think it did, because it was just like you could change the button kind of.*

Later Eleanor created velocity-time graphs that were shown by the computer screen with the following display (see Figure 4).

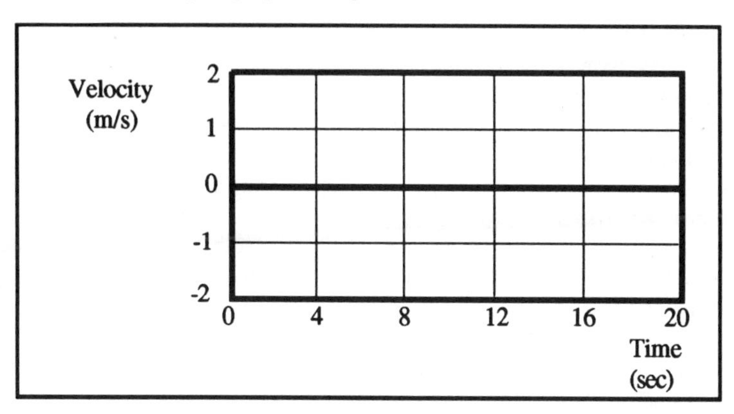

Figure 4.

After producing several velocity-time graphs, Tracey and Eleanor were talking about the velocity graphs. Observing the computer screen, Eleanor pointed to the "1" and "–1" on the vertical scale of velocities:

Eleanor: *They are like the same because they are up and down.*
Tracey: *What do you mean, that they are up and down?*
Eleanor: *[1] is going away from it [the motion detector] and [–1 is] coming toward it [the motion detector]. Coming towards it comes below.*

Then Tracey asked Eleanor about the horizontal lines displayed on the velocity graph display. Eleanor, pointing to the zero velocity line, said:

Eleanor: *The zero line is kind of stop if it stays on that line ... see, on this one [velocity-time graph, as opposed to distance-time graph] like, to change direction it [the velocity graph] shows that it has to stop. Like it has to go through this one [zero line] to change direction.*

The continuity of the velocity curve moved Eleanor to anticipate that a change of direction requires a "stop" of the train. Something that was necessary *on the graph* led her to envision a necessity for the physical movement of the train.

I am not advocating that functional graphs should always be continuous. I think that students can learn from both discrete and continuous graphs, depending on the situation and the tools, and that none of them is more basic or simple. More generally, my suggestion is to adopt a broader approach to functions than the pointwise and discrete focuses. There is a whole realm of experiences involving variation, qualitative analysis, and continuity that can enormously enrich students' learning about functions. This broadening of perspective seems to me a key to help students to envision what is possible and plausible in a situation, independently of computational procedures.

2. REAL WORLD PROBLEMS

My reaction to Heid's chapter (this volume) focuses on the notion of "real world problems." My motivation for this focus is twofold: (a) the use of real world problems is one of the most central elements in the *Computer-Intensive Algebra* curriculum, and (b) a critical analysis of the notion of real world problem is relevant for mathematics education in general. The accumulated experience of Heid and her collaborators over the years, designing and implementing activities for the learning of algebra that involve "realistic situations," might be a unique and rich source of insights. The chapter, however, seems to take for granted the use and nature of real world problems. This reaction is basically an attempt to spell out implicit assumptions and raise questions around them.

The plan is to elaborate on the following three different expectations for the contributions of realistic situations to the learning of algebra and then to explore issues surrounding them:

- Real world problems enable students to approach new mathematical concepts using ideas and situations that are familiar to them.
- Real world problems suggest to students that the formal definitions adopted in algebra are natural and reasonable.
- Real world problems offer fruitful contexts for students' learning to deal with complexity.

I avoid a discussion on the definition of what a real world problem is because I suspect that it would be sterile. Rather, I will repeatedly refer to the three examples of real world problems that Heid included in her chapter and add some examples from my own experience.

There is a fourth expectation commonly held for the contribution of real world problems that I am not addressing here, namely, that real world problems help students to realize that algebra is useful and relate to actual issues of social, financial, or political relevance. In connection to this expectation, there are a number of questions that Heid's chapter elicits, such as: If we think that algebra relates to factual issues of political or personal relevance, why invent problems? Wouldn't it be better to use situations that appear in newspaper articles or actual personal stories? Also, what are the issues that students are supposed to deal with in problems like the *Mountain Resort* or the *All Stars Game*, besides familiarizing themselves with a set of linear and quadratic equations? A recent elaboration on these important themes can be found in Lesh, Hoover, and Kelly (1993).

The following sections focus on each one of the three expectations cited above. My analysis centers on examples of problems and how students approached them.

2.1. *Real World Problems Enable Students to Approach New Mathematical Concepts Using Ideas and Situations that are Familiar to them*

This thesis reflects a vision of cultivating algebra learning in contexts that maximize the use of resources students can develop from their life experience. Most educators would agree that the richness of a learning activity is nourished by the background of intuitions and tools that students can bring to their understanding of the situation. Because students live in the real world, it is often assumed that real world problems will enrich the domain of possibilities that students can envision. Examining the chapter's examples of real world problems, one wonders whether this is necessarily true. Are students familiar with the management of a hotel or with the organization of all-star games? Do they have intuitions as to what are reasonable profits and costs for a special winter weekend at a resort hotel in the mountains?

In order to elaborate deeper questions, I will describe a conversation that I had with my son, Damian, on the problem of how many people can sit around a table of varying length. My son had just finished 7th grade and had not yet taken algebra. I asked him what general formula could tell the number of people who can sit at a table formed by a sequence of unit-tables:

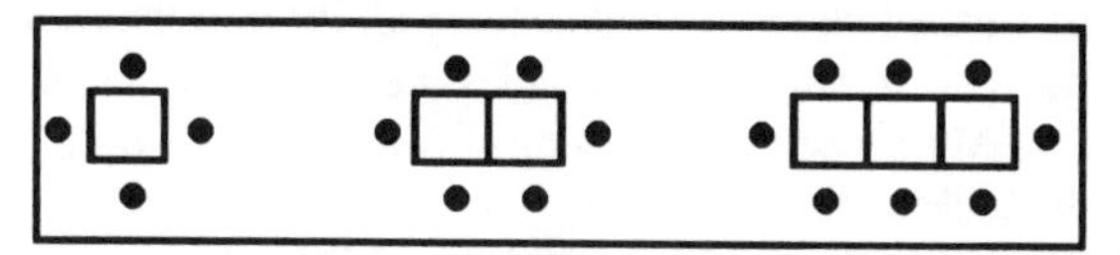

We counted that the single table seats four people, the double one six people, the triple one eight people, and so forth. In order to generalize, Damian created the following table in which the column to the left indicated the number of unit-tables, whereas the column to the right is the number that needs to be added to the one on the left to obtain the amount of available places to sit (e.g., if there are 2 tables, the number of available places to sit is 2+4; he recorded this idea as 2 = 4):

2 = 4
3 = 5
4 = 6
5 = 7
6 = 8
7 = 9
8 = 10
9 = 11
10 =

Damian commented that the number to the right is always two more than the number of tables. Trying to make sense of this pattern he drew a couple of long rectangles standing for a sequence of 10 and 100 tables respectively:

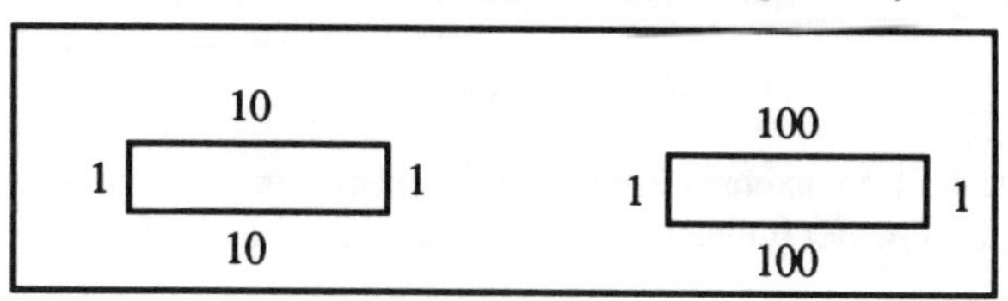

As he drew the rectangles, he said that the number of sitting places on the long sides is equal to the number of tables and wrote the following expression where "each side" meant "each end":

Table × 2 + how many can sit on each side

Damian wanted to show that, for a single line of tables, the "how many can sit on each side" is 2, whereas for a double line of tables it is 4:

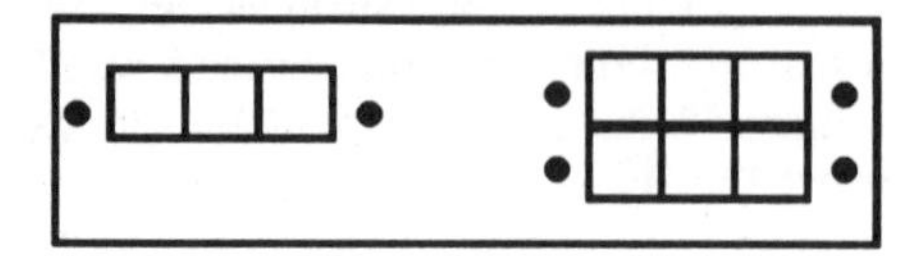

I asked him to use "P" for the number of people and "T" for the number of tables. He then wrote the following expressions:

$T \times 2 + 2 = P$ $T \times 2 + 4 = P$ (for the double line of tables).

At this point I asked him to imagine having all the unit-tables separated. It was immediately clear to him that, if all the tables are separated, the number of people who can sit is T×4. My subsequent question was how many sitting places would be lost by joining the tables. His first impulse was to create the following table of data. The column on the left is the number of tables, and the column on the right ("D") is the number of people displaced (losing their sitting places) by joining the tables:

T	D
1	0
2	2
3	4
4	6
5	8
6	10

Looking at this table, Damian tried to find number patterns. He was particularly interested in the fact that the second line is the only one with equal values. One of patterns that he mentioned was that each number on the right doubles the previous one on the left. I asked him how he would express such a pattern in a formula. He wrote:

D = 2(last #).

The ensuing conversation was on alternative ways to express the "last #" factor. It occurred to him that T-1 was a good alternative:

$$D = 2(T-1).$$

Then we had an exchange on how to check whether the latest expression "works," that is, whether it matches correct numeric values for D and T. After we felt satisfied with a few particular cases, Damian represented the process of joining the tables with the following expression that subtracts, from T×4, the sitting places that disappear when the tables are joined:

$$P = T\times 4 - 2(T-1).$$

I asked Damian how he would show that P = T×4 - 2(T-1) and P = T×2 + 2 tell the same thing. He said that they had to be equivalent and seemed to feel that the way in which we had obtained them proved such a conclusion.

This conversation with Damian illustrates, I think, what a child who feels resourceful might do in such a situation. Within our interaction, Damian could find multiple perspectives to deal with the problem of the sitting places; he generated his own tables of data, identified diverse number patterns, explored formulaic ways of expressing them, created visual representations, articulated word descriptions of the situation, felt numerous expectations as to what kinds of results make sense (e.g., the longer the table, the more people who can sit; the number of people who can sit on the extremes is constant, etc.), and constructed linear equations to model the situation. It is possible to argue that the problem of the table/people is less "real world" than the one of owning a restaurant or of bacterial population (e.g., its numbers are not messy). However, the argument can be reversed. Instead of starting with a general scheme of what constitutes a real world problem, one could say that one of the aspects of "reality" that matters to us, is the flourishing of students' resourcefulness. From this point of view the situation of the table/people might be experienced by the student as more real--tangible, full of significant details--than the one of owning a restaurant. The main question that I want to formulate as a mode of conclusion is this: Where is the reality of a "real world problem"? How do we account for the fact that what is experienced as real by someone may not be real for someone else?

2.2. *Real World Problems Suggest to Students that the Formal Definitions Adopted in Algebra are Natural and Reasonable*

One of the real world problems included in Heid's chapter intends to illustrate how students "develop properties of exponential functions through consideration of their meaning in applied settings," such as the property

$$a^{-n} = 1/a^n$$

where a is the population of bacteria that doubles every day. There is an important tradition within mathematics education that seeks to design realistic contexts in which the formulation of general and formal properties can be recognized as sensible, and even necessary. One the best known examples concerns the rules to combine signs in addition and multiplication of integers. Electric charges, mailmen delivering notes of credit and debit, switches, and elevators are some of the themes that have been used to make a case for the rules with signs. Ball (1990) described how she used in her 3rd grade classroom the metaphors of an elevator and money transactions to discuss the meaning of additive expressions. The children could interpret many additive expressions, but not others, such as 3-(-6). Ball experienced a tension between the boundaries of children's interpretations in the elevator context and the mathematical meanings of the number sign. Perhaps there is a more general tension between the "reality" of the context and how faithfully it can suggest the consistency of formal systems. Schwarz, Kohn, and Resnick (1992) designed a computer microworld with simulated trains for learning about the sign rules. As they tried to reflect accurately 14 general rules that define the use of signs in addition and multiplication, the simulation seemed to lose its resemblance to what happens with real trains. Is this an issue? If students are supposed to develop properties out of considering meanings in an applied setting, what happens when the applied setting becomes an unrealistic artifact of the property that it is intended to teach?

With some colleagues, we have recently discussed scenarios for children's conversations about the rules for operating with signs. We wondered what would happen if we inverted the question, that is, if instead of suggesting the correct rules by designing ad hoc contexts, we asked them what might the addition of signed numbers mean, in whatever context they want to use. Jan Mokros asked this question to Erica, her 10-year-old daughter. Erica started with the number line and thought of adding and subtracting signed numbers as taking out or putting in the signed number on the number line itself:

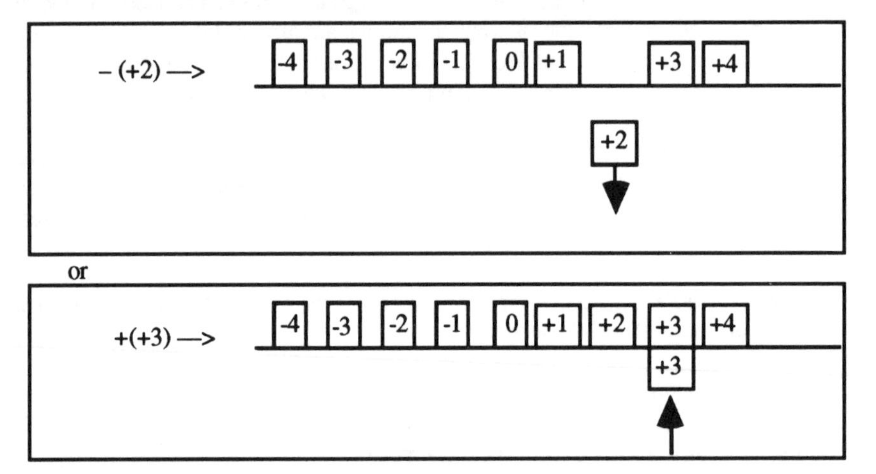

Subsequently, Erica thought that to figure out the result, one would have to balance everything out: Each number and its additive inverse disappear (e.g., the +1 card goes away with the -1 card, etc.). The net result, that is, the number card that is left on the table, is the one to be used; for example, – (+2) . –2 or –(–4) . +4. Erica devised a consistent scheme for reducing a pair of contiguous signs--where the first one means the action of taking out or putting in, and the second one means location on the number line--into a single sign. Her scheme was unexpected and new to us.

The story with Erica suggests that what counts might be not what we judge as being real and compelling, but how we facilitate the students' mathematical encounter with what is real and compelling to them. For Erica, the use of the number line as a context may have been more fruitful than the more "realistic" ones of the elevator or the electric charges.

In order to deepen this analysis, I wanted to explore the problem of bacteria growth and the property $a^{-n} = 1/_{a^n}$ with a student, Ana, who was 15 years old and had completed 9th grade. Ana had been taught in school about positive and negative exponents. I began asking her about the powers of 2. Ana showed me, easily, that:

$$2^4 = 16$$
$$2^3 = 8$$
$$2^2 = 4$$
$$2^1 = 2$$

Then I asked Ana about the zero and negative powers of 2. She completed the list in the following way:

$$2^4 = 16$$
$$2^3 = 8$$
$$2^2 = 4$$
$$2^1 = 2$$
$$2^0 = 2$$
$$2^{-1} = -2$$
$$2^{-2} = 4$$
$$2^{-3} = -8$$

Ana explained to me that one has to multiply 2 by itself as many times as the number that is "on top." In the case of negative exponents, she reckoned that the number that one has to multiply by itself is –2. I said that I understood how she had come up with the values for the negative powers of 2, but not with $2^0 = 2$. Ana said that 2^0 is "like not multiplying, so you are left with 2."

I drew a binary tree and we talked about the correspondence between the powers of 2 and the numbers of dots in each level:

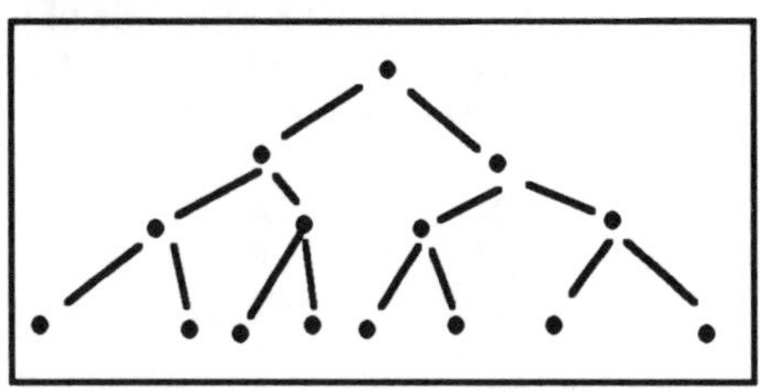

Ana saw that the correspondence suggested that $2^0 = 1$, but she was unsure about this result because "everything starts with 2 dots," that is, 2^0 is not really part of the series because its real beginning is 2^1. My understanding of her intuition is that she perceived the initial dot of the binary tree as uninformative and external. It did not tell that the sequence was about powers of 2, nor indicate binary bifurcation.

Then I posed for her the problem of the bacterial population that doubles every day starting with 500. We elaborated the following table:

N	Day
500	1
1000	2
2000	3
4000	4
8000	5

I asked Ana to think of a general equation that would give N for any Day. First, Ana tried to use proportions and checked the following correspondence:

Day	N
1	500
3	1500

She commented, "no, it does not work," because on Day 3, N had to be 2000. Ana pointed out that each N is the previous one times 2, and then she started to investigate the use of powers of 2. Ana wrote:

$$2^3 \times 500 = 4000$$
$$2^4 \times 500 = 8000$$

Reflecting on these results as compared to the former table, Ana concluded that one has to use an exponent one less than the Day. Subsequently she wrote the general expression:

$$2^{D-1} \times 500.$$

After trying out several values of D, she felt satisfied that her formula "worked" because it produced the correct values of N. Regarding the case of Day 1, she wrote:

$$N = 2^0 \times 500$$
$$N = 500$$

using, possibly influenced by our previous conversation on the binary tree, the fact that 2^0 is equal to 1. I asked Ana about the bacteria population on the day before the first one. It was clear to her that, following the posited rule of population growth, on the day before Day 1 the population must have been 250. She also concluded that such a day should be referred to as "0" and added it to the table:

N	Day
250	0
500	1
1000	2
2000	3
4000	4
8000	5

Ana tried to check whether the general equation worked for Day 0 too. She wrote:

$$N = 2^{-1} \times 500$$
$$N = -2 \times 500$$
$$N = -1000$$

Ana commented that her formula was not good for the day before Day 1. I asked her what should 2^{-1} be equal to for her formula to hold on Day 0. My question did not make sense to her. She said that multiplying 500 by a power of 2 would always give a number greater than 500, and so it could not produce 250. She concluded that the formula cannot be extrapolated to the past. I felt stuck. I had a strong feeling that

I had to respect her conclusion. There was something mathematically important and profound in her sense that one does not arbitrarily redefine mathematical operations, such as powers of 2, to accommodate results in an equation. I thought that such an attitude was in many ways much more important than knowing the true value of 2^{-1}. I felt relieved that our time was over and we agreed to continue on another day.

Two days later, I met Ana again. I asked her to reconstruct her table for the powers of 2; she wrote:

$$\begin{aligned}
2^5 &= 32\\
2^4 &= 16\\
2^3 &= 8\\
2^2 &= 4\\
2^1 &= 1\\
2^0 &= 2\\
2^{-1} &= -2\\
2^{-2} &= 4\\
2^{-3} &= -8
\end{aligned}$$

Then I asked Ana to think of an alternative approach: Imagine that one wants the powers of 2 to be such that anytime one "moves up" in the table the number doubles, and anytime one "moves down" the number halves. She deemed that, in that case, the table should be:

$$\begin{aligned}
2^5 &= 32\\
2^4 &= 16\\
2^3 &= 8\\
2^2 &= 4\\
2^1 &= 2\\
2^0 &= 1\\
2^{-1} &= .5\\
2^{-2} &= .25\\
2^{-3} &= .125
\end{aligned}$$

We then had a long conversation about this way of defining powers of 2 (as reflected in her "doubling-halfing" table). I drew two number lines, so that I could move my finger marking "n" and she could move hers marking "2^n":

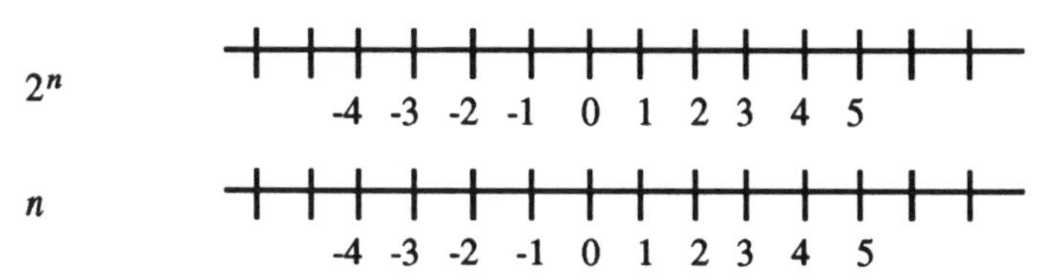

As I displaced my finger toward the left, Ana moved hers closer to zero but without passing it to the negative side. We talked about the fact that our fingers were never side by side. She commented that our fingers were closest for $n = 0$ and $n = 1$, because to the right of 1 or to the left of 0 they moved farther apart. I asked Ana how one could figure out directly 2^{-2}, knowing that $2^2 = 4$, that is, without halving repeatedly. She tried several cases noticing that:

$$2^5 = 32 \qquad 2^{-5} = 1/32$$
$$2^6 = 64 \qquad 2^{-6} = 1/64$$

From these cases Ana wrote a generalization:

$$2^{-n} = 1/2^n$$

Then I proposed to revisit the problem of the bacteria growth. After recalling what we had done in the first session, Ana deemed that, using the new scheme for the definition of the powers of 2, her formula worked for the days before Day 1. Close to the end of the session, with an expression of perplexity in her face, she asked me: "But this is not how they [powers of 2] really are, right?" She reacted with surprise when I told her that the "official" definition of powers of 2 is according to the "doubling-halfing" table that she had generated. We talked about Ana's sense that the definition of powers of 2 had to be grounded in the multiplication of 2 by itself, an element that seemed to her absent for the negative exponents in that table.

Looking back at the role of the *Bacteria Growth* problem in the conversation with Ana, it seems to me that it was a helpful element for our exchanges. But I do not think that its importance lay in highlighting a realistic situation. I believe that the problem could equally have been about alien creatures reproducing in another planet or a strange book that doubles the number of letters in each page. Its contribution was to evoke in Ana and me a shared language (days, number of bacteria, doubling, halving, etc.) with which we could talk about the subtleties of a functional situation.

More generally, the tales of Erica and Ana suggest to me that properties, such as rules of signs or negative exponents, reflect broad perspectives about the nature of mathematics, its uses, and inner relationships. Students come to reflect about them with a rich background of personal stories and expectations. Defining those properties or inferring how they play in applied settings are just small pieces within a wide and rich landscape. What matters about problems like the one on bacterial growth, I think, is how they become a part of conversations and attitudes framing the discussion of formal properties.

2.3. *Real World Problems offer Fruitful Contexts for Students Learning to deal with Complexity*

The problems posed in school mathematics often have the flavor of being extreme oversimplifications that seem at odds with an essential aspect of our experience of the real: feeling intimacy with complexity. It is commonly postulated that, by posing real world problems, students are more likely to overcome the sense of uselessness and artificiality that pervades their path throughout mathematics education, fostering the perception that mathematics offers empowering means to deal with complexity. The point that I want to elaborate in this section is that complexity is not a quality dwelling in the problem but in the experience of the problem solver. To articulate the implications of this point of view, I will relate another conversation with Damian about a simple problem and how it motivated an excursion into complexity.

I asked Damian about the problem of the staircase, namely, how many blocks are needed to construct a staircase like the following one, up to an "n" number of levels:

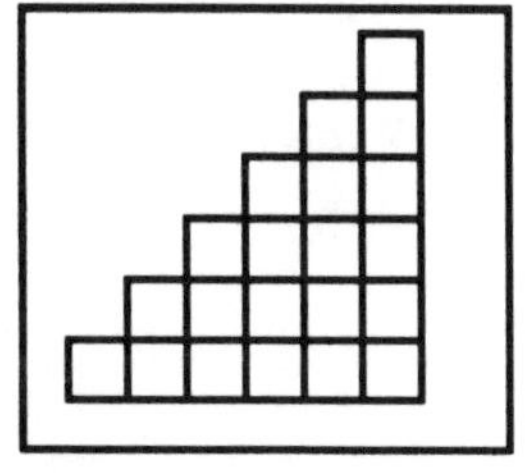

At first he could not find where to begin. I suggested he consider the square of blocks encompassing the staircase:

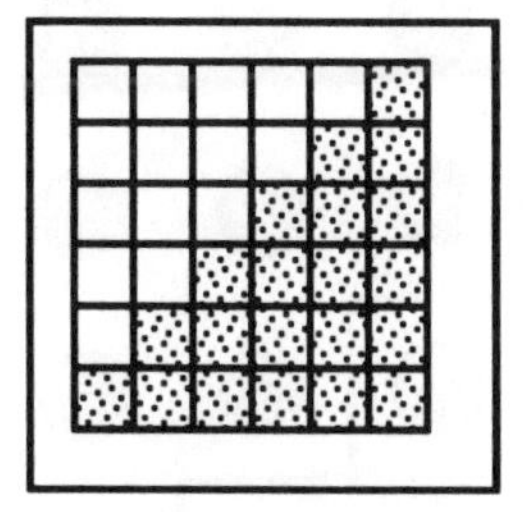

Damian knew that to construct the square he needed $n \times n$ blocks. The problem became that of taking away those that are not part of the staircase. At one point, he thought that, after taking out the diagonal, half of the blocks were inside and half outside of the staircase:

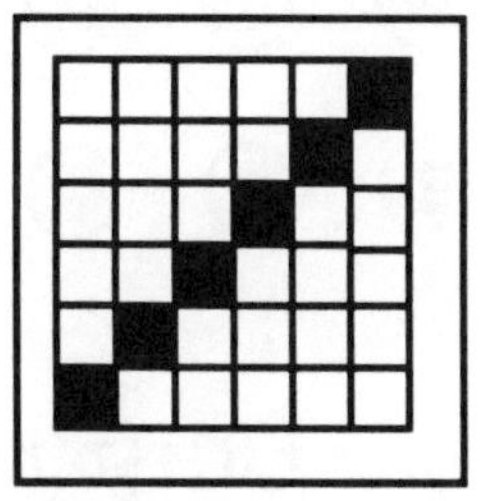

He counted that the diagonal had 6 blocks, the same as "n." Damian expressed the idea of taking out the diagonal and leaving half of them with this expression:

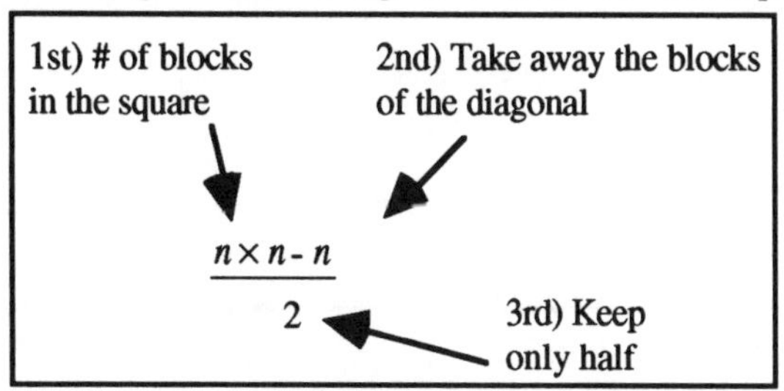

To complete the staircase, he then added the blocks of the diagonal and arrived at a final expression, that he checked for several values of n:

$$\frac{n \times n - n}{2} + n$$

I showed and discussed with Damian two other solutions (see Figures 5 and 6) that are included in *Proofs Without Words* (Nelsen, 1993).

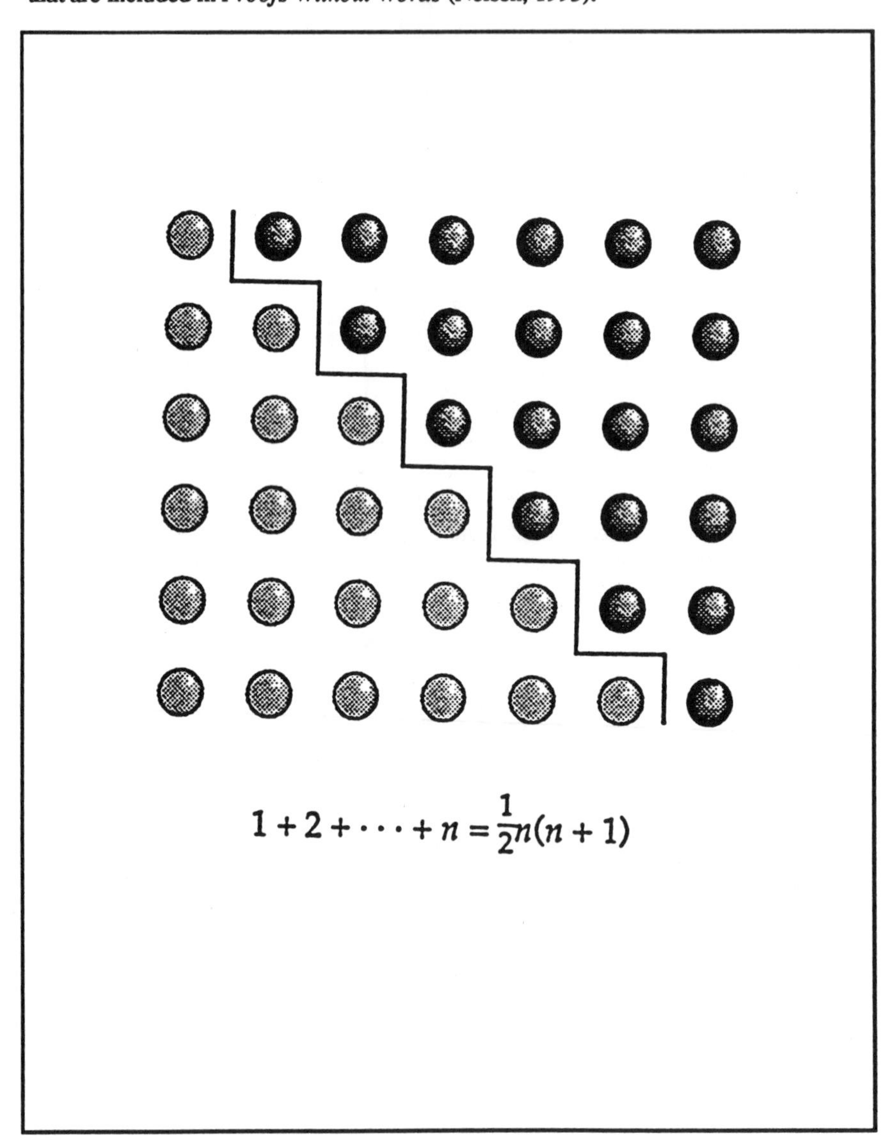

Figure 5.

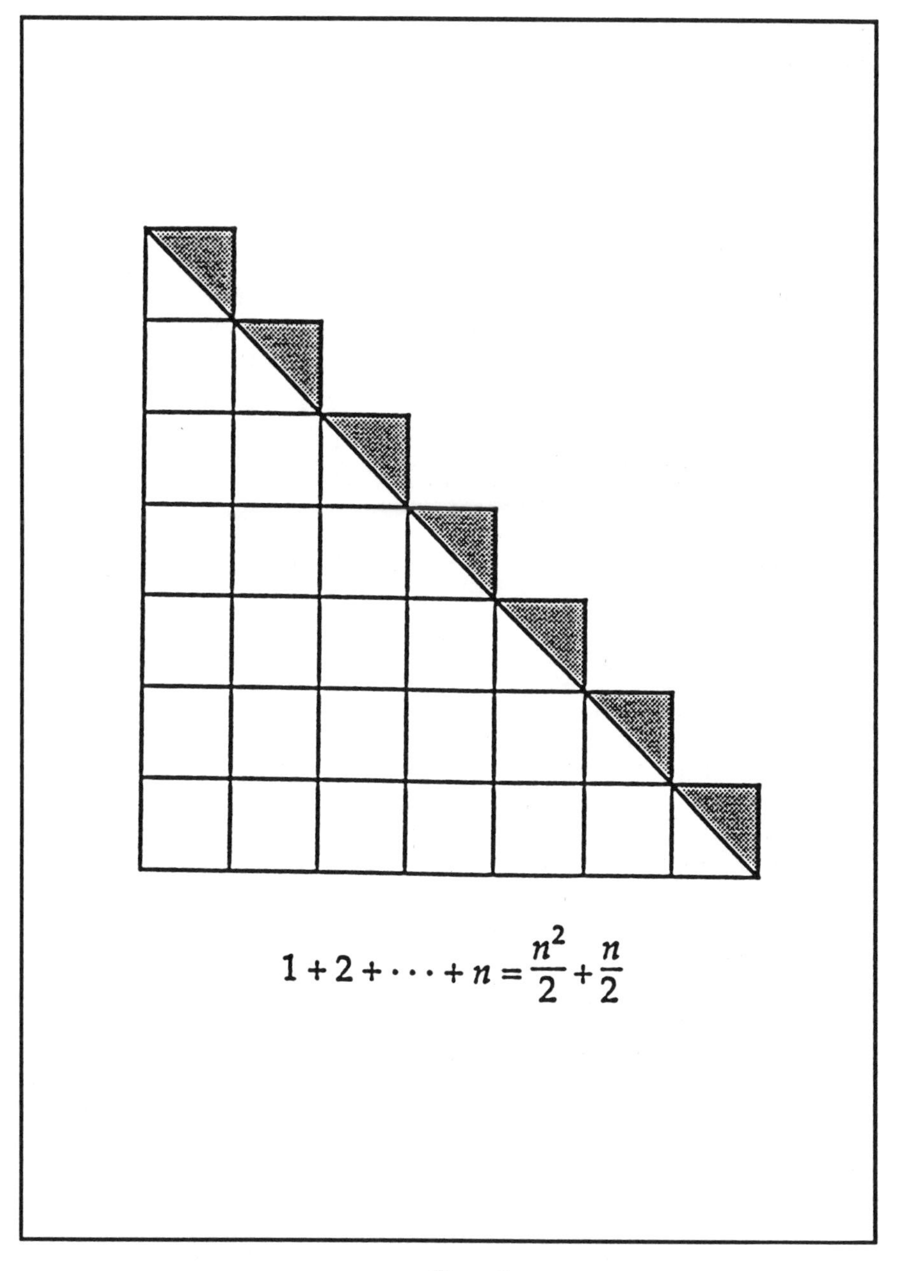

Figure 6.

Since it had become very clear that the problem of the staircase was equivalent to the sum of the first n whole numbers, I wanted to tell Damian the famous story of Gauss (as I know it), when he was a child, and the teacher asked for the sum of the first 10 whole numbers. To talk about Gauss' solution, I intended to write the following:

$$\begin{array}{r} 1 + 2 + 3 + 4 + 5 + 6 + 7 + 8 + 9 + 10 = 55 \\ \underline{10 + 9 + 8 + 7 + 6 + 5 + 4 + 3 + 2 + 1} = 55 \\ 11 + 11 + 11 + 11 + 11 + 11 + 11 + 11 + 11 + 11 = 110 \end{array}$$

But as soon as I had completed the first line, Damian said with a tone of insight, "Oh, I know," and marked the following lines as he said that one could add 1 and 9, 2 and 8, 3 and 7, ..., and then the one that is left in the middle:

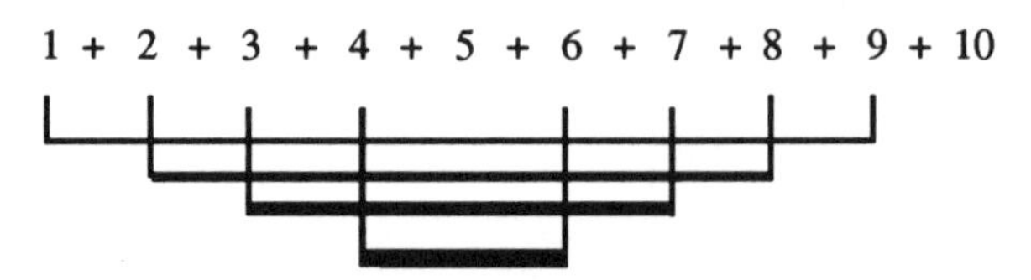

I asked him how this could be expressed by a formula. He wrote, avoiding the use of specific numbers: Add the first with the one before to the last, the second with the one second to the last, ...

Sensing that this long sentence could not be a formula, Damian said that he knew how to do it but not how to show it with a formula. Talking about the fact that each addition always gives 10, he counted that 10 was being added 5 times, that is, half of itself times itself, an idea that he expressed with the following expression:

$$n \times n/2$$

Then he wanted to add the number in the middle, 5, which was half of the last one. His final expression was the following:

$$n \times n/2 + n/2$$

He commented that this procedure works only if the last number, n, is even; otherwise, if the last number were odd, there would not be a number left in the middle (like 5 for the case of 10) to add at the end. Damian was surprised to see that his expression was the same as the one derived in Figure 6. Since the solution represented in Figure 6 works for any n, even or odd, the issue that arose in the ensuing conversation was why the formula that he had obtained can be used for odd integers too. My sense is that we could not find an explanation satisfactory to us.

This conversation suggests that whether the staircase problem is simple or complex is not intrinsic to the problem itself, but to what one does with it. During the exchange, Damian and I shared a growing sense of complexity and richness. This story suggests to me that the roots of complexity, at least the ones that matter the most for mathematics education, are in what students bring to their strivings to making sense of the problem situations. Complexity, rather than being an exogenous factor defined by the problem, emerges from the qualities that surround the students' experience of the problem.

One's stance with respect to complexity has practical implications as to what kind of problems one poses to students. For instance, in Heid's examples the students are provided with the "model," that is, with the functions to be used as a model of the situation. The idea is that real world situations are too complex for the students' own construction of appropriate models. Such an enforced stipulation of the model may not always be necessary, however, if one rethinks where the complexity is and dismisses generic taxonomies separating simple from complex problems in favor of deepening our understanding of how complexity can flourish in the student. With Damian, Erica, and Ana, I have tried to exemplify students who constructed models on their own, including linear, quadratic, and exponential functions. This may have been an important aspect of their feeling of intimacy with complexity.

My intention in this section of the chapter is not to criticize the *Computer-Intensive Algebra* curriculum, nor to discuss whether real world problems are good or bad for algebra education, but to revise some of the assumptions that are often invoked in their use. I think that the four expectations that I mentioned in the introduction to this section reflect legitimate and important goals: to foster mathematics learning by enriching students' resources and intuitions, to relate the definition of mathematical properties to broader ideas, to learn mathematics in the analysis of the complex, and to perceive social and personal relevance in the use of mathematical ideas. The theme that I recognize emerging throughout my examples and comments is that in order to achieve these goals it may be necessary to free ourselves from our stereotypes of what a real world problem is, and open our minds to explore what is real in the experience of the learner. Such a conceptual shift forces us to confront the fact that what is real to the student and to us--what is felt as significant, full of interconnected details, tangible, and accessible from multiple points of view--may break the mold of common stereotypes.

The distinction between real and unreal world problems is parallel to another distinction very common in the literature, and whose treatment is equally misleading, I believe, that between contextualized and decontextualized problems. Often problems are characterized as being decontextualized because they are just about numbers (as opposed to quantities or measures of specific things), as if all the rich background of ideas and experiences that students develop around numbers could not offer a context. I see the origin of this mistaken notion in the assumption that the context belongs to the formulation of the problem, ignoring that real contexts are to be found in the experience of the problem solvers.

PART VI

SYNTHESIS
AND
DIRECTIONS FOR FUTURE RESEARCH

CHAPTER 21

BACKWARDS AND FORWARDS: REFLECTIONS ON DIFFERENT APPROACHES TO ALGEBRA

DAVID WHEELER

This chapter is based on my closing remarks at the colloquium that gave rise to this book. But before I get to the main purpose of the chapter, I would like to make a comment triggered by one of the contributions to our closing discussion. Lesley Lee gave a passionate, and very moving, plea that we share the algebra culture more widely. Induction to it, she feels, shouldn't be restricted to a minority, depriving whole chunks of the general population of the sort of intense pleasure, of fun, of achievement, that all of us were probably fortunate enough to experience.[1] While I empathize very strongly with the general sentiment, I also feel sure that there will always be people who will resist attempts to get them to join the algebra culture. I don't deplore their resistance; I expect it, even welcome it. There are many other worthy human activities for people to enjoy and profit from. A person can have a very good and useful life without knowing any algebra at all--any mathematics, even--just as I have managed to get along without any Latin and Greek, and without ever painting in oils, playing football, or learning to meditate.

Educators are often encouraged to make prescriptive statements about what all people *ought* to know and be able to do, and they wouldn't be human if they didn't tend to find the strong cases close to their own interests and specialisms. But I'm sure it's healthier to acknowledge that we are not dealing with imperatives here. We are too inclined to confuse the question of what may be important for *every individual* in our society to know with the question of what knowledge is important for our society *as a whole*--the knowledge that must be deliberately preserved and built on for everyone's benefit. Educators can help to answer these questions, but they will lose credibility if they make exactly the same response to both.

The meaning I give to the claim of "algebra for all" is that everyone should have a real chance to be introduced to the algebra culture and allowed to enter it *if they want to*. We don't yet know how to achieve this level of opportunity and we will have to work extraordinarily hard to find out, but it is unnecessary and unwise to invest in the hope that everyone will immediately run to take advantage of the offer.

This colloquium has been conspicuous for the extent of its attention to the history of algebra, so I shall begin with a few remarks about that, and follow it with remarks about the nature of algebra. Then, mainly for my own amusement, I take each of the four "approaches" to "beginning algebra" laid down for our consideration, and guess what the respective motivations for learners using these approaches might be. Then I take a slightly more serious look at the four approaches and begin to consider "the four questions" proposed in the brief we were all given.

I was asked to attempt to make a synthesis of what has taken place at this colloquium--a difficult task for anyone, and one I feel quite unsuited for. The best I can do is offer some reflections "after the event" and, since they *are* after the event, they can't help being to a large extent influenced by what has transpired. If I hold my

N. Bednarz et al. (eds.), Approaches to Algebra, 317-325.

mirror up to the experiences of the past few days, maybe this can serve as some sort of a summary and a focus for further discussions.

1. ALGEBRA IN HISTORY

As soon as one starts considering the lessons that can be learned from history, one can't help wishing that all mathematics educators had a basic knowledge of the history of mathematics--and regretting that most of them don't. Although it gives us no prescriptions for action, history is invaluable for giving us a perspective, for alerting us to difficulties and possibilities.

The first lesson history tells us is that algebra took an extraordinarily long time to develop. Most accounts of its origins make the full development stretch over at least 1000 years; some suggest it started earlier and took even longer. Interpretations of the specific historical events vary, but the weight of the evidence suggests that the total historical process was complex, subtle, and non-intuitive. It seems reasonable to suppose that the time was needed to meet and overcome some extremely difficult challenges.

Teresa Rojano reminded us that algebra is in some sense an *extension* of arithmetic, picking up some of the questions that arithmetic throws up but can't handle by itself. Putting it more strongly, algebra is a *completion* of arithmetic: Arithmetic appears to need the real numbers for its trouble-free functioning, and these could not be fully developed without the aid of algebra. Alan Bell recalled for us that there is a multiplicity of algebras, not just a single one. That phenomenon suggests that algebra, in spite of the specifics of its historical origins, isn't restricted to the world of numbers and may therefore not be inextricably tied to arithmetic.

It seems very natural for us to relate elementary algebra back to arithmetic, not nearly so natural to relate it to geometry; yet one of the lessons of history is that algebra might have been built on geometrical foundations. For a time, particularly during the period when Islamic mathematics was flourishing, that was the direction the development appeared as if it would go, and the possibility was still open at the time of the Renaissance. The evidence suggests to some people that "beginning algebra" for students today could, in principle, start from their geometrical knowledge and intuitions. There is little sign, though, that this alternative will be worked out practically in a detailed teaching program--it would make too dramatic a break with pedagogical tradition and cause all sorts of adjustment difficulties in other programs in the school curriculum.

Another lesson of history is perhaps even more unexpected: There are a number of instances that suggest that it isn't easy--maybe not even possible--to integrate elementary arithmetic, algebra, and geometry into one harmonious mathematical whole. The Greeks dealt with the discovery of irrational numbers by putting those numbers aside as in some way improper mathematical objects. The great achievement of the 19th century, the arithmetization of analysis, found no place in its program for geometrical intuition. One might say that in the first case geometry "won" and elbowed out the aberrant arithmetical entities, whereas in the other it "lost" and was banished from the further development of one of the most powerful intellectual tools ever invented. However one interprets these historico-mathematical events, they seem to demonstrate the existence of some deep difficulty, as of a psychological splitting, that prevents arithmetic, algebra, and geometry from cohabiting in a completely non-contradictory fashion.

2. WHAT IS ALGEBRA?

A difficulty in defining algebra is that when one thinks one has got hold of its essence, other aspects occur to one and have to be made room for. It always seems to comprise rather more than any simple story suggests.

- Algebra is a *symbolic system*--symbols are the signs by which we often recognize the presence of algebra--but it's also more than a symbolic system.
- Algebra is a *calculus*--among its primary elementary uses is the computation of numerical solutions to problems--but it's also more than a calculus.
- Algebra is a *representational system*--it usually plays a large role in the mathematization of situations and experiences--but it's also more than a representational system.

During the meeting, someone suggested that algebra begins as soon as a symbol or a set of symbols is chosen to stand for an object or a situation. I think that this is not enough to warrant the use of the name algebra and that the representation, to be truly algebraic, must be endowed with a *structure* and a corresponding *dynamics*. Indeed, I suggest that although algebra has both static and dynamic aspects, its essence--at an elementary level, at least--is dynamic and transformational. Elementary algebra is *action*: collecting like terms, factoring, expanding, solving equations, simplifying expressions, summing sequences, drawing graphs, and so on. Elementary algebra appears to be an extraordinary collection of transformational procedures, a grand rococo fantasia on the theme of substituting one arrangement of symbols for another. Everything algebraic seems fluid and unstable. An algebraic statement, as soon as it is written down, seems to have an inner urge to become something else. But one also learns that there are strong constraints on this fluidity, constraints in the form of structural properties arising out of the nature of the mathematical objects--numbers, for example--that the algebra manipulates.

The difficulty of breaking algebra down into its basic components "to see what it's made of" is that one may end up with a list of actions--for example, *classifying, doing and undoing*, along with the others in the list that John Mason provided--in which all the mathematical content appears to have been stripped away. The actions in Mason's list, as Jacques Lefebvre pointed out, are just as basic to biology as to algebra. Well, one can tackle the definitional problem in two broadly different ways, each of which has some merit. One can take the line, as Mason does, that these basic operations are, indeed, truly algebraic, and together constitute algebraic thinking, in spite of the absence of mathematical-looking symbols. This then implies that, yes, biology and most other intellectual endeavors rest on a foundation of mental activity that is algebraic in character, just as mathematics itself does. The other line acknowledges the fundamental importance of these very general mental operations, but won't allow itself to call them algebra until they have been clothed in an appropriate symbolic form. One option doesn't seem necessarily any more "correct" than the other: Both positions are tenable. But there can be some confusion if people don't clearly announce their choice.

Algebraic activity, and mathematical activity in general, are not univocal forms of mental action: They occur in different modes, or at different levels, and there is no compulsion on anyone to stay in one mode or at one level of activity. Indeed it is a mark of true mathematical competence to be able to switch from one to another at will. Algorithms are so prevalent and so powerful in elementary arithmetic and algebra that there is always a temptation to reduce the question of competence to that of possessing a kit bag of algorithms and knowing which to use when. But even very elementary mathematical competence is more than that. One learns a "universal" algorithm for multiplying two natural numbers, for instance, and there is some implication that the algorithm is all that one needs since it covers all possible cases. But one is either blind or foolish to use the algorithm to multiply 500 by 2000. Without instruction in formal equation-solving methods, students show that they can solve a simple equation by using a test-and-adjust strategy. Even after they have been taught the usual "universal" methods for solving linear, quadratic, and simultaneous equations, the test-and-adjust strategy is still available and, with calculator at hand, may be a more sensible way to find an approximate solution quickly to some particular equation. Increasing one's mathematical competence isn't a matter of replacing inadequate or limited techniques with better or more universal ones, but of continually adding to one's repertoire so that one has more and more technical options to choose from in any situation.

3. INTERLUDE

People say that the central pedagogical problem in teaching algebra is finding convincing intrinsic reasons for the value of the study. (Extrinsic reasons, such as "required for high school graduation" or "required for entry to a differential and integral calculus course, which in turn is required for ...," and so on up the ladder, are plentiful enough.) But perhaps in no school subject is the teacher so likely to be plagued with questions of the order, "What's the use of this?" My own response in the classroom has usually been to try to divert the attack by suggesting that most other school subjects, including physical education and athletics, have very little immediate and obvious utility either--but I'm not proud of this tactic. Once it *has* raised its head, the issue of intrinsic motivation is tough for the teacher to deal with. The trap awaiting anyone trying to justify the value of algebra is that of oversimplification.

Consider, as an exercise, justifications separately based on the four viewpoints on algebra represented in this book: Generalization, Functions, Problem Solving, and Modeling.

- "Algebra is essentially about generalization." Learning algebra would then seem to be learning to do more of what people already do far too much of. Very young children already know how to generalize, and over-enthusiastic generalization is the root of human pride and prejudice and many other dangerous things.
- "Algebra is essentially about functions." This turns history upside-down and has the effect of throwing scorn on all the mathematics developed before the advent of the calculus or independently of it. (This claim can be alternatively expressed as "calculus is the real thing"--or "the hegemony of the calculus.")

- "Algebra is essentially about problem solving." This accords with the vector of history, and is in the main the approach codified in school textbooks, but in practice almost all the problems that this approach considers can be solved by arithmetic and don't need the weapons of algebra anyway.
- "Algebra is essentially about modeling." This emphasizes the important role algebra plays in mathematization, but it sidetracks the proper understanding of working with signs and significations by the undue stress it places on the three R's: "reality," "representation," and "relevance."

The above can perhaps be regarded as rather poor and sketchy jokes, which may or may not have some kernel of sense to them. But they don't contribute to the main business of this chapter, and to that I must now return.

4. FOUR APPROACHES

The four topic approaches to "beginning algebra" proposed for the consideration of this book have been referred to a number of times in the previous chapters (as well as in the section above), and I find little to say that has not already been said. One cannot doubt that the four topics can claim a place in all but the most basic mathematics curriculum, and each makes an important contribution to mathematical literacy. I was very taken with a remark made during the conference to the effect that "all generalization is a justification." I have yet to clear up in my mind to what extent I agree with this, but if I assume that there is something in it, it certainly strengthens the case for the "generalization" component of the curriculum. Justification involves arguing for, giving reasons, and anything seems worth encouraging that entails more of this rational activity in the mathematics classroom as a counterbalance to all the automating and routinizing that tends to dominate the scene.

Of course, like the other topics, "generalization" is an extremely broad term, covering far more than those aspects that would be seized on if "beginning algebra" were actually to begin here. The proponents of this approach don't seem to me to have made clear yet how the start they recommend would eventually lead to students possessing a fully functioning symbolic system. I can't help feeling that this particular suggestion, despite its interest, remains at the level of an idea rather than a workable program.

I could say the same about "modeling," only more so. Basing algebra on the concept of models seems to me like approaching history through the study of political trends or biology through evolutionary theories: The only way the approach can possibly be made to work with novices is by trivializing it. The reasons why this approach is talked about so frequently now are not mathematical or pedagogical but ideological. The modeling approach has the advantage of seeming "down to earth," overtly linking "abstract" mathematics to experiential stuff in the "real" world; in this sense it's non-elitist, and a-historical too, so it has the potential to free school mathematics from thc constraints of the Eurocentric mold.

The other two topic approaches can be worked out in a program. "Problem solving" must still be allowed the edge: It has the sanction of history and a century of teaching tradition behind it. It connects algebra directly to what some people regard as the main purpose of mathematics--the solving of well-formulated problems. The approach through "functions" has been made a practical possibility by the arrival of the computer. We have yet to find out whether an induction to algebra by this route will make more sense to students than the traditional path.

For understandable reasons--the almost total lack of work on it--the possibility of approaching "beginning algebra" from the direction of geometry has been largely underplayed here. The flush of current enthusiasm with the power of the computer will probably make this alternative no more likely to be developed in the immediate future, though there seems a real opportunity here to investigate an approach that might, in principle, anchor the symbolisms of algebra in the concrete tactile world of geometrical shapes.

5. BIG IDEAS

What are the ideas that we hope students of algebra will meet, by whatever path they take? I think this is a good question to put, and to keep on putting, though it's hard to answer (and harder still to decide what counts as a good answer). But learning algebra "because it is there" definitely isn't a good reason, and "because it's required for something else" isn't entirely satisfactory either.

Just to start us off, I'll put forward two suggestions. The first relates to the omnipresent "*x*," the feature of algebra that everyone notices. The idea, or the awareness, that I'd like students to acquire is that:

an as-yet-unknown number
a general number
a variable

} can be symbolized and operated upon "as if" it was a number.

The second "big one" is the idea, or the awareness, that algebra introduces one to a set of tools--tables, graphs, formulas, equations, arrays, identities, functional relations, and so on--that are related to each other sometimes almost interchangeably and together constitute a substantial technology that can be used to discover and invent things.

6. THE FOUR QUESTIONS

The conference organizers put four questions to us--which have varied a little in their formulation at different times. I will make a few brief comments about each before concluding.

6.1. *What are the Essential Characteristics of Algebraic Thinking?*

I tend to think this question has not yet been worked on enough, so it calls for too much speculation, though it's an excellent long-term research question. We've seen that there is no consensus on the attempt to differentiate algebraic thinking from mathematical thinking in general, or on the attempt to reduce the essential content of algebraic thinking to a set of very elementary operations. There's not much doubt in my mind that we can't wait until this question is decently answered before setting about experimenting with some new curricula. In the meantime, I think my "big ideas" (just above) could provide a reasonably uncontroversial starting assumption.

6.2. *What can Algebra Learning build on?*

The usual approach to algebra assumes familiarity with the behavior of the integers, and there is no particular reason to abandon this particular foundation--unless and until the experiment of starting from geometry is tried. Indeed, as soon as students have acquired generalized procedures for dealing with and producing integers, a certain amount of algebra is already implicit in their arithmetical competence. The second foundation that can also be assumed is students' experience of working with a language. Algebra is a symbolic system that has a number of the properties of a language and although there are many differences between algebraic and natural language, there are some similarities that can be exploited (though please note the qualification under the next heading). And, thirdly, everyone of sufficient maturity has mastery of some elementary mental operations, as Piaget's work has shown--mental operations such as classifying, substituting, arranging, inverting, and so on--that have obvious analogs in the symbolic operations of elementary algebra.

6.3. *What are the Obstacles to Algebra Learning?*

An interesting question, which I don't know how to answer, is whether it is possible to distinguish the learning obstacles that are intrinsic to the nature of the subject matter, the algebraic content itself, from others that are artifacts of the traditional approach to the teaching of algebra. In a practical sense, the distinction seems impossible to draw because, for almost everyone, algebra is identical with "school algebra." Most people don't have access to any school-free source of information about what elementary algebra "is really like." The relatively small proportion of people who have gone beyond elementary algebra and obtained a perspective on it probably construct *their* picture of algebra out of the accidents of history. The testimony of working mathematicians can be sought, of course, on the grounds that, as users, they should be closer to apprehending the essence of algebra, but a surprising number of those haven't reflected on the matter and can't produce an articulate answer, and some are not at all interested in the question.

This question of the four is probably the one we are best equipped to answer at this time. Research has added substantially to our awareness of the character of students' errors, for example, so that we are closer to a better understanding of what it is about algebraic activity that causes students to make mistakes. I can't contribute any fresh specifics to this discussion, but I will make two rather general--meta-level--observations.

One very general obstacle, it seems to me, can be regarded as the flip side to one of the points made in the answer to the previous question. Algebra, as a language, employs many of the words and symbols that students have already met in arithmetic. This commonality has its positive side--it permits students virtually immediate access to simple "meaningful" algebraic statements--but a negative side too--it encourages the assumption that the common words and symbols bear exactly the same significance as they do in arithmetic, and this is not so in any but the most trivial contexts. What happens to one's interpretation of the plus sign, for instance, when it is now placed between two symbols which *cannot* be combined and replaced by another symbol? We saw the "new math" movement try to tackle this kind of problem--the need to "unlearn" and adapt associations between signifiers and their signifieds--by introducing a variety of temporary notations, but that only added to the impression of algebra as formalist gobbledy-gook.

Another obstacle has to do with the matter of "noise," in the cybernetic sense. Pedagogical devices of any sort introduce an element of noise into the classroom--that is, "stuff" that isn't meant to be assimilated because it's not part of what the device is intended to teach, but it's there because it's part of the "embodiment" that makes the "real" content easier to grasp than it otherwise would be. You can tell how suspect this procedure is by the number of quote marks I've had to use in the preceding sentence; nevertheless, even if I've caricatured it, my description does draw attention to the distinct possibility that a pedagogical device can be too noisy, too full of distractions, to achieve its objective. I had that feeling very strongly about Dienes' pedagogy, which emphasized a particularly busy approach to the embodiment of mathematical concepts in a variety of settings. Current developments using computers to display multiple representations give me the same feeling of foreboding. The presentation is too noisy by half. (So is *Sesame Street.*) How can students possibly see the wood for the trees?

But, of course, one can't have a wood without any trees. The metaphor of cybernetic noise may be misleading because it suggests that information can be transmitted noise-free--and almost certainly it cannot. There is a good deal of talk at the present time about the need to re-contextualize the mathematics that is taught in the classroom, largely on the grounds that knowledge-in-context is meaningful and knowledge-out-of-context is not. Although one can't disagree with the broad truth of this assertion, in practice it is often put forward in connection with a very limited view of what constitutes a meaningful context for mathematics--as if the only context worth considering is that of everyday social practice. Since mathematics influences social practices very little, in fact, this choice of context severely limits the kind of mathematics that can be recontextualized in this way. My tentative opinion is that this strategy merely exchanges one form of pedagogical impoverishment for another.

I am sure we need fewer unsupported assertions and more research in this whole area. Let's look carefully at the relationship between context and meaning in algebra, at alternative contexts for introducing algebraic ideas and techniques, at the relationships between context and (cybernetic) noise, at how much noise can be tolerated, and so on. And we mustn't forget that algebra is a special case: It is *intrinsically* very general and very abstract and context-free. The reason why algebra works so efficiently as a calculus is that it treats data impartially and is not concerned with the "meaning" inherent in the context in which the data are situated.

6.4. *How shall We Choose which Approach to take to "Beginning Algebra"?*

I want to make only one comment on the answer to this question, and that is to point out that every choice has a price, and that the price must be paid. The four choices we have before us are not independent alternatives. Although presented as choices for a starting-point, they are really somewhat complementary objectives for the teaching of algebra: That is, we want students of algebra to come to know how to use it to solve problems, to model situations, to handle functions, and to make generalizations. Choosing one of these as a starting point affects how the others can be reached. If we opt for "functions" as a starting point (and many curriculum developers seems to be running blindly in this direction at the moment--I'll refrain from recalling the appropriate Biblical image), then we postpone the attainment of "problem solving" and make that more difficult to reach. And vice-versa, of course (though this is roughly the situation of traditional school algebra). So, in making our choice, let us not forget to factor in the cost.

7. DISCUSSION

Further discussion of the four approaches at the colloquium brought up a number of issues:

- The separation into four approaches to "beginning algebra" is artificial; all four components are needed in any algebra program.
- Each approach represents a spectrum of possibilities: for example, there can be a number of quite different ways of starting algebra while placing an emphasis on functions.
- Some other possible approaches have probably been omitted.
- Do we know precisely what algebraic content can be developed easily from each starting point, and how the shift can be made from one set of concepts to another?
- Is there indeed a unity--that we call algebra--behind the apparent diversity of the approaches?
- Different criteria can be used in choosing an approach; for example: Which approach provides students with the best set of tools for dealing with quantitative situations? Which approach makes more sense for a majority of high school students?

There was general agreement among the group that the various approaches had not yet been adequately researched, and that there was not enough hard evidence to use in evaluating the merits of the alternative proposals for the teaching of algebra now being developed. The division into four approaches (which the organizers had selected as representing the principal trends in current research and program development), though indeed artificial, had been useful for the purposes of analysis. The participants felt they now had a better idea of the nature of the challenge of finding a meaningful approach to "beginning algebra" and an enriched perspective on the possibilities.

NOTES

1 Lee's chapter, in this book, tells more about her own induction to the algebra culture.

REFERENCES

Ainley, J. (1987). Telling questions, *Mathematics Teaching*, 118, 24-26.

Al-Khwarizmi. (1831). *The algebra of Mohammed ben Musa* (F. Rosen, Trans. & Ed.). London: Oriental Translation Fund.

Arzarello, F. (1991a). Procedural and relational aspects of algebraic thinking. In F. Furinghetti (Ed.), *Proceedings of the 15th International Conference for the Psychology of Mathematics Education* (Vol. I, pp. 80-87). Assisi, Italy: PME Program Committee.

Arzarello, F. (1991b). Pre-algebraic problem solving. In J. P. Ponte, J. F. Matos, J. M. Matos, & D. Fernandes (Eds.), *Mathematical problem solving and new information technologies* (NATO ASI Series F, Vol. 89, pp. 155-166). Berlin: Springer-Verlag.

Ball, D. L. (1990, April). *With an eye on the mathematical horizon: Dilemmas of teaching*. Paper presented at the 1990 annual meeting of the American Educational Research Association, Boston, MA.

Barnes, M. (1991). *Investigating change: An introduction to calculus for Australian schools.* Carlton, South Australia: Curriculum Corporation.

Bashmakova, I. G., & Slavutin, E. I. (1977). "Genesis triangularum" de François Viète et ses recherches dans l'analyse indéterminée. *Archives for the History of Exact Sciences, 16*, 289-306.

Bednarz, N., & Janvier, B. (1991). Illustrations de problèmes mathématiques complexes mettant en jeu un changement ou une séquence de changements par des enfants du primaire. In F. Furinghetti (Ed.), *Proceedings of the 15th International Conference for the Psychology of Mathematics Education* (Vol. I, pp. 112-119). Assisi, Italy: PME Program Committee.

Bednarz, N., & Janvier, B. (1994). The emergence and development of algebra in a problem solving context: An analysis of problems. In J. P. da Ponte & J. F. Matos (Eds.), *Proceedings of the 18th International Conference for the Psychology of Mathematics Education* (Vol. II, pp. 64-71). Lisbon, Portugal: PME Program Committee.

Bednarz, N., & Janvier, B. (1995). L'enseignement de l'algèbre au secondaire, une caractérisation du scénario actuel et des problèmes qu'il pose aux élèves. In A. Daife et al. (Eds.), *Actes du colloque sur la didactique des mathématiques et la formation des enseignants* (pp. 21-40). Marrakech, Maroc: École Normale Supérieure de Marrakech.

Bednarz, N., Radford, L., Janvier, B., & Lepage, A. (1992). Arithmetical and algebraic thinking in problem solving. In W. Geeslin & K. Graham (Eds.), *Proceedings of the 16th International Conference for the Psychology of Mathematics Education* (Vol. I, pp. 65-72). Durham, NH: PME Program Committee.

Bednarz, N., Schmidt, S., & Janvier, B. (1989). *Problèmes de reconstruction d'une transformation arithmétique* (FCAR Report). Montréal, Canada: Université du Québec à Montréal, CIRADE.

Bell, A. (1964a). Timetables, calendars and clocks. In T. J. Fletcher (Ed.), *Some lessons in mathematics*. Cambridge, UK: Cambridge University Press.

Bell, A. (1964b). *Algebraic structures*. London: Allen & Unwin.

Bell, A. (1988). Algebra--Choices in curriculum design. In A. Borbas (Ed.), *Proceedings of the 12th International Conference for the Psychology of Mathematics Education* (Vol. I, pp. 147-153). Veszprém, Hungary: OOK.

Bell, A. (1995). Purpose in school algebra. In C. Kieran (Ed.), New perspectives on school algebra: Papers and discussions of the ICME-7 algebra working group (special issue). *Journal of Mathematical Behavior, 14,* 41-73.

Bell, A., Hart, M., Love, E., & Swan, M. (1980). *Algebra.* Nottingham, UK: Shell Centre for Mathematical Education, South Notts Project.

Bell, A., Malone, J., & Taylor, P. (1988). *Algebra: An exploratory teaching experiment.* Nottingham, UK: Shell Centre for Mathematical Education; Perth, Australia: Curtin University, SMEC.

Boers-van Oosterum, M. A. M. (1990). Understanding of variables and their uses acquired by students in traditional and Computer-Intensive Algebra. (Doctoral dissertation, University of Maryland, 1990). *Dissertation Abstracts International, 51A,* 1538.

Boileau, A., & Garançon, M. (1987). *CARAPACE* [Computer software]. Montréal, Canada: Université du Québec à Montréal, Département de Mathématiques.

Booth, L. R. (1984). *Algebra: Children's strategies and errors.* Windsor, UK: NFER-Nelson.

Boyer, C. B. (1956). History of analytic geometry. *The Scriba Mathematica Studies* (no. 6 & 7). New York: Scripta Mathematica.

Boyer, C. B., & Merzbach, U. C. (1991). *A history of mathematics* (2nd ed.). New York: John Wiley & Sons.

Brookes, B. (1976). Philosophy and action in education: When is a problem? *ATM Supplement, 19,* 11-13.

Brousseau, G. (1989). Les obstacles épistémologiques et la didactique des mathématiques. In N. Bednarz & C. Garnier (Eds.), *Construction des savoirs: Obstacles et conflits* (pp. 41-63). Montréal, Canada: Agence d'Arc.

Brown, J., & Burton, R. (1978). Diagnostic models for procedural bugs in basic mathematical skills. *Cognitive Science, 2,* 155-192.

Brown, M. (1981). Place value and decimals. In K. M. Hart (Ed.), *Children's understanding of mathematics: 11-16* (pp. 48-65). London: John Murray.

Brown, S., & Walter, M. (1983). *The art of problem posing.* Philadelphia, PA: Franklin Press.

Burkhardt, H. (1981). *The real world and mathematics.* London: Blackie & Son.

Capponi, B., & Balacheff, N. (1989). Multiplan calcul algébrique. *Educational Studies in Mathematics, 20,* 147-176.

Cardano, G. (1968). *The great art or the rules of algebra by Girolamo Cardano* (T. R. Witmer, Trans. & Ed.). Cambridge, MA: MIT Press. (Original work published 1545)

Carraher, T. N., Carraher, D. W., & Schliemann, A. D. (1985). Mathematics in the streets and in schools. *British Journal of Developmental Psychology, 3,* 21-29.

Charbonneau, L., & Lefebvre, J. (1991). *Une lecture de Viète, l'introduction à l'art analytique, cinq livres des Zététiques* (Technical report). Montréal, Canada: Université du Québec à Montréal, CIRADE.

Charbonneau, L., & Lefebvre, J. (1992). L'introduction à l'art analytique (1591) de François Viète: Programme et méthode de l'algèbre nouvelle. In I. Kleiner, H. Grant, & A. Schnitzer (Eds.), *Proceedings of the 17th Annual Meeting of the Canadian Society for the History and Philosophy of Mathematics* (pp. 103-116). Toronto, Canada: HPM Program Committee.

Chazan, D., & Bethell, S. C. (1994). Sketching graphs of an independent quantity: Difficulties in learning to make stylized, conventional "pictures." In J. P. da Ponte & J. F. Matos (Eds.), *Proceedings of the 18th International Conference for the Psychology of Mathematics Education* (Vol. II, pp. 176-184). Lisbon, Portugal: PME Program Committee.

Chevallard, Y. (1985). *La transposition didactique*. Grenoble, France: La Pensée Sauvage.

Chevallard, Y. (1989). *Arithmétique, algèbre, modélisation, étapes d'une recherche* (Vol. 16). Aix-Marseille, France: IREM.

Christiansen, B., & Walther, G. (1986). Task and activity. In B. Christiansen, A. G. Howson, & M. Otte (Eds.), *Perspectives on mathematics education* (pp. 243-307). Dordrecht, The Netherlands: Reidel.

Chuquet, N. (1880). La triparty en la science des nombres par Maistre Nicolas Chuquet parisien (A. Marre, Ed.). *Bullettino di Bibliografia E Di Storia Delle Scienza Mathematiche E Fisichei, 13*, 593-659, 693-814.

Chuquet, N. (1979). *La géométrie, première géométrie algébrique en langue française (1484)* (Introduction, text, & notes by H. l'Huillier). Paris: Librairie Philosophique J. Vrin.

Clagett, M. (1968). *Nicole Oresme and the medieval geometry of qualities and motions*. Madison: University of Wisconsin Press.

Clement, J. (1982). Algebra word problem solutions: Thought processes underlying a common misconception. *Journal for Research in Mathematics Education, 14*, 16-30.

Clement, J., Lochhead, J., & Monk, G. (1981). Translation difficulties in learning mathematics. *American Mathematical Monthly, 88*, 289-290.

Colín, J., & Rojano, T. (1991). Bombelli, la sincopación del álgebra y la resolución de ecuaciones. *L'Educazione Matematica, Anno XII, Serie III* (Vol. 2, No. 2). Cagliari, Italy: Centro di ricerca e sperimentazione dell'educazione matematica.

Collis, K. F. (1972). *A study of concrete and formal operations in school mathematics*. Unpublished doctoral dissertation, University of Newcastle, Australia.

Collis, K. F. (1974). *Cognitive development and mathematics learning*. Paper presented at the Psychology of Mathematics Workshop, Chelsea College, London.

Confrey, J. (1992a). Using computers to promote students' inventions of the function concept. In S. Malcolm, L. Roberts, & K. Sheingold (Eds.), *This year in school science, 1991* (pp. 131-161). Washington, DC: American Association for the Advancement of Science.

Confrey, J. (1992b). *Function Probe* [Computer software]. Santa Barbara, CA: Intellimation.

Courant, R. (1981). Reminiscences from Hilbert's Gottingen. *Mathematical Intelligencer, 3*, 154-164.

Crawford, J. T., Dean, J. E., & Jackson, W. A. (1954). *A new algebra for high schools*. New York: Macmillan.

Davis, P., & Hersh, R. (1981). *The mathematical experience*. Boston: Birkhauser.

Davydov, G. (1990). Types of generalization in instruction. In J. Kilpatrick (Ed.), *Soviet studies in mathematics education* (J. Teller, Trans.). Reston, VA: National Council of Teachers of Mathematics.

de Lange, J. (1987). *Mathematics insight and meaning*. Utrecht, The Netherlands: Rijksuniversiteit, OW&OC.

Descartes, R. (1954). *The geometry of René Descartes* (D. E. Smith & M. L. Latham, Trans.). New York: Dover. (Original work published 1637).

diSessa, A. A., Hammer, D., Sherin, B., & Kolpakowski, T. (1991). Inventing graphing: Meta-representational expertise in children. *Journal of Mathematical Behavior, 10,* 117-160.

Dolmans, F. (1982). Magnitudes in mathematical education. In G. van Barneveld & H. Krabbendam (Eds.), *Proceedings of Conference on Functions* (pp. 56-61). Enschede, The Netherlands: National Institute for Curriculum Development.

D'Ooge, M. L. (Trans.). (1926). *Nicomachus of Gerasa. Introduction to arithmetic. With studies in Greek arithmetic* (F. E Robbins & L. C. Karpinski, Eds.). New York: Macmillan.

Dubinsky, E., & Levin, P. (1986). Reflective abstraction and mathematics education: The genetic decomposition of induction and compactness. *Journal of Mathematical Behavior, 5,* 55-92.

Dugdale, S. (1993). Functions and graphs--Perspectives on student thinking. In T. A. Romberg, E. Fennema, & T. P. Carpenter (Eds.), *Integrating research on the graphical representation of functions* (pp. 101-130). Hillsdale, NJ: Lawrence Erlbaum.

Engelder, J. (1991). *A comparison of thinking skills elicited in textbooks used in Computer-Intensive Algebra and traditional algebra.* Unpublished master's paper, Pennsylvania State University, University Park, PA.

Euclid. (1956). *The thirteen books of the Elements.* New York: Dover. (Original work published 1925)

Eves, H. (1983). *Great moments in mathematics before 1650* (Dolciani Mathematical Expositions, No. 5). Washington, DC: Mathematical Association of America.

Fauvel, J., & Gray, J. (1987). *The history of mathematics: A reader.* London: Macmillan.

Ferrier, R. D. (1980). *Two exegetical treatises of François Viète, translated, annotated, and explained.* Unpublished doctoral dissertation, Indiana University, Bloomington.

Fey, J. T. (1989). School algebra for the year 2000. In S. Wagner & C. Kieran (Eds.), *Research issues in the learning and teaching of algebra* (pp. 199-213). Reston, VA: National Council of Teachers of Mathematics; Hillsdale, NJ: Lawrence Erlbaum.

Fey, J. T., Heid, M. K., Good, R., Sheets, C., Blume, G., & Zbiek, R. M. (1991). *Computer-Intensive Algebra.* College Park: University of Maryland; University Park: Pennsylvania State University.

Fikrat, L. (1994). *La notion de fonction et ses représentations.* Unpublished master's thesis, Université du Québec à Montréal.

Filloy, E., & Rojano, T. (1984a). From an arithmetical to an algebraic thought. In J. M. Moser (Ed.), *Proceedings of the 6th Annual Meeting of the North American Chapter of the International Group for the Psychology of Mathematics Education* (pp. 51-56). Madison: University of Wisconsin.

Filloy, E., & Rojano, T. (1984b). La aparición del lenguaje aritmetico-algebraico. *L'Educazione Matemática, Anno V* (N. 3; pp. 278-306). Cagliari, Italy: Centro di Recerca e sperimentazione dell'educazione matemática.

Filloy, E., & Rojano, T. (1989). Solving equations: The transition from arithmetic to algebra. *For the Learning of Mathematics, 9*(2), 19-25.

Filloy, E., & Rubio, G. (1991). Unknown and variable in analytical methods for solving word arithmetic/algebraic problems. In R. G. Underhill (Ed.), *Proceedings of the 13th Annual Meeting of the North American Chapter of the International Group for the Psychology of Mathematics Education* (Vol. 1, pp. 64-69). Blacksburg, VA: PME-NA Program Committee.

Filloy, E., & Rubio, G. (1993). Didactic models, cognition and competence in the solution of arithmetic and algebra word problems. In I. Hirabayashi, N. Nohda, K. Shigematsu, & F.-L. Lin (Eds.), *Proceedings of the 17th International Conference for the Psychology of Mathematics Education,* (Vol. I, pp. 154-161). Tsukuba, Japan: PME Program Committee.

Fischbein, E., Tirosh, D., & Hess, P. (1979). The intuition of infinity. *Educational Studies in Mathematics, 10,* 3-40.

Fleck, L. (1981). *Genesis and development of a scientific fact.* (T. Trenn & R. Merton, Eds.; F. Bradley & T. Trenn, Trans.). Chicago: University of Chicago Press.

Flegg, G., Hay, C., & Moss, B. (1985). *Nicolas Chuquet, renaissance mathematician. A study with extensive translation of Chuquet's mathematical manuscript completed in 1484.* Dordrecht, The Netherlands: Reidel.

Foucault, M. (1981). *Las palabras y las cosas.* Madrid: Siglo XXI Editores. (Original work published 1966)

Freudenthal, H. (1968). Why to teach mathematics so as to be useful. *Educational Studies in Mathematics, 1,* 3-8.

Freudenthal, H. (1982). Variables and functions. In G. van Barneveld & H. Krabbendam (Eds.), *Proceedings of Conference on Functions* (pp. 7-20). Enschede, The Netherlands: National Institute for Curriculum Development.

Freudenthal, H. (1983). *Didactical phenomenology of mathematical structures.* Dordrecht, The Netherlands: Reidel.

Gallardo, A. (1994). *El estatus de los números negativos en la resolution de ecuaciones algebraicas.* Unpublished doctoral dissertation, CINVESTAV del Institutuo Politécnico Nacional, México.

Galvin, W. P., & Bell, A. W. (1977). *Aspects of difficulties in the solution of problems involving the formation of equations.* Nottingham, UK: Shell Centre for Mathematical Education.

Gandz, S. (1938). The origin and development of the quadratic equations in Babylonian, Greek and early Arabic Algebra. *Orisis, 3,* 405-557.

Garançon, M., Kieran, C., & Boileau, A. (1990). Introducing algebra: A functional approach in a computer environment. In G. Booker, P. Cobb, & T. N. de Mendicuti (Eds.), *Proceedings of the 14th International Conference for the Psychology of Mathematics Education* (Vol. 2, 51-58). Oaxtepec, México: PME Program Committee.

Garançon, M., Kieran, C., & Boileau, A. (1993). Using a discrete computer graphing environment in algebra problem solving: Notions of infinity/continuity. In I. Hirabayashi, N. Nohda, K. Shigematsu, & F.-L. Lin (Eds.), *Proceedings of the 17th International Conference for the Psychology of Mathematics Education* (Vol. II, pp. 25-32). Tsukuba, Japan: PME Program Committee.

Gattegno, C. (1978). *What we owe children: The subordination of teaching to learning.* London: Routledge & Kegan Paul.

Gattegno, C. (1990). *The science of education.* New York: Educational Solutions.

Gillings, R. (1982). *Mathematics in the time of the pharoahs.* New York: Dover.

Giordan, A. (1989). Vers un modèle didactique d'apprentissage allostérique. In N. Bednarz & C. Garnier (Eds.), *Construction des savoirs: Obstacles et conflits* (pp. 240-257). Montréal, Canada: Agence d'Arc.

Goldenberg, E. P. (1988). Mathematics, metaphors, and human factors: Mathematical, technical, and pedagogical challenges in the educational use of graphical representations of functions. *Journal of Mathematical Behavior, 7,* 135-173.

Goldenberg, P., Lewis, P., & O'Keefe, J. (1992). Dynamic representation and the development of a process understanding of function. In G. Harel & E. Dubinsky (Eds.), *The concept of function: Aspects of epistemology and pedagogy* (MAA Notes, Vol. 25, pp. 235-260). Washington, DC: Mathematical Association of America.

Griffin, P., & Mason, J. (1990). Walls and windows: A study of seeing using Routh's theorem and associated results. *Mathematical Gazette, 74,* 260-269.

Grouws, D. A. (Ed.). (1992). *Handbook of research on mathematics teaching and learning*. New York: Macmillan.

Hall, R., Kibler, D., Wenger, E., & Truxaw, C. (1989). Exploring the episodic structure of algebra story problem solving. *Cognition and Instruction, 6,* 223 283.

Halmos, P. (1975). The teaching of problem solving. *American Mathematical Monthly, 82,* 466-470.

Harel, G., & Dubinsky, E. (Eds.). (1992). *The concept of function: Aspects of epistemology and pedagogy* (MAA Notes, Vol. 25). Washington, DC: Mathematical Association of America.

Harel, G., & Tall, D. (1991). The general, the abstract, and the generic in advanced mathematics. *For the Learning of Mathematics, 11*(1), 38-42.

Harper, E. (1987). Ghosts of Diophantus. *Educational Studies in Mathematics, 18,* 75-90.

Healy, L., & Sutherland, R. (1991). *Exploring mathematics with spreadsheets.* London: Blackwell.

Heath, T. L. (1964). *Diophantus of Alexandria. A study in the history of Greek algebra* (2nd ed.). New York: Dover. (Original work published 1910)

Heid, M. K. (1987). *"Algebra with Computers" in XX High School: A description and an evaluation of student performance and attitudes* (Report submitted to the XX Area Schools, Board of Education). Unpublished manuscript.

Heid, M. K. (1992). *Final report: Computer-Intensive Curriculum for Secondary School Algebra* (Report submitted to the National Science Foundation, NSF Project Number MDR 8751499). Washington, DC: NSF.

Heid, M. K., Sheets, C. S., & Matras, M. (1990). Computer-enhanced algebra: New roles and challenges for teachers and students. In T. Cooney & C. Hirsch (Eds.), *Teaching and learning mathematics in the 1990's* (1990 Yearbook of the National Council of Teachers of Mathematics, pp. 194-204). Reston, VA: NCTM.

Heid, M. K., Sheets, C., Matras, M., & Menasian, J. (1988, April). *Classroom and computer lab interaction in a computer-intensive environment.* Paper presented at the annual meeting of the American Educational Research Association, New Orleans, LA.

Heid, M. K., & Zbiek, R. M. (1993). Nature of understanding of mathematical modeling by beginning algebra students engaged in a technology-intensive conceptually based algebra course. In J. R. Becker & B. J. Pence (Eds.), *Proceedings of the 15th Annual Meeting of the North American Chapter of the International Group for the Psychology of Mathematics Education* (Vol. 1, pp. 128-134). Pacific Grove, CA: PME-NA Program Committee.

Herscovics, N.,·& Linchevski, L. (1991). Pre-algebraic thinking: Range of equations and informal solution processes used by seventh graders prior to any instruction. In F. Furinghetti (Eds.), *Proceedings of the 15th International Conference for the Psychology of Mathematics Education* (Vol. II, pp. 173-180). Assisi, Italy: PME Program Committee.

Hewitt, D. (in preparation). *The principle of economy in the learning and teaching of mathematics.* Doctoral thesis, Open University, Milton Keynes.

Hiebert, J. (1988). A theory of developing confidence with written mathematical symbols. *Educational Studies in Mathematics, 19*, 333-355.

Holt, J. (1964). *How children fail.* Harmondsworth, UK: Penguin.

Høyrup, J. (1985). *Babylonian algebra from the viewpoint of geometrical heuristics. An investigation of terminology, methods and patterns of thought.* Roskilde, Denmark: Roskilde University Centre.

Høyrup, J. (1986). Al-Khwarizmi, Ibn Turk and the Liber Mensurationum: On the origins of Islamic algebra. *Erdem, 2*, 445-484.

Høyrup, J. (1987). *Algebra and naive geometry. An investigation of some basic aspects of old Babylonian mathematical thought.* Roskilde, Denmark: Roskilde University Centre.

Høyrup, J. (1990). Dynamis, the Babylonians and Theaetetus 147c7-148d7. *Historia Mathematica, 17*, 201-222.

Hugues, B. (1981). *Jordanus de Nemore: De Numeris Datis.* Berkeley: University of California Press.

Janvier, B., & Bednarz, N. (1993). The arithmetic-algebra transition in problem solving: Continuities and discontinuities. In J. R. Becker & B. J. Pence (Eds.), *Proceedings of the 15th Annual Meeting of the North American Chapter of the International Group for the Psychology of Mathematics Education* (Vol. II, 19-25). Pacific Grove, CA: PME-NA Program Committee.

Janvier, C. (1978). *The interpretation of complex Cartesian graphs representing situations.* Unpublished doctoral dissertation, University of Nottingham.

Janvier, C. (1993). Les graphes cartésiens: Des traductions aux chroniques. In J. Baillé & S. Maury (Eds.), Les représentations graphiques dans l'enseignement et la formation (special issue). *Les sciences de l'éducation pour l'ère nouvelle, 1-3*, 17-37.

Janvier, C., Charbonneau, L., & René de Cotret, S. (1989). Obstacles épistémologiques à la notion de variable: Perspectives historiques. In N. Bednarz & C. Garnier (Eds.), *Construction des savoirs: Obstacles et conflits* (pp. 64-75). Montréal, Canada: Agence d'Arc.

Kang, W., & Kilpatrick, J. (1992). Didactic transposition in mathematics textbooks. *For the Learning of Mathematics, 12*(1), 2-7.

Kaput, J. (in press). Democratizing access to calculus: New routes using old roots. In A. Schoenfeld (Ed.), *Mathematical thinking and problem solving.* Hillsdale: Lawrence Erlbaum.

Kaput, J., & Sims-Knight, J. (1983). Errors in translations to algebraic equations: Roots and implications. In M. Behr & G. Bright (Eds.), Mathematical learning problems of the post secondary student (special issue). *Focus on Learning Problems in Mathematics, 5*, 63-78.

Katz, V. (1993). *The history of mathematics. An introduction*. New York: Harper Collins.

Kerslake, D. (1981). Graphs. In K. M. Hart (Ed.), *Children's understanding of mathematics: 11-16* (pp. 120-136). London: John Murray.

Kieran, C. (1981). Concepts associated with the equality symbol. *Educational Studies in Mathematics, 12*, 317-326.

Kieran, C. (1989). The early learning of algebra: A structural perspective. In S. Wagner & C. Kieran (Eds.), *Research issues in the learning and teaching of algebra* (pp. 33-56). Reston, VA: National Council of Teachers of Mathematics; Hillsdale, NJ: Lawrence Erlbaum.

Kieran, C. (1992). The learning and teaching of school algebra. In D. A. Grouws (Ed.), *Handbook of research on mathematics teaching and learning* (pp. 390-419). New York: Macmillan.

Kieran, C. (1994). A functional approach to the Introduction of algebra: Some pros and cons. In J. P. da Ponte & J. F. Matos (Eds.), *Proceedings of the 18th International Conference for the Psychology of Mathematics Education* (Plenary address, Vol. 1, pp. 157-175). Lisbon, Portugal: PME Program Committee.

Kieran, C., Boileau, A., & Garançon, M. (1989). Processes of mathematization in algebra problem solving within a computer environment: A functional approach. In C. A. Maher, G. A. Goldin, & R. B. Davis (Eds.), *Proceedings of the 11th Annual Meeting of the North American Chapter of the International Group for the Psychology of Mathematics Education* (pp. 26-34). New Brunswick, NJ: PME-NA Program Committee.

Kieran, C., Boileau, A., & Garançon, M. (1992). *Multiple solutions to problems: The role of non-linear functions and graphical representations as catalysts in changing students' beliefs* (Research report). Montréal, Canada: Université du Québec à Montréal, Département de Mathématiques.

Kieran, C., Garançon, M., Boileau, A., & Pelletier, M. (1988). Numerical approaches to algebraic problem solving in a computer environment. In M. J. Behr, C. B. Lacampagne, & M. M. Wheeler (Eds.), *Proceedings of the 10th Annual Meeting of the North American Chapter of the International Group for the Psychology of Mathematics Education* (pp. 141-149). DeKalb, IL: PME-NA Program Committee.

Kieran, C., Garançon, M., Lee, L., & Boileau, A. (1993). Technology in the learning of functions: Process to object? In J. R. Becker & B. J. Pence (Eds.), *Proceedings of the 15th Annual Meeting of the North American Chapter of the International Group for the Psychology of Mathematics Education* (Vol. 1, pp. 91-99). San José, CA: PME-NA Program Committee.

Kieren, T., & Pirie, S. (1992). The answer determines the question: Interventions and the growth of mathematical understanding. In W. Geeslin & K. Graham (Eds.), *Proceedings of the 16th International Conference for the Psychology of Mathematics Education* (Vol. II, pp. 1-8). Durham, NH: PME Program Committee.

Klein, J. (1968). *Greek mathematical thought and the origin of algebra*. Cambridge, MA: MIT Press.

Kline, M. (1972). *Mathematical thought from ancient to modern times*. Oxford, UK: Oxford University Press.

Krabbendam, H. (1982). The non-quantitative way of describing relations and the role of graphs: Some experiments. In G. van Barneveld & H. Krabbendam (Eds.), *Proceedings of the Conference on Functions* (pp. 125-146). Enschede, The Netherlands: National Institute of Curriculum Development.

Küchemann, D. E. (1981). Algebra. In K. M. Hart (Ed.), *Children's understanding of mathematics, 11-16* (pp. 102-119). London: John Murray.

Labov, W. (1972). *Language in the inner city: Studies in black English vernacular.* Philadelphia: University of Pennsylvania Press.

L'arithmétique des écoles. (1927). Montréal, Canada: Les Clercs de Saint-Viateur.

Lee, L., & Wheeler, D. (1987). *Algebraic thinking in high school students: Their conceptions of generalisation and justification* (Research report). Montréal, Canada: Concordia University, Mathematics Department.

Lee, L., & Wheeler, D. (1989). The arithmetic connection. *Educational Studies in Mathematics, 20,* 41-54.

Lee, L., & Wheeler, D. (1990). *Algebraic thinking in adult students: A comparison with high school students--Part 2. Interview behavior* (Research report). Montréal, Canada: Concordia University, Mathematics Department.

Lesh, R., Hoover, M., & Kelly, A. E. (1993). Equity, assessment, and thinking mathematically: Principles for the design of model-eliciting activities. In I. Wirszup & R. Streit (Eds.), *Developments in school mathematics education around the world* (Vol. 3). Reston, VA: National Council of Teachers of Mathematics.

Lins, R. C. (1992). *A framework for understanding what algebraic thinking is.* Unpublished doctoral dissertation, University of Nottingham.

Lobato, J., Gamoran, M., & Magidson, S. (1993). *Linear functions, technology, and the real world: An Algebra 1 replacement unit.* Unpublished manuscript, University of California, Berkeley, CA.

Lochhead, J., & Mestre, J. P. (1988). From words to algebra: Mending misconceptions. In A. F. Coxford (Ed.), *The ideas of algebra, K-12* (1988 Yearbook of the National Council of Teachers of Mathematics, pp. 127-135). Reston, VA: NCTM.

Luckow, A. (1993). *The evolution of the concept of algebraic expression in a problem solving computer environment: A process-object perspective.* Unpublished doctoral dissertation, Université du Québec à Montréal.

MacGregor, M., & Stacey, K. (1993). Seeing a pattern and writing a rule. In I. Hirabayashi, N. Nohda, K. Shigematsu, & F.-L. Lin (Eds.), *Proceedings of the 17th International Conference for the Psychology of Mathematics Education* (Vol. 1, pp. 181-188), Tsukuba, Japan: PME Program Committee.

MacLane, S. (1986). *Mathematics: Form and function.* New York: Springer-Verlag.

Mahoney, M. S. (1968). Another look at Greek geometrical analysis, *Archives for the History of Exact Sciences, 5,* 318-348.

Mahoney, M. S. (1971). Die Anfäng der algebraischen Denkweise im 17 Jahrhundert. *Rete, 1,* 15-31.

Mahoney, M. S. (1972). Babylonian algebra: Form vs. content, *Studies in History and Philosophy of Science, 1,* 369-380.

Mahoney, M. S. (1973). *The mathematical career of Pierre de Fermat (1601-1665).* Princeton, NJ: Princeton University Press.

Mahoney, M. S. (1980). The beginnings of algebraic thought in the seventeenth century. In S Gaukroger (Ed.), *Descartes, philosophy, mathematics and physics* (pp. 141-155). Sussex, UK: Harvester Press.

Margolinas, C. (1991). Interrelations between different levels of didactic analysis about elementary algebra. In F. Furinghetti (Ed.), *Proceedings of the 15th International Conference for the Psychology of Mathematics Education* (Vol. II, pp. 381-388). Assisi, Italy: PME Program Committee.

Mason, J. (1980). When is a symbol symbolic? *For the Learning of Mathematics, 1*(2), 8-12.

Mason, J. (1984). What do we really want students to learn? *Teaching At A Distance, 25*, 4-11.

Mason, J. (1986). Tensions. *Mathematics Teaching, 114*, 28-31.

Mason, J. (1988). *Expressing generality*. Milton Keynes, UK: Open University Press.

Mason, J. (1989). Mathematical abstraction seen as a delicate shift of attention. *For the Learning of Mathematics, 9*(2), 2-8.

Mason, J. (1991a). Epistemological foundations for frameworks which stimulate noticing. In R. G. Underhill (Ed.), *Proceedings of the 13th Annual Meeting of the North American Chapter of the International Group for the Psychology of Mathematics Education* (Vol. 2, pp. 36-42). Blacksburg, VA: PME-NA Program Committee.

Mason, J. (1991b). *Supporting primary mathematics: Algebra*. Milton Keynes, UK: Open University Press.

Mason, J. (1992). Frames for teaching. In E. Love & J. Mason (Eds.), *Teaching mathematics: Action and awareness* (pp. 29-53). Milton Keynes, UK: Open University Press.

Mason, J., Burton, L., & Stacey, K. (1984). *Thinking mathematically*. London: Addison Wesley.

Mason, J., Graham, A., Pimm, D., & Gowar, N. (1985). *Routes to/roots of algebra*. Milton Keynes, UK: Open University Press.

Mason, J., & Pimm, D. (1984). Generic examples: Seeing the general in the particular. *Journal of Educational Studies, 15*, 277-289.

Matras, M. (1989). The effects of curricula on students' ability to analyze and solve problems in algebra (Doctoral dissertation, University of Maryland, 1988). *Dissertation Abstracts International, 49*, 1/26.

Maturana, U. (1978). Biology of language: The epistemology of reality. In G. A. Miller & E. Lennenberg (Eds.), *Psychology and biology of language and thought: Essays in honor of Eric Lennenberg*. New York: Academic Press.

Mazzinghi, M. A. D. (1967). *Trattato di Fioretti* (nella trascelta a cura di Mo. Benedetto, secondo la lezione del Codice L, IV, 21 [Sec. XV] della Biblioteca degl'Intronati di Siena, e con introduzione di Gino Arrighi). Pisa, Italy: Domus Galileana.

Menger, K. (1979). *Selected papers in logic and foundations, didactics, economics*. Dordrecht, The Netherlands: Reidel.

Monagan, M. (1992). Exact eigenvalue and eigenvector computation in Maple. *MapleTech, 8*, 24-32.

Movshovits-Hadar, N. (1988). School mathematics theorems--An endless source of surprise. *For The Learning of Mathematics, 8*(3), 34-40.

Nelsen, R. B. (1993). *Proofs without words*. Washington, DC: Mathematical Association of America.

Nemirovsky, R. (1992). *On the basic forms of variation* (Unpublished manuscript). Cambridge, MA: TERC.

Nemirovsky, R. (1993). *Motion, flow, and contours: The experience of continuous change*. Unpublished doctoral dissertation, Harvard University.

Nemirovsky, R., & Rubin, A. (1991). It makes sense if you think about how the graphs work. But in reality... In F. Furinghetti (Ed.), *Proceedings of the 15th International Conference for the Psychology of Mathematics Education* (Vol. III, pp. 57-64). Assisi, Italy: PME Program Committee.

Neugebauer, O. (1935-37). *Mathematische Keilschrift-Text* (3 vols.). Berlin: Springer-Verlag.

Neugebauer, O. (1957). *Les sciences exactes dans l'antiquité* (P. Souffrin, Trans.). Paris: Actes du Sud.

Neugebauer, O., & Sachs, A. (1945). *Mathematical cuneiform texts* (Vol. 29). New Haven, CT: American Oriental Society.

Ochs, E., Taylor, C., Rudolph, D., & Smith, R. (1992). Storytelling as a theory-building activity. *Discourse Processes, 15*(1), 37-72.

Open University. (1982). *Developing mathematical thinking*. Milton Keynes, UK: author.

Orton, A., & Orton, J. (1994). Students' perception and use of pattern and generalization. In J. P. da Ponte & J. F. Matos (Eds.), *Proceedings of the 18th International Conference for the Psychology of Mathematics Education* (Vol. III, pp. 407-414). Lisbon, Portugal: PME Program Committee.

Paradís, J., Miralles, J., & Malet, A. (1989). *La génesis del algebra simbólica. Vol II: El álgebra en el período renacentista.* Barcelona, España: PPU.

Parshall, K. H. (1988). The art of algebra from Al-Kwarizmi to Viète: A study in the natural selection of ideas. *History of Science, 26,* 129-164.

Peel, E. A. (1971). Psychological and educational research bearing on school mathematics. In W. Servais & T. Varga (Eds.), *Teaching school mathematics.* Harmondsworth, UK: Penguin Education for UNESCO.

Penrose, R. (1991). *The emperor's new mind.* Oxford, UK: Oxford University Press.

Peterson, C., & McCabe, A. (1991). Linking children's connective use and narrative structure. In A. McCabe & C. Peterson (Eds.), *Developing narrative structure* (pp. 29-53). Hillsdale, NJ: Lawrence Erlbaum.

Pimm, D. (1987). *Speaking mathematically*. London: Hodder & Stoughton.

Poirier, L., & Bednarz, N. (1991). Etude des modèles implicites mis en oeuvre par les enfants lors de la résolution de problèmes complexes mettant en jeu une reconstruction d'une transformation arithmétique. In F. Furinghetti (Ed.), *Proceedings of the 15th International Conference for the Psychology of Mathematics Education* (Vol. I, pp. 185-190). Assisi, Italy: PME Program Committee.

Polya, G. (1966a). *Mathematical discovery.* New York: John Wiley & Sons.

Polya, G. (1966b). *On teaching problem solving. The role of axiomatics and problem solving in mathematics.* Washington, DC: Ginn.

Ptolemy. (1984). *Ptolemy's Almagest* (G. J. Toomer, Trans. & Annotations). London: Duckworth.

Puig, L., & Cerdán, F. (1990). Acerca del carácter aritmético o algebraico de los problemas verbales. In E. Filloy & T. Rojano (Eds.), *Memorias del Segundo Simposio Internacional sobre Investigación en Educación Matemática* (pp. 35-48). Cuernavaca, México: PNFAPM.

Pycior, H. M. (1984). Internalism, externalism and beyond: 19th century British algebra. *Historia Mathematica, 2,* 424-441.

Radford, L. (1992). Diophante et l'algèbre pré-symbolique. *Bulletin de l'Association Mathématique du Québec, 31/32*(4/1), 73-80.

Radford, L. (1995a). L'émergence et le développement conceptuel de l'algèbre (III siècle - XIV siècle). In F. Lalonde & F. Jaboeuf (Eds.), *Actes de la première Université d'été européenne--Histoire et Épistémologie dans l'éducation Mathématique* (pp. 69-83). Montpellier, France: IREM.

Radford, L (1995b). Before the other unknowns were invented: Didactic inquiries on the methods and problems of medieval Italian algebra. *For the Learning of Mathematics, 15*(3), 28-38.

Radford, L., & Berges, V. (1988). Explicaciones y procesos de resolucion de problemas de tipo logico en un contexto aritmético en adolescentes. *Memorias de la segunda reunion centroamericana y del Caribe sobre Formacion de Profesores e Investigacion en Matematica Educativa* (pp. 31-51). Guatemala: Universidad de San Carlos.

Regiomontanus (1967). *On triangle, De triangulis omnimodis* (B. Hugues, Trans.). Madison: University of Wisconsin. (Original work published 1533)

René de Cotret, S. (1986). *Histoire de la notion de fonction: Frontière épistémologique et conséquences didactiques.* Unpublished master's thesis, Université du Québec à Montréal.

Rey, A. (1948). *L'apogée de la science technique grecque. L'essor de la mathématique.* Paris: Éditions Albin Michel.

Ricoeur, P. (1981). *Hermeneutics and the human sciences.* Cambridge, UK: Cambridge University Press.

Riesz, F., & Nagy, B. (1968). *Leçons d'analyse fonctionnelle* (2nd ed.). Budapest: Académie des Sciences de Hongrie, Akadémiai Kiadó.

Ritter, F. (1895). Analyse des oeuvres de Viète. *Revue occidentale philosophique, sociale et politique* (2e série), *10* , 354-415.

Robbins, F. (1929). P. Mich. 620: A series of arithmetical problems. *Classical Philology, 24*, 321-329.

Rojano, T. (1985). *De la aritmética al algebra (un estudio clínico con niños de 11-12 años de edad).* Unpublished doctoral dissertation, CINVESTAV del Institutuo Politécnico Nacional, México.

Rojano, T., & Sutherland, R. (1991a). La sintaxis algebraica en el proyecto viético. *Memorias del Tercer Simposio Internacional sobre Investigación en Educación Matemática.* Valencia, España: Universidad de Valencia, PNFAPM.

Rojano, T., & Sutherland, R. (1991b). Symbolising and solving algebra word problems: The potential of a spreadsheet environment. In F. Furinghetti (Ed.), *Proceedings of the 15th International Conference for the Psychology of Mathematics Education* (Vol. III, pp. 207-213.). Assisi, Italy: PME Program Committee.

Rojano, T., & Sutherland, R. (1992). *A new approach to algebra: Results from a study with 15 year old algebra-resistant pupils.* Paper presented at Cuarto Simposio Internacional sobre Investigación en Educación Matemática/ PNFAPM, Ciudad Juárez, México.

Rojano, T., & Sutherland, R. (1993). Towards an algebra approach: The role of spreadsheets. In I. Hirabayashi, N. Nohda, K. Shigematsu, & F. Lin (Eds.), *Proceedings of the 17th International Conference for the Psychology of Mathematics Education* (Vol. I, pp. 189-196). Tsukuba, Japan: PME Program Committee.

Romberg, T. A., Fennema, E., & Carpenter, T. P. (Eds.). (1993). *Integrating research on the graphical representation of function.* Hillsdale, NJ: Lawrence Erlbaum.

Rouche, N. (1992). *Le sens de la mesure. Des grandeurs aux nombres rationnels.* Bruxelles: Didier-Hatier.

Rubio, G. (1990). Algebra word problems: A numerical approach for its resolution (a teaching experiment in the classroom). In G. Booker, P. Cobb, & T. N. de Mendicuti (Eds.), *Proceedings of the 14th International Conference for the Psychology of Mathematics Education* (Vol. II, pp. 125-132). Oaxtepec, México: PME Program Committee.

Rubio, G. (1994). *Modelos didacticos para resolver problemas verbales aritmetico-algebraicos: Tesis teoricas y observacion empirica.* Unpublished doctoral dissertation, CINVESTAV del Institutuo Politécnico Nacional, México.

Schmidt, S. (1994). *Passage de l'arithmétique à l'algèbre et inversement de l'algèbre à l'arithmétique chez les futurs enseignants dans un contexte de résolution de problème.* Unpublished doctoral dissertation, Université du Québec à Montréal.

Schoenfeld, A., & Arcavi, A. (1988). On the meaning of variable. *Mathematics Teacher, 81*, 420-427.

Schwartz, J., & Yerushalmy, M. (1992). Getting students to function in and with algebra. In G. Harel & E. Dubinsky (Eds.), *The concept of function: Aspects of epistemology and pedagogy* (MAA Notes, Vol. 25, pp. 261-289). Washington, DC: Mathematical Association of America.

Schwarz, B. B., Kohn, A. S., & Resnick, L. B. (1992, April). *Positive about negatives.* Paper presented at the 1992 annual meeting of the American Educational Research Association, San Francisco, CA.

Seeger, F. (1989). *Davydov's theory of generalization: Theoretical thinking and representation in learning and teaching algebra* (Occasional Paper 117). Bielefeld, Germany: Universitat Bielefeld, Institut fur Didaktik der Mathematik.

Sesiano, J. (1982). *Books IV to VII of Diophantus' Arithmetica, in the Arabic translation attributed to Qusta ibn Luqa.* New York: Springer-Verlag.

Sfard, A. (1987). Two conceptions of mathematical notions. In J. C. Bergeron, N. Herscovics, & C. Kieran (Eds.), *Proceedings of the 11th International Conference for the Psychology of Mathematics Education* (Vol. III, pp. 162-169). Montréal, Canada: PME Program Committee.

Sfard, A. (1991). On the dual nature of mathematical conceptions: Reflections on processes and objects as different sides of the same coin. *Educational Studies in Mathematics, 22,* 1-36.

Sfard, A. (1992). Operational origins of mathematical notions and the quandary of reification: The case of function. In G. Harel & E. Dubinsky (Eds.), *The concept of function: Aspects of epistemology and pedagogy* (MAA Notes, Vol. 25, pp. 59-84). Washington, DC: Mathematical Association of America.

Sfard, A. (1995). The development of algebra--Confronting historical and psychological perspectives. In C. Kieran (Ed.), New perspectives on school algebra: Papers and discussions of the ICME-7 algebra working group (special issue). *Journal of Mathematical Behavior, 14,* 15-39.

Sfard, A., & Linchevski, L. (1994). The gains and the pitfalls of reification--The case of algebra. *Educational Studies in Mathematics, 26,* 191-228.

Sheets, C. (1993). Effects of computer learning and problem solving tools on the development of secondary school students' understanding of mathematical functions. (Doctoral dissertation, University of Maryland, 1993). *Dissertation Abstracts International, 54A,* 1/14. (University Microfilms No. aac 93-2/495)

Shuard, H., & Neill, H. (1977). *From graphs to calculus.* Glasgow: Blackie.

Sierpinska, A. (1992). On understanding the notion of function. In G. Harel & E. Dubinsky (Eds.), *The concept of function: Aspects of epistemology and pedagogy* (MAA Notes, Volume 25, pp. 25-58). Washington, DC: Mathematical Association of America.

Sleeman, D. H. (1986). Introductory algebra: A case study of students' misconceptions. *Journal of Mathematical Behavior, 5,* 25-52.

Soloway, E., Lochhead, J., & Clement, J. (1982). Does computer programming enhance problem solving ability? Some positive evidence on algebra word problems. In R. J. Seidel, R. E. Anderson, & B. Hunter (Eds.), *Computer literacy* (pp. 171-185). New York: Academic Press.

Stacey, K. (1989). Finding and using patterns in linear generalising problems. *Educational Studies in Mathematics, 20,* 147-164.

Sutherland, R. (1989). Providing a computer-based framework for algebraic thinking. *Educational Studies in Mathematics, 20,* 317-344.

Sutherland, R. (1992a). Some unanswered research questions on the teaching and learning of algebra. *For the Learning of Mathematics, 11*(3), 40-46.

Sutherland, R. (1992b). *Thinking algebraically: Pupil models developed in Logo and a spreadsheet environment.* London: University of London, Institute of Education.

Sutherland, R., & Rojano, T. (1993). A spreadsheet approach to solving algebra problems. *Journal of Mathematical Behavior, 12,* 353-383.

Swan, M. B. (1985) *The language of functions and graphs.* Nottingham, UK: Shell Centre for Mathematical Education.

Tahta, D. (Ed.). (1972). *A Boolean anthology: Selected writings of Mary Boole.* Derby, UK: ATM.

Tahta, D. (1980). About geometry. *For the Learning of Mathematics, 1*(1), 2-9.

Tall, D., & Thomas, M. (1991). Algebraic thinking with a computer. *Educational Studies in Mathematics, 22,* 125-147.

Tall, D., & Vinner, S. (1981). Concept image and concept definition in mathematics with particular reference to limits and continuity. *Educational Studies in Mathematics, 12,* 151-169.

Tannery, P. (1893). *Diophanti Alexandrini opera omnia cum Graecis commentariis.* Lipsiae, Germany: Teubneri.

Thureau-Dangin, F. (1938a). *Textes mathématiques Babyloniens* (Ex Oriente Lux, I). Leiden, The Netherlands: E. J. Brill.

Thureau-Dangin, F. (1938b). La méthode de fausse position et l'origine de l'algèbre. *Revue d'Assyriologie et d'archéologie Orientale, 35,* 71-77.

Tierney, C., Weinberg, A., & Nemirovsky, R. (1992). Telling stories about plant growth: Fourth grade students interpret graphs. In W. Geeslin & K. Graham (Eds.), *Proceedings of the 16th International Conference for the Psychology of Mathematics Education* (Vol. III, pp. 66-73). Durham, NH: PME Program Committee.

Tierney, C., Weinberg, A., & Nemirovsky, R. (1994) *Graphs: Changes over time. Curricular unit for grade 4.* Palo Alto, CA: Dale Seymour.

Treutlein, P. (1879). Die deutsche Coss. *Abhandlungen zur Geschichte der Mathematik, 2,* 1-124.

Tropfke, J. (1933). *Geschichte der elementar-mathematik* (Vol. II, Chap. A, 3rd ed.). Berlin.

Unguru, S. (1975). On the need to rewrite the history of Greek mathematics, *Archives for the History of Exact Sciences, 15,* 69-113.

Ursini, S. (1992). *An exploratory study of pupils' capability to discriminate and connect different characterizations of the variable in a Logo environment.* Unpublished doctoral dissertation, University of London, Institute of Education.

Usiskin, Z. (1988). Conceptions of school algebra and uses of variables. In A. F. Coxford (Ed.), *The ideas of algebra, K-12* (1988 Yearbook of the National Council of Teachers of Mathematics, pp. 8-19). Reston, VA: NCTM.

van Barneveld, G., & Krabbendam, H. (Eds.). (1982). *Proceedings of the Conference on Functions.* Enschede, The Netherlands: National Institute of Curriculum Development.

van der Waerden, B. L. (1961). *Science awakening* (A. Dresden, Trans.). Oxford, UK: Oxford University Press.

van der Waerden, B. L. (1976). Defence of a "shocking" point of view. *Archives for the History of Exact Sciences, 15,* 199-210.

van der Waerden, B. L. (1980). *A history of algebra: From Al-Khwarizmi to Emmy Noether.* New York: Springer-Verlag.

van der Waerden, B. L. (1983). *Geometry and algebra in ancient civilizations.* Berlin: Springer-Verlag.

Van Egmond, W. (1980). Practical mathematics in the Italian Renaissance (A catalog of Italian Abbacus Manuscripts and Printed Books to 1600). *Suplemento agli Annali dell'Instituto e Museo di Storia della Scienza, 1980--Fascicolo 1.* Firenze, Italy: Instituto e Museo di Storia della Scienza.

Van Egmond, W. (1988). How algebra came to France. In C. Hay (Ed.), *Mathematics from manuscript to print, 1300-1600* (pp. 127-144). Oxford, UK: Clarendon Press.

van Lehn, K. (1982). Bugs are not enough: Empirical studies of bugs, impasses, and repairs in procedural skills. *Journal of Mathematical Behavior, 3*(2), 3-71.

Ver Eecke, P. (1959). *Diophante d'Alexandrie. Les six livres arithmétiques et le livre des nombres polygones.* Paris: Blanchard.

Vergnaud, G. (1981). Quelques orientations théoriques et méthodologiques des recherches françaises en didactique des mathématiques, In C. Comité & G. Vergnaud (Eds.), *Proceedings of the Fifth International Conference for the Psychology of Mathematics Education* (Vol 2, pp. 7-17). Grenoble, France: IMAG.

Vergnaud, G. (1982). A classification of cognitive tasks and operations of thought involved in addition and subtraction problems. In T. P. Carpenter, J. M. Moser, & T. A. Romberg (Eds.), *Addition and subtraction: A cognitive perspective* (pp. 39-59). Hillsdale, NJ: Lawrence Erlbaum.

Vergnaud, G. (1989). L'obstacle des nombres négatifs et l'introduction à l'algèbre. In N. Bednarz & C. Garnier (Eds.), *Construction des savoirs: Obstacles et conflits* (pp. 76-83). Montréal, Canada: Agence d'Arc.

Viète, F. (1970). *Opera mathematica* (Van Schooten, Ed.). New York: Georg Ohms Verlag. (Original work published 1646)

Viète, F. (1983). *The analytic art, nine studies in algebra, geometry and trigonometry from the Opus Restitutae Mathematicae Analyseos, seu Algebrâ Novâ by François Viète* (T. R. Witmer, Trans.). Kent, OH: Kent State University Press.

Vygotsky, L. (1978). *Mind in society: The development of the higher psychological processes.* Cambridge, MA: Harvard University Press.

Whitehead, A. N. (1947). *Essays in science and philosophy.* New York: Philosophical Library.

Yerushalmy, M., & Schternberg, B. (1992). *The algebra sketchbook* [Computer software]. Pleasantville, NY: Sunburst.

Zbiek, R. M., & Heid, M. K. (1989). *Use of tools, representations, and strategies by beginning algebra students: Preferences and influences.* Unpublished manuscript. (Available from M. K. Heid, 171 Chambers Bldg., Penn State University, University Park, PA 16803)

Zbiek, R. M., Hess, L., & Rodriguez, P. (1989). *Assessing eighth graders' understanding of algebra by focusing on the concept of variable.* Unpublished manuscript. (Available from R. M. Zbiek, University of Iowa, 289 Lindquist Center North, Iowa City, IA 52246-4449)

AUTHOR AFFILIATIONS

Nadine Bednarz
CIRADE
Université du Québec à Montréal
C.P. 8888, succursale Centre-ville
Montréal, Québec
Canada
H3C 3P8

Alan Bell
Shell Centre for Mathematical Education
University Park
Nottingham NG7 2RD
United Kingdom

André Boileau
Département de Mathématiques
Université du Québec à Montréal
C.P. 8888, succursale Centre-ville
Montréal, Québec
Canada
H3C 3P8

Louis Charbonneau
Département de Mathématiques
Université du Québec à Montréal
C.P. 8888, succursale Centre-ville
Montréal, Québec
Canada
H3C 3P8

Maurice Garançon
Département de Mathématiques
Université du Québec à Montréal
C.P. 8888, succursale Centre-ville
Montréal, Québec
Canada
H3C 3P8

M. Kathleen Heid
Department of Curriculum and Instruction
The Pennsylvania State University
University Park, PA 16803
USA

Bernadette Janvier
CIRADE
Université du Québec à Montréal
C.P. 8888, succursale Centre-ville
Montréal, Québec
Canada
H3C 3P8

Claude Janvier
CIRADE
Université du Québec à Montréal
C.P. 8888, succursale Centre-ville
Montréal, Québec
Canada
H3C 3P8

Carolyn Kieran
Département de Mathématiques
Université du Québec à Montréal
C.P. 8888, succursale Centre-ville
Montréal, Québec
Canada
H3C 3P8

Lesley Lee
Département de Mathématiques
Université du Québec à Montréal
C.P. 8888, succursale Centre-ville
Montréal, Québec
Canada
H3C 3P8

Jacques Lefebvre
Département de Mathématiques
Université du Québec à Montréal
C.P. 8888, succursale Centre-ville
Montréal, Québec
Canada
H3C 3P8

John Mason
Mathematics Faculty
Open University
Milton Keynes MK7 6AA
United Kingdom

Ricardo Nemirovsky
TERC
2067 Massachusetts Avenue
Cambridge, MA 02140
USA

Luis Radford
École des sciences de l'éducation
Université Laurentienne
Sudbury, Ontario
Canada
P3E 2C6

Teresa Rojano
Dep. de Matemática Educativa
Centro de Investigación y Estudios Avanzados del IPN
Nicolás San Juan 1421
Col. del Valle
México City D.F., c.p. 03100
México

David Wheeler
205-1230 Haro Street
Vancouver, British Columbia
Canada
V6E 4J9

Mathematics Education Library

Managing Editor: A.J. Bishop, Melbourne, Australia

1. H. Freudenthal: *Didactical Phenomenology of Mathematical Structures.* 1983
   ISBN 90-277-1535-1; Pb 90-277-2261-7
2. B. Christiansen, A. G. Howson and M. Otte (eds.): *Perspectives on Mathematics Education.* Papers submitted by Members of the Bacomet Group. 1986. ISBN 90-277-1929-2; Pb 90-277-2118-1
3. A. Treffers: *Three Dimensions.* A Model of Goal and Theory Description in Mathematics Instruction – The Wiskobas Project. 1987 ISBN 90-277-2165-3
4. S. Mellin-Olsen: *The Politics of Mathematics Education.* 1987
   ISBN 90-277-2350-8
5. E. Fischbein: *Intuition in Science and Mathematics.* An Educational Approach. 1987 ISBN 90-277-2506-3
6. A.J. Bishop: *Mathematical Enculturation.* A Cultural Perspective on Mathematics Education. 1988
   ISBN 90-277-2646-9; Pb (1991) 0-7923-1270-8
7. E. von Glasersfeld (ed.): *Radical Constructivism in Mathematics Education.* 1991 ISBN 0-7923-1257-0
8. L. Streefland: *Fractions in Realistic Mathematics Education.* A Paradigm of Developmental Research. 1991 ISBN 0-7923-1282-1
9. H. Freudenthal: *Revisiting Mathematics Education.* China Lectures. 1991
   ISBN 0-7923-1299-6
10. A.J. Bishop, S. Mellin-Olsen and J. van Dormolen (eds.): *Mathematical Knowledge: Its Growth Through Teaching.* 1991 ISBN 0-7923-1344-5
11. D. Tall (ed.): *Advanced Mathematical Thinking.* 1991 ISBN 0-7923-1456-5
12. R. Kapadia and M. Borovcnik (eds.): *Chance Encounters: Probability in Education.* 1991 ISBN 0-7923-1474-3
13. R. Biehler, R.W. Scholz, R. Sträßer and B. Winkelmann (eds.): *Didactics of Mathematics as a Scientific Discipline.* 1994 ISBN 0-7923-2613-X
14. S. Lerman (ed.): *Cultural Perspectives on the Mathematics Classroom.* 1994
    ISBN 0-7923-2931-7
15. O. Skovsmose: *Towards a Philosophy of Critical Mathematics Education.* 1994 ISBN 0-7923-2932-5
16. H. Mansfield, N.A. Pateman and N. Bednarz (eds.): *Mathematics for Tomorrow's Young Children.* International Perspectives on Curriculum. 1996
    ISBN 0-7923-3998-3
17. R. Noss and C. Hoyles: *Windows on Mathematical Meanings.* Learning Cultures and Computers. 1996 ISBN 0-7923-4073-6; Pb 0-7923-4074-4

Mathematics Education Library

Managing Editor: A.J. Bishop, Melbourne, Australia

18. N. Bednarz, C. Kieran and L. Lee (eds.): *Approaches to Algebra.* Perspectives for Research and Teaching. 1996
ISBN 0-7923-4145-7; Pb 0-7923-4168-6

KLUWER ACADEMIC PUBLISHERS – DORDRECHT / BOSTON / LONDON